T0205664

BIOTECHNOLOGY OF MICROORGANISMS

Diversity, Improvement, and Application of
Microbes for Food Processing, Healthcare,
Environmental Safety, and Agriculture

Innovations in Biotechnology Series

BIOTECHNOLOGY OF MICROORGANISMS

Diversity, Improvement, and Application of Microbes for Food Processing, Healthcare, Environmental Safety, and Agriculture

Edited by

Jeyabalan Sangeetha, PhD
Devarajan Thangadurai, PhD
Somboon Tanasupawat, PhD
Pradnya Pralhad Kanekar, PhD

APPLE
ACADEMIC
PRESS

Apple Academic Press Inc. | Apple Academic Press Inc.
3333 Mistwell Crescent | 1265 Goldenrod Circle NE
Oakville, ON L6L 0A2 | Palm Bay, Florida 32905
Canada USA | USA

Library and Archives Canada Cataloguing in Publication

Title: Biotechnology of microorganisms : diversity, improvement, and application of microbes for food processing, healthcare, environmental safety, and agriculture / edited by Jeyabalan Sangeetha, Devarajan Thangadurai, Somboom Tanasupawat, Pradnya Pralhad Kanekar.

Names: Sangeetha, Jeyabalan, editor. | Thangadurai, Devarajan, editor. | Tanasupawat, Somboom, editor. | Kanekar, Pradnya Pralhad, editor.

Description: Series statement: Innovations in biotechnology ; volume 2 | Includes bibliographical references and index.

Identifiers: Canadiana (print) 20190092688 | Canadiana (ebook) 20190092831 | ISBN 9781771887472 (hardcover) | ISBN 9780429434112 (eBook)

Subjects: LCSH: Microbial biotechnology.

Classification: LCC TP248.27.M53 B56 2019 | DDC 660.6/2—dc23

CIP data on file with US Library of Congress

Apple Academic Press also publishes its books in a variety of electronic formats. Some content that appears in print may not be available in electronic format. For information about Apple Academic Press products, visit our website at **www.appleacademicpress.com** and the CRC Press website at **www.crcpress.com**

ABOUT THE EDITORS

Jeyabalan Sangeetha, PhD

Jeyabalan Sangeetha, PhD, is an Assistant Professor in the Central University of Kerala at Kasaragod, South India. She earned her BSc in Microbiology and PhD in Environmental Science from Bharathidasan University, Tiruchirappalli, Tamil Nadu, India. She holds an MSc in Environmental Science from Bharathiar University, Coimbatore, Tamil Nadu, India. She is the recipient of a Tamil Nadu Government Scholarship and a Rajiv Gandhi National Fellowship of the University Grants Commission, Government of India for her doctoral studies. She served as a Dr. D.S. Kothari Postdoctoral Fellow and a UGC Postdoctoral Fellow at Karnatak University, Dharwad, South India during 2012–2016 with funding from the University Grants Commission, Government of India, New Delhi. Her research interests are in environmental toxicology, environmental microbiology, and environmental biotechnology, and her scientific/community leadership included serving as editor of an international journal, *Acta Biologica Indica.*

Devarajan Thangadurai, PhD

Devarajan Thangadurai, PhD, is a Senior Assistant Professor at Karnatak University in South India, President of the International Society for Applied Biotechnology, and General Secretary for the Association for the Advancement of Biodiversity Science. In addition, Dr. Thangadurai is Editor-in-Chief of two international journals, *Biotechnology, Bioinformatics and Bioengineering* and *Acta Biologica Indica.* He received his PhD in Botany from Sri Krishnadevaraya University in South India. During 2002–2004, he worked as a CSIR Senior Research Fellow with funding from the Ministry of Science and Technology, Government of India. He served as a Postdoctoral Fellow at the University of Madeira, Portugal, University of Delhi, India and ICAR National Research Centre for Banana, India. He is the recipient of the Best Young Scientist Award with a Gold Medal from Acharya Nagarjuna University, India and the VGST-SMYSR Young Scientist Award of the Government of Karnataka, Republic of India. He has authored and edited 19 books, including *Genetic Resources and Biotechnology* (3 vols.), *Genes, Genomes and Genomics* (2 vols.), *Mycorrhizal Biotechnology, Genomics and Proteomics, Industrial Biotechnology,* and *Environmental Biotechnology* with publishers of national and international reputation. He has also visited Portugal, Indonesia, Ukraine, Nepal, Sri Lanka, Bangladesh, Thailand, Oman, Vietnam, Malaysia, China, Georgia, United Arab Emirates, Myanmar, Egypt, Italy, and Russia for academic work, scientific meetings, and international collaborations.

Somboon Tanasupawat, PhD

Somboon Tanasupawat, PhD, is a Professor at the Department of Biochemistry and Microbiology, Faculty of Pharmaceutical Sciences, Chulalongkorn University, Bangkok, Thailand. He received his MSc in Microbiology (Kasetsart University, Thailand), Diploma in Microbiology and Biotechnology (Osaka University, Japan), and PhD in Agricultural Chemistry (Tokyo University of Agriculture, Japan). He has published more than 180 research and review articles on microbial taxonomy and applied microbiology. He received a Scientist Award (Nagai Award) from the Nagai Foundation, Tokyo. He has authored several book chapters with publishers of international reputations. In addition, he is an editorial board member of the journals *BioMed Research International, Annals of Microbiology,* and *Thai Journal of Pharmaceutical Sciences.*

Pradnya Pralhad Kanekar, PhD

Pradnya Pralhad Kanekar, PhD, is a CSIR Emeritus Scientist at the Modern College of Arts, Science and Commerce, Pune, India. She received her PhD in Microbiology from the University of Pune in 1980. She joined the MACS-Agharkar Research Institute (MACS-ARI) during 1974 and headed the Microbiology Department of MACS-ARI from 1988 followed by Microbial Sciences Division of MACS-ARI till 2011. She also headed MACS-ARI as Acting Director during 2009–2010. Her research areas include extremophilic microorganisms, microbial diversity, microbial biotechnology, bioremediation, and biodegradation of toxic organic chemopollutants. She has been a Fellow of the Academy of Environmental Biology since 1990 and received a Life Fellow Award, Archana Gold Medal and Citation in 1995 for her contributions to the development of microbial technology for industrial wastewater treatment. She is also a Fellow of the Maharashtra Academy of Sciences since 2009 and an Honorary Fellow of the Society for Applied Biotechnology since 2011 and was honored with a Women Bioscientist Award in 2011. She served as President, Pune Chapter of the Association of Food Scientists and Technologists of India (AFSTI) during 1988–1990, President, Pune Unit of the Association of Microbiologists of India (AMI) in 2009, and Convener, Pune Branch of the Indian Women Scientists Association (IWSA) during 1999–2005. She has 80 research papers published in peer-reviewed journals, five Indian patents granted, 11 book chapters published, 25 papers published in proceedings of national and international conferences, one proceedings of the national conference, and over 125 papers presented at international and national conferences to her credit. She has successfully implemented two International (Indo-Swiss and Indo-US) projects; 12 projects sponsored by CSIR, DBT, DST, MoEF, DRDO, Government of India; and four projects by industries. She has visited Russia, England, Switzerland, France, Czechoslovakia, Austria and USA for project meetings and international conferences.

INNOVATIONS IN BIOTECHNOLOGY

SERIES EDITOR

Devarajan Thangadurai, PhD
Assistant Professor, Karnatak University, Dharwad, South India

BOOKS IN THE SERIES

Volume 1: Fundamentals of Molecular Mycology
Devarajan Thangadurai, PhD, Jeyabalan Sangeetha, PhD, and
Muniswamy David, PhD

Volume 2: Biotechnology of Microorganisms: Diversity, Improvement,
and Application of Microbes for Food Processing, Healthcare,
Environmental Safety, and Agriculture
Edited by Jeyabalan Sangeetha, PhD, Devarajan Thangadurai, PhD,
Somboon Tanasupawat, PhD, and Pradnya Pralhad Kanekar, PhD

Volume 3: Phycobiotechnology: Biodiversity and Biotechnology of
Algae and Algal Products for Food, Feed, and Fuel
Edited by Jeyabalan Sangeetha, PhD, Devarajan Thangadurai, PhD,
Sanniyasi Elumalai, PhD, and Shivasharana Chandrabanda Thimmappa,
PhD

CONTENTS

CONTRIBUTORS

Antony Alex Kennedy Ajilda
Department of Microbiology, Science Campus, Alagappa University, Karaikudi, Tamil Nadu – 630003, India

Subathra Devi Chandrasekaran
School of Biosciences and Technology, VIT University, Vellore, Tamil Nadu, 632014, India

Muniswamy David
Department of Zoology, Karnatak University, Dharwad, 580003, Karnataka, India

Ravichandra Hospet
Department of Botany, Karnatak University, Dharwad, 580003, Karnataka, India

Nasser El-Din Ibrahim
Department of Biology, University of Waterloo, ON, N2L 3G1, Canada

Shrinivas Jadhav
Department of Zoology, Karnatak University, Dharwad, 580003, Karnataka, India

Jyotsna Jotshi
formerly Microbial Sciences Division, Agharkar Research Institute, Pune, Maharashtra 411004, India

Pradnya Pralhad Kanekar
Department of Biotechnology, Modern College of Arts, Science, and Commerce, Shivajinagar, Pune 411005, Maharashtra, India

Sagar Pralhad Kanekar
Department of Biotechnology, Modern College of Arts, Science and Commerce, Ganeshkhind, Pune 411007, Maharashtra, India

Anupreet Kaur
Department of Biotechnology, University Institute of Engineering and Technology, Punjab University, Chandigarh 160014, India

Jaspreet Kaur
Department of Biotechnology, University Institute of Engineering and Technology, Punjab University, Chandigarh 160014, India

Snehal Omkar Kulkarni
Department of Biotechnology, Modern College of Arts, Science and Commerce, Shivajinagar, Pune 411005, Maharashtra, India

Kesen Ma
Department of Biology, University of Waterloo, ON, N2L 3G1, Canada

Essam Abdellatif Makky
Faculty of Industrial Sciences and Technology, University Malaysia Pahang, Gambang, 26300 Kuantan, Pahang, Malaysia

Taweesak Malimas
Department of Biochemistry and Microbiology, Faculty of Pharmaceutical Sciences,
Chulalongkorn University, Bangkok 10330, Thailand

Abhishek Channayya Mundaragi
Department of Botany, Karnatak University, Dharwad, 580003, Karnataka, India

Subir Kumar Nandy
Technical University of Denmark, Søltofts Plads, 2800 Kgs, Lyngby, Denmark

Nagarajan Padmini
Department of Microbiology, Science Campus, Alagappa University, Karaikudi,
Tamil Nadu–630003, India

Nittaya Pitiwittayakul
Department of Agricultural Technology and Environment, Faculty of Sciences and Liberal Arts,
Rajamangala University of Technology Isan, Nakhon Ratchasima Campus,
Nakhon Ratchasima 30000, Thailand

Purushotham Prathima
Department of Botany, Karnatak University, Dharwad, 580003, Karnataka, India

Jeyabalan Sangeetha
Department of Environmental Science, Central University of Kerala, Kasaragod 671316, Kerala, India

Gopal Selvakumar
Department of Microbiology, Science Campus, Alagappa University, Karaikudi,
Tamil Nadu–630003, India

Jemimah Naine Selvakumar
School of Biosciences and Technology, VIT University, Vellore, Tamil Nadu, 632014, India

Weilan Shao
School of Environment, Jiangsu University, Zhenjiang, China

Natesan Sivakumar
School of Biotechnology, Madurai Kamaraj University, Madurai, Tamil Nadu–625021, India

Khomsan Supong
Department of Applied Science and Biotechnology, Faculty of Agro-Industrial Technology,
Rajamangala University of Technology, Twan-ok Chantaburi Campus, Chantaburi 22210, Thailand

Somboon Tanasupawat
Department of Biochemistry and Microbiology, Faculty of Pharmaceutical Sciences,
Chulalongkorn University, Bangkok 10330, Thailand

Devarajan Thangadurai
Department of Botany, Karnatak University, Dharwad, 580003, Karnataka, India

Shivasharana Chandrabanda Thimmappa
Department of Microbiology and Biotechnology, Karnatak University, Dharwad, 580003, Karnataka,
India

Mohanasrinivasan Vaithilingam
School of Biosciences and Technology, VIT University, Vellore, Tamil Nadu, 632014, India

Huong Thi Lan Vu
Faculty of Biology and Biotechnology, University of Science, Vietnam National University-HCM City, Ho Chi Minh City, Vietnam

Yuzo Yamada
Department of Applied Biological Chemistry, Faculty of Agriculture, Shizuoka University, Suruga-Ku, Shizuoka 422-8529, Japan

Pattaraporn Yukphan
BIOTEC Culture Collection (BCC), National Center for Genetic Engineering and Biotechnology (BIOTEC), National Science and Technology Development Agency (NSTDA), 113 Thailand Science Park, Pathumthani 12120, Thailand

Mashitah Mohd Yusoff
Faculty of Industrial Sciences and Technology, University Malaysia Pahang, Gambang, 26300 Kuantan, Pahang, Malaysia

ABBREVIATIONS

%	percentage
°C	Degree Celsius
μA	microampere
μM	micromolar
^{124}I-PTH-G	^{124}I-phenolphthalein-glucuronide probe
^{18}F-FEAnGA-Me	1-O-(4-(2-fluoroethyl-carbamoyloxymethyl)– 2-nitrophenyl)-O-β-D-glucopyronuronate methyl ester
2,5DKG	2,5-diketo-D-gluconate
2D	2-dimensional
2KGA	2-keto-D-gluconate
4KA	4-keto-D-arabonate
4KAS	4-keto-D-arabonate synthase
5KGA	5-keto-D-gluconate
A MFC	microbial fuel cell using *Acetobacter aceti*
A	Ampere
AAB	acetic acid bacteria
ADH	alcohol dehydrogenase
AFLP	amplified fragment length polymorphism
Ag	silver
AG-MFC	microbial fuel cell using mixed culture cell (*Acetobacter aceti* and *Gluconobacter roseus*)
AgNPs	Silver nanoparticles
AIDs	antiinflammatory drugs
ALDH	aldehyde dehydrogenase
AMF	arbuscular mycorrhizal fungi
As	arsenic
As^{3+}	arsenites
As^{5+}	arsenate
ATCC	American type culture collection
ATP	adenosine triphosphate
ATPS	aqueous two-phase system
Au	gold

AuCN	gold cyanide
bat	bacterio-opsin activator
BC	bacterial cellulose
BDEPT	bacteria directed enzyme prodrug therapy
BHIB	brain heart infusion broth
Bi	bismuth
blp	bacterioopsin linked product
BOD	biological oxygen demand
bop	bacterio-opsin
bp	base pairs
BR	bacteriorhodopsin
brp	bacterio-opsin related protein
BSSC	bio-sensitized solar cell
Ca^{2+}	calcium ion
CAGR	compound annual growth rate
CALB	*Candida antarctica* lipase B
CaMV	cauliflower mosaic virus
Cd	cadmium
cDNA	complementary DNA
CDW	cell dry weight
Ce	cerium
CFU	colony forming units
cm	Centimeter
Co	Cobalt
CO^{2+}	Cobalt ion
COS-SSTR	compressed oxygen supply-sealed and stirred tank reactor
Cr	chromium
CrO	chromium (II) oxide
CS	chondroitin sulfate
CSLM	confocal scanning laser microscope
CSTRs	continuous stirred tank reactors
CTS	carpal tunnel syndrome
Cu	copper
Cu^{2+}	cupric ion
DAPI	4′,6-diamidino–2-phenylindole
DEAE	diethylaminoethyl cellulose
d-GA	d-glyceric acid

DHA	dihydroxyacetone
DMAPP	dimethylallyl diphosphate
DNA	deoxyribonucleic acid
DP	drop plate
E. coli	*Escherichia coli*
EC	enzyme commission
ECM	extracellular matrix
EDTA	ethylene diamine tetraacetic acid
EIA	enzyme immunoassay
EPS	extracellular polymeric substances
ER	endoplasmic reticulum
ERIC-PCR	enterobacterial repetitive intergenic consensus sequence-based PCR
EU	European Union
FACITs	fibril-associated collagens with interrupted triple helices
FBA	FITC-biotin-avidin
FCM	flow cytometry
FDGlcU	fluorescein di-β-D-glucuronide probe
Fe	iron
FHL	*Fusarium heterosporum* lipase
FITC-TrapG	fluorescein isothiocyanate linked to TrapG
g/l	gram per liter
Ga	gallium
GADH	gluconate dehydrogenase
GAG	glycosaminoglycan
GalNAc	N-acetylgalactosamine
GFP	green fluorescent protein
GH	glycoside hydrolase
GI	gastrointestinal
GlcA	glucuronic acid
GLDH	glycerol dehydrogenase
G-MFC	microbial fuel cell using *Gluconobacter roseus*
GMO	genetically modified organisms
GRAS	generally recognized as safe
GUSB	glucuronidase, beta
H/D	height to diameter ratio
HCD	high cell density
HCl	hydrochloric acid

Hg	mercury
Hg^{2+}	mercuric ion
HNQ	2-hydroxy-1,4-naphthoquinone
HS	Hestrin-Schramm medium
HTP	high throughput technology
HVAC	heating, ventilating and air-conditioning
IAA	indole-3-acetic acid
IC_{50}	half maximal inhibitory concentration
IFA	immunofluorescence assay
IME	intron-mediated enhancement
IPP	isopentenyl diphosphate
ITO	indium tin oxide
ITS	internally transcribed spacer
Kb	kilobyte
kDa	kilo Dalton
$k_L a$	volumetric oxygen mass transfer coefficient
l	liter
LacA	lactobionic acid
LED	light emitting diode
LPS	lipopolysaccharide
Lys	lysine
MACITs	membrane-associated collagens with interrupted triple helices
MALDI-TOF	matrix-assisted laser desorption ionization time-of-flight
MBRT	methylene blue dye reduction test
MDR	multiple drug resistant
MEP	methylerythritol phosphate pathway
MFCs	microbial fuel cells
mg	milligram
mg/l	milligram per liter
Mg^{2+}	magnesium ion
MIC	minimum inhibitory concentration
ml	milliliter
mM	millimolar
MMP	matrix metalloprotease
Mn	manganese
Mn^{2+}	manganese ion
MPEW	mixture of pre-hydrolysates and ethanol-fermented waste liquid

MPS	mucopolysaccharidosis type
mRNA	messenger RNA
MRSA	methicillin-resistant *Staphylococcus aureus*
MS	mass spectrometry
MSSA	methicillin-sensitive *Staphylococcus aureus*
MULTIPLEXINs	multiple triple helix domains and interruptions
NA	nonanionic acid
NaCl	sodium chloride
NAD	nicotinamide adenine dinucleotide
NADP	nicotinamide adenine dinucleotide phosphate
NC-IUBMB	Nomenclature Committee of the International Union of Biochemistry and Molecular Biology
Ni	nickel
Ni^{2+}	nickel ion
NIR-TrapG	near-infrared dye (IR-820) linked to TrapG
nm	nanometer
nM	nanomolar
nmol	nanomole
NNI	National Nanotechnological Initiative
NO	nitric oxide
NPs	nanoparticles
NSAIDs	nonsteroidal anti-inflammatory drugs
NsPNP	nanosuspension polymeric nanoparticle
OCV	open circuit voltage
ORF	open reading frame
OTR	oxygen transfer rate
Pb	lead
Pb^{2+}	lead ion
PBR	packed bed reactor
PCR	polymerase chain reaction
PDMS	polydimethylsiloxane
PEG	polyethylene glycol
PET	positron emission tomography
PFRs	plug flow reactors
pH	potential of hydrogen
PHA	polyhydroxyalkanoate
PKS	polyketide synthase
PLA	polylactic acid

PLA2s	phospholipases A2
PLD	phospholipase D
PM	purple membrane
PPII	polyproline II
PQQ	pyrroloquinoline quinone
Pt	platinum
PUF	polyurethane foam
PUFAs	polyunsaturated fatty acids
Q	ubiquinone Q
Rep-PCR	repetitive extragenic palindromic-PCR
RML	*Rhizomucor miehei* lipase
RNA	ribonucleic acid
ROL	*Rhizopus oryzae* lipase
rpm	revolutions per minute
SAMBRs	submerged anaerobic membrane bioreactors
Sb	antimony
Sb^{3+}	antimony (III) ion
SBW	soya bean whey
SCP	single cell protein
SDS-PAGE	sodium dodecyl sulfate-polyacrylamide gel electrophoresis
SGU	sucrose density gradient ultracentrifugation
SKDH	shikimate dehydrogenase
SLN	solid lipid nanoparticles
SMF	submerged fermentation
Sn	tin
SP	spread plate
sp.	species
SRB	sulfate-reducing bacteria
SSF	solid-state fermentation
STRs	stirred tank reactors
TCA	tricarboxylic acid
Te	tellurium
TI^+	titanium ion
TiO_2	titanium dioxide
Tl	thallium
TLL	*Thermomyces lanuginosus* lipase
U	uranium

UDP	uridine diphosphate
UDPG	UDP-α-D-glucose
UDPGA	UDP-glucuronic acid
UGDH	UDP-glucose dehydrogenase
UGT	uridine 5′-diphospho-glucuronosyltransferase
UV	ultraviolet-A
V	vanadium
var.	variety
VLP	virus-like proteins
vvm	vessel volume per minute
W	Watt
w/v	weight by volume
WHO	World Health Organization
Zn	zinc
Zn^{2+}	zinc ion
μg	microgram

PREFACE

Growth in biotechnology rose to a new high after the advent of genetic engineering to expand the potential of microbes, namely bacteria, viruses, yeast cells, and other fungi. These miniature organisms act as biochemical factories. Obtaining the complete genome sequence of these microbes provides crucial information about their biology and their biological capabilities, which can be modified for beneficial purposes. The findings from these microorganisms and improvement of these microorganisms describe the branch of microbial biotechnology.

Knowledge in microbial biotechnology is realizing novel stature through the determination of genomic sequences of hundreds of microorganisms and the invention of new technologies, such as genomics, transcriptomics, and proteomics, to control this deluge of data. These genomic data are now exploited in many applications, ranging from those in medicine to agriculture. This is an epoch of explosive increase of analysis and manipulation of microbial genomes. Microbial biotechnology has led to breakthroughs such as improved vaccines, better disease-diagnostic tools, improved microbial agents for biological control of plant and animal pests, modifications of plant and animal pathogens for reduced virulence, development of new industrial catalysts and fermentation organisms, and new microbial agents for bioremediation of soil and water.

This book provides an exciting interdisciplinary journey through the rapidly changing backdrop of the invention in microbial biotechnology by professional researchers, scientists, and experts in the field. This book has been written in simple and very clear language. The tables and figures included in this book make the content very easy to understand by the readers at any level. This book will serve as good reference material for students, scientists, researchers, and academicians of microbiology and biotechnology.

Chapters 1 and 2 discuss in detail the traditional CFU method for quantitative population analysis of pure and mixed culture cultivations and advanced microbial cultivation strategies in bioreactor designing and industrial operation. The third chapter reviews the diversity of acetic acid producing bacteria in Asia and its applications in various fields. Various microbial secondary metabolites such as microbial enzymes, microbial proteins and its applications in agriculture, food, health, and environment are discussed from Chapters 4 through 9. The microbial potential on biosorption and bioaccumulation of heavy metals, application of molecular tools for the bioremediation of heavy metal pollution, and ecological restoration of the polluted ecosystem through microbial genetic resources are discussed in Chapters 10 and 11. Chapter 12 discusses the advancement of microbial application in nanotechnology for future sustainable development elaborately.

The chapters in this book were prepared and reviewed by many well-known professionals in their respective fields. Editing and publishing of this book received massive support from experts. We are exceedingly grateful to all these people for their valuable suggestions, contributions, and guidance to finish this book successfully.

We also wish to thank Sandy Jones Sickels, Vice President, and Ashish Kumar, Publisher and President, Apple Academic Press, Inc., for quality production and humble communications to publish this book.

We recognize consistent support and constant encouragement from our family members. Finally, we would like to express our gratitude to every individual who directly or indirectly was involved and contributed support in preparing this book.

Jeyabalan Sangeetha, PhD
Devarajan Thangadurai, PhD
Somboon Tanasupawat, PhD
Pradnya Pralhad Kanekar, PhD

CHAPTER 1

COMPARATIVE STUDIES ON CFU DETERMINATION BY PURE AND/OR MIXED CULTURE CULTIVATIONS

SUBIR KUMAR NANDY

Technical University of Denmark, Søltofts Plads, 2800 Kgs, Lyngby, Denmark

1.1 INTRODUCTION

Existing methodologies used in the determination of metabolic active cell count (MACC) are either time consuming, laborious, error-prone or expensive. Studies of any culture require an easy and fast methodology for quantification of MACC of different types of microorganisms. Thus, it is necessary to combine all probable methodology for quantification of MACC in the biological studies. Further, such a platform should be required to compare rapid and precise methodologies for viable cell quantification so that it could be easier to evaluate the dynamic response of any culture. Thus, the overall objective of this chapter is to develop one platform for a quantitative method to resolve MACC in biological cultures. Furthermore, these quantification processes are crucial in the pharmaceutical and food industries for epidemiologic investigation.

The study of the above phenomenon requires quantification of MACC in pure and mixed cultures. Mixed cultures are usually used in several different processes involving bioremediation and food processing. Quantification of MACC of mixed cultures is monotonous. Quantification of more than one culture may be helpful in characterizing contamination in fermentation processes. Existing methodologies used in the determination of MACC are either time consuming or expensive.

Thus, it is necessary to generate a process to quantify MACC in mixed cultures. The method should not only yield a total cell count of all the microorganisms, but also yield the MACC of individual organism present in the culture having more than one microorganism.

1.2   METHODS FOR VIABLE CELL COUNT

1.2.1   PLATE COUNT TECHNIQUES

This normal method is used to estimate cell culturability. Here, agar plates are used to obtain CFU (colony forming units).

1.2.1.1   SPREAD PLATE TECHNIQUE

The total number of bacteria in the experimental flask can be quantified by using spread plate (SP) techniques. In this process sample is diluted several times depending on the growth of organisms and a fraction is transferred to an agar plate. The bacteria are evenly spread over the surface of agar by SP method. The procedure is carried out in a laminar hood to minimize contamination. The colonies are grown overnight in the incubator with specific temperature depending on the type of strain to be quantified. All the colonies are counted, and total bacteria are calculated depending on dilution. Finally, the total number of bacteria is determined based on a serial dilution and plating them serially. The most important advantage of this process is that only metabolically active cells grow on the agar (Kirkpatrick et al., 2001).

The precaution of this method is that all the plates and tips should be autoclaved for less chance of contamination. Sterilized plates are also available now, and that saves the time for sterilization. The bent glass rod is used to spread the culture over the plates. The glass rod should be decontaminated by dipping into ethanol and holding in Bunsen burner to avoid contamination. Finally, the plate should be counted when colonies are in the range of 40–200. The total number of colonies can be calculated by multiplying the average colonies with the dilution factor in terms of CFU/ml. The major advantage of the method is its simplicity, with the disadvantage being that the method is a time-consuming process.

1.2.1.2   STREAK PLATE TECHNIQUE

In this method, the microorganism is streaked on the agar using sterilized loop. This method is used to separate pure colony from a mixed culture or from contamination and cannot be used for determining bacterial count. The colony separation procedure is also depended on the dilution of the broth.

1.2.2   STAINING PROCESSES

1.2.2.1   DYE EXCLUSION METHOD

One of the widely used techniques is the dye exclusion test. This method is based on specific dyes such as methylene blue, trypan blue and acridine orange (Sampson et al., 1924). This procedure works where dead cells are stained, but active metabolic cells remain unstained when exposed to the dye. The advantage of this process is that this assay is easy and fast.

This process is not an accurate method (error-prone technique). Cells are not stained while culture was stored for a long time at low temperatures (Hauschka et al., 1959). Other disadvantages are (a) effective concentration of trypan blue is based on the cell suspension; and (b) maintaining the exact condition of sample preparation was difficult (Pappenheimer et al., 1917).

1.2.2.2   GRAM POSITIVE AND GRAM NEGATIVE STAINING

1.2.2.2.1   Slide Gram Stain

Gram staining method is not suggested for the whole broth quantification procedure. This method is used to separate gram positive and gram negative bacteria (Gram, 1884) very well. The exact mechanism of the gram stain is not clearly understood and due to the errors in staining the procedure is not effective (Beveridge and Davies, 1983; Davies et al., 1983). The disadvantage of this process is that it cannot detect CFU at low concentrations (<1000 live cells).

1.2.2.2.2   Filter Gram Stain

Microfiltration has become a popular procedure for the concentration and enumeration of bacteria. This is a rapid and a sensitive procedure than the normal slide gram stain method to differentiate bacteria utilizing a poly-carbonate membrane filter, crystal violet, iodine, 95% ethanol, and 6% carbol fuchsin that can be completed in 60 to 90 s. This method is useful to detect less number of bacteria (~ 100 cells) but requires modification in the protocol for specific systems (Romero et al., 1988).

1.2.3   MICROSCOPIC PROCEDURE

1.2.3.1   GENERAL APPROACH

A microscopic approach can be used to calculate live cells directly within a very short time, but this approach is not accurate. The culture was stained and fixed on slides to quantify under a microscope (Zhang and Shen, 2006). Few researchers used hemocytometer to quantify the microorganisms.

1.2.3.2   IMMUNOFLUORESCENCE AND EPIFLUORESCENCE ASSAY

These assays are used for labeling antibodies and antigens with fluorescent dyes (Zhang and Shen, 2006). Direct epifluorescence counting was also a suitable method for enumeration of total bacteria in environmental samples (Kepner et al., 1994). Dunne et al. (1987) use of immunofluorescence have been outmoded by the development of recombinant proteins containing fluorescent protein domains, e.g., a green fluorescent protein (GFP).

1.2.3.3   CONFOCAL SCANNING LASER MICROSCOPE ASSAY

Confocal scanning laser microscope (CSLM) is a technique for obtaining high-resolution optical images. Depending on the fluorescence proper-ties of the used dyes, there is a subtle improvement in lateral resolution compared to conventional microscopes. This is a fast method compared to other available techniques to detect the active metabolic cells from heat-killed bacteria than standard SP method. This CSLM method is specially used in the dairy industry to check microbial or contamination load in milk (Pettipher et al., 1980). This method also quantifies bacteria rapidly by image analysis (Caldwell et al., 1992).

1.2.3.4 FLUORESCENCE MICROSCOPIC METHOD

In this process cells inoculated on the chamber, slides were fixed in methanol at $-20°C$ for 10 min for cyclin B1 detection. The cells for microtubule detection were fixed with a 10% formaldehyde solution containing 0.5% Triton X–100, 1 mM $MgCl_2$ at room temperature for 10 min and further fixed with the same solution without Triton X–100 for 5 min. The fixed cells were washed with cold PBS and incubated overnight with a monoclonal anti-cyclin B1 (Pharmingen) or anti-β-tubulin antibody (Sigma) in 1% BSA/PBS at 4°C. They were washed and incubated with a FITC-conjugated anti-mouse IgG antibody (Pharmingen) in 1% BSA/PBS in the dark at room temperature for 1 h. The cells were then washed and stained with 0.2 µg/ml Hoechst 33258 (bis-benzimide trihydrochloride, Sigma) at room temperature for 10 min. The slide was washed and mounted in glycerol. The cells were viewed on a fluorescence microscope using epi-illumination (BX60, Olympus Optics, Tokyo, Japan) and photographed on Tmax film (Kodak, ASA 400). This assay is also a rapid process for direct assessment of cell viability (Kepner and Pratt, 1994). Very few researchers were used LIVE/DEAD BacLight Viability Kit (Molecular Probes Inc., Eugene, OR) to differentiate live/dead bacteria based on plasma membrane permeability (Virta, 1998).

1.2.4 DROP PLATE COUNT METHOD

Metabolically active cells present in a known volume can be determined by the drop plate (DP) method. Though the DP method is not optimized, recent studies have shown some advantages over traditional SP method. Colony counting in DP method is very accurate and faster but depending on various parameters such as dilutions which may vary from laboratory to laboratory or technician to technician. DP method is a quick and easy process where drops can be dispensed onto an agar plate compared to SP technique where the sample volume spread on the agar. Few researchers use 10-fold dilutions, others use two-fold, or few laboratories use a total volume of 0.1 ml on a plate, others plate 0.2 ml, but all these factors use while practicing the DP method.

This assay is used for an active metabolic cell which is faster than SP technique (Herigstad et al., 2001). In this assay, the culture drops on agar in a little volume, and it will absorb quickly. Several researchers want to demonstrate this assay for bacterial counts (Herigstad et al., 2001). In this method, the small drops are separated on agar, and after incubation of the plates, colonies within the drops are counted, and finally, overall bacteria count based on the flask volume. This method is not useful for swarming type bacteria such as *Proteus mirabilis* and *Proteus vulgaris*. This method is a labor-intensive procedure.

1.2.5 INSTRUMENTATION PROCESSES

PCR and Flow Cytometry (FCM) can also yield valuable information regarding viability (Davey and Kell, 1996; Decre et al., 1998). FCM is a means of measuring certain physical and chemical characteristics of cells or particles as they pass in a fluid stream by a beam of laser light. The term "flow cytometry" derives from the measurement (meter) of single cells (cyto) as they flow past a series of detectors. Flow sorting extends FCM by using electrical or mechanical means to divert and collect cells with one or more measured characteristics falling within a range or ranges of values set by the user. The major applications of FCM include the analysis of cell cycle, apoptosis, necrosis, multicolor analysis, cell sorting, functional analysis, and stem cell analysis. Oravcova et al. (2008) show that real-time PCR based method has been used to enumerate *Listeria monocytogenes* in food samples. Callister et al. (2003) demonstrated a laser integrated microarray scanner was used to quantify and compare the biomass of *Burkholderia cepacia* G4 alone in the presence of phenol degrading community of microorganisms. This method has a detection limit up to 10^3–10^4 cells ml^{-1}. Therefore, these methods are accurate but very expensive. These methods are discussed in detail below.

1.2.6 POLYMERASE CHAIN REACTION (PCR)

Polymerase chain reaction (PCR) is the amplification of a nucleic acid target sequence (Saiki et al., 1985). The targeted sequences can be a specific gene, repetitive areas in the sequence or arbitrary sequences. However, when the PCR is based on the amplification of a specific portion of the DNA, the targeted DNA sequence, except for arbitrary PCR, must be known for the synthesis of the oligonucleotides. The PCR-based techniques have also been developed for the screening of genetically modified organisms (Holst-Jensen et al., 2003). Post-PCR detection methods vary from gel electrophoresis, hybridization analysis, and usage of specific nucleic acid probes. In some cases, probes simplify the detection of the PCR product, in a similar way as gel electrophoresis, while in other cases it can further discriminate for only certain bacteria. By using fluorometric or colorimetric labeled probes on PCR products, further detection and specification of species and strains can be performed on membranes or microwells (Mandrell and Wachtel, 1999; O'Connor et al., 2000; O'Sullivan et al., 2000; Grennan et al., 2001).

16S rRNA gene is a favorable PCR amplification target by universal or species-strain specific primers for identification and phylogenic purposes since it is universally distributed among bacteria and it contains enough variations amongst strains and species within the DNA sequence (Weisburg et al., 1991). The availability of whole genome to small subunit ribosomal RNA gene sequences, such as 16S rRNA, data is constantly increasing, and public-domain databases have been established, such as Ribosomal Data Base Project (http://rdp.cme.msu.edu/) and named like National Center for Biotechnology Information Blast Library (http://www.ncbi.nlm.nih.gov/blast/Blast.cgi). These database libraries can be applied for identification of cultured and uncultured microorganisms from environmental, clinical, and food samples by comparing the 16S rRNA gene sequences in the databases to those of the unknown microorganism (Drancourt et al., 2000). However, 16S rRNA gene sequence analyses have shown limited variability within strains of a bacteria species (Woese, 1987; Bottger, 1989; Olsen and Woese, 1993). This is especially evident amongst homogenous groups, such as the cereulide-producing *B. cereus*. For example, it has been shown that the 16S rRNA gene sequences of thirteen cereulide-producing *B. cereus* strains were identical to each other as well as to the 16S rRNA gene sequences of *Bacillus anthracis* strains Ames, Sterne, and NC 08234–02. Therefore, 16S rRNA gene sequence analysis for homology is not always capable to completely identify an unknown organism.

1.2.6.1   VARIATIONS OF PCR

Different variations of PCR have been advanced to fulfill simultaneous detection of more than one bacteria, quantification, and differentiation of viable bacterial cells from samples. Simultaneous bacterial detection can be performed by multiplex PCR, which uses several different primers targeted for specific genes of each bacterial strain (Yaron and Matthews, 2002; Touron et al., 2005). Many multiplex PCR systems have been developed for the differentiation of multiple species belonging to single genera and for differentiation of mixed bacterial pathogens (Settanni and Corsetti, 2007). Conventional PCR is not able to indicate if the bacterial cells are viable or dead and therefore reverse transcriptase PCR (RT-PCR) was established for specifically detecting viable cells. In RT-PCR, reverse transcriptase enzyme that can use messenger RNA as a template for synthesizing single-stranded DNA in the 5′ to 3′ direction (Lazcka et al., 2007; Rodríguez-Lázarro et al., 2007). The technique is sensitive and requires no pre-enrichment steps, which decreases the time of analysis (Deishing and Thompson, 2004). Another advantage of the RT-PCR is the detection of VNC cells that are not detected with culturing. For the quantitative detection of bacteria in a sample, quantitative PCR can be used (Monis and Giglio, 2006). This technology is based on the monitoring of the formation of PCR product simultaneously as the reaction occurs by using fluorescent probes or dyes that are sequence-specific or nonspecific. Molecular beacons are an example of a sequence-specific fluorescent probe which undergoes a fluorogenic conformational change when hybridizing to their target (Tyagi and Kramer, 1996). SYBER Green I and SYBER gold are nonspecific dyes that bind to the double-stranded DNA of the PCR product (Glynn et al., 2006). Only fluorescent probes labeled with different reporter dyes are used to identify multiple amplicons within the same reaction mixture while double-stranded DNA dyes are limited to a single product per reaction (Robertson and Nicholson, 2005).

1.2.6.2   LIMITATIONS OF NUCLEIC ACID AMPLIFICATION METHODS

The acceptance and application of nucleic acid amplification methods in routine detection of foodborne pathogens have been limited due to the standardization and validation of PCR protocols. The protocols need to be synchronized between laboratories so that the PCR results are reliable and reproducible when performed in different locations or times (Malorny et al., 2003). However, this is problematic because many of the aspects that can affect PCR are difficult to control. These are the quality of the DNA template, the environment (humidity, chemical, and microbiological cleanliness, temperature), the equipment, personal practice, and the reaction conditions and the reaction materials (Malorny et al., 2003). In addition, the food itself is a difficult matrix since it can contain substances that affect the PCR reaction. Food can contain substances that can degrade the target nucleic acid sequence or inhibits the enzyme activity in the PCR, which can give false negative results (Glynn et al., 2006). Also, PCR does not confirm the presence of toxins in the food, but only the genetic potential of a microorganism to produce them (de Boer and Beumer, 1999). Although the technique of PCR is simple and quick, the technique still needs improvement for bacterial investigations of different foods and for standardized protocols in public health laboratories.

1.2.7 FUTURE METHODS IN DETECTION AND QUANTIFICATION OF BACTERIA

Constantly more rapid and easier techniques are being developed for simultaneous detection and identification of bacteria from biological experiments. These techniques are based on the similar concepts as the phenotypic and genotypic methods, but they include new instrumentation or set-ups for the analysis of the samples. Although FCM, DNA microarrays and biosensors will be discussed in this chapter, other bacterial detection systems have been developed based on instruments that are traditionally used in the field of chemistry, such as Reflectance Spectroscopy (Rahman et al., 2006), Fourier Transform Raman Spectroscopy (Yang and Irudayaraj, 2003) and Mass Spectrometry (Mandrell and Wachtel, 1999).

1.2.7.1 FLOW CYTOMETRY METHOD

Although FCM was discovered in the late 1960s, its applicability in the field of microbiology has not yet been fully reached. It has been extensively applied to mammalian cells and chromosomes, such as cell cycle analysis and medical diagnostic studies (Steen, 2000; Longobardi, 2001).

However, its potential in microbiology is continuously investigated, and therefore it can still be considered a developing technique in the field of detection and identification of bacteria.

1.2.7.1.1   Instrumentation of FCM

Steen (2000) described FCM simply as a fluorescent microscope with cells flowing through the focus. The instrument composes of four main elements: a light source, fluid lines, and controls (fluidics), an electronic network, and a computer (Longobardi, 2001). Optical lenses shape the laser beam (light source) into a focused light that illuminates the samples one at a time, such as cells. As the cells pass through the illumination spot in the FCM, sheath fluid surrounds them to assure the uniformity of the alignment between the cells and the laser beam. At the analysis point, lenses collect the light signal from the cells by focusing it on photodetectors (photodiodes and photomultipliers) that convert the light signal to an electrical signal. The electrical signal is then converted from analog-to-digital and analyzed accordingly using standard analysis computer software.

In an FCM, there are multiple photodetectors that measure simultaneously different aspects from a single cell (Longobardi, 2001). The forward-angle photodiode is located directly at the analysis point. An obscuration bar is in front of this detector to only allow the light detection that has been bent by passing through the cell. This signal is called forward scatter or forward-angle light scatter which is related to the size or volume and the cell refractive index (Davey and Kell, 1996). To the right angles of the illuminating beam, three or more photodetectors are located that detect any light that is deflected to the side from the analysis point. Since photodetectors measure all colors of light, filters must be in front of them to specify the light which each one measures. One of the photodetectors registers the illuminating light that has been bounced 90° from the surface of the analyzed cell. This signal is called side scatter light which reflects the granularity of a particle (Davey and Kell, 1996). The other photomultipliers are there to detect other colors of light that might be emitted by the cell due to endogenous fluorescent compounds or to staining by fluorescent dyes which allow the study of surface proteins, intracellular proteins and DNA (Nebe-von-Caron et al., 2000; Brehm-Stecher and Johnson, 2004).

1.2.7.1.2 FCM in Microbiology

The size of bacteria is one of the obstacles that have slowed down the utilization of FCM in bacteriological investigations. The diameter of a bacterial cell is around one-tenth of that of a mammalian blood cell resulting in a smaller surface area for staining, and the DNA content of a bacterial cell, such as *E. coli*, is about 10^3 times that of diploid human cell (Longobardi, 2001). Therefore, the bacterial cells require sensitive instruments and bright fluorescent dyes (Steen, 2000). The first microbiologists studied the nucleic acid and protein amounts in bacterial cells during different growth stages by FCM (Bailey et al., 1977; Paau et al., 1977; Steen, 2000). FCM studies in bacteriology have become more common by improving the sensitivity of FCM. FCM has developed into an intriguing tool for microbiology research due to its capability to simultaneously measure multiple aspects of a homogenous or heterogeneous sample, such as cell detection, cell counting, and cellular structure analysis (Brehm-Stecher and Johnson, 2004).

Microbial identification by FCM is also used in phenotypic and genotypic methods described earlier in the chapter. The size and granularity of the cells are indicated by the light scatter which is used in differentiating cells, for example, yeast from bacterial cells (Malacrinò et al., 2001). The autofluorescent properties of cells, such as the presence of photosynthetic pigments, have been used in identifying and classifying algae (Troussellier et al., 1993). FCM has been utilized in serological discrimination of bacteria, fungi, viruses, and parasites (Álvarez-Barrientos et al., 2000).

In molecular biology, FCM can be used for DNA fragment analysis of bacteria (Huang et al., 1999) and viruses (Ferris et al., 2005), and detection of clones and mutants by reporters encoded by genes, such as *LUX* and *GFP* (Huang et al., 1996; Link et al., 2007). In food microbiology (Ueckert et al., 1995), FCM is advantageous since it is used in differentiating between dead, viable, and VNC cells by using fluorescent dyes (Breeuwer and Abee, 2000) that indicate the membrane integrity, membrane potential, respiration, intracellular pH, and enzyme activity of cells, directly from food, such as milk, juice, wine, vegetable products, and ground beef (Laplace-Builhé et al., 1993; Gunasekera et al., 2000; Malacrinò et al., 2001; Yamaguchi et al., 2003).

1.2.7.2 DNA MICROARRAYS

DNA microarrays consist of a solid surface (glass, silicon, nylon substrates) to which many probes, DNA fragments or oligonucleotides, are immobilized that will hybridize to fluorescently labeled DNA (target) from the sample (Call, 2003). The target can be genomic DNA isolated from the sample or an amplified PCR product.

Genomic microarrays and oligonucleotides arrays are two different types of DNA microarrays. In genomic DNA microarrays, the probes are complete genes or their fragments from a strain of a microorganism, while in oligonucleotides microarrays the target DNA hybridizes to 18 to 70 nucleotides long oligos. Although both types of microarrays can be used in the detection of pathogens, commonly oligonucleotides microarrays are used in the detection of either genomic DNA directly, or PCR amplified a portion of the genomic DNA, such as rRNA genes or virulence genes (Kostrzynska and Bachand, 2006).

Microarrays have been developed for identification of food-borne bacterial pathogens belonging to *Bacillus* spp., *C. jejuni*, *E. coli*, *L. mono-cytogenes*, *S. enterica* (Call et al., 2003; Chiang et al., 2006; Garaizar et al., 2006; Sergeev et al., 2006; Eom et al., 2007) and for discrimination from multiple different pathogens and their virulence factors (Sergeev et al., 2006; Wang et al., 2007) in case of food outbreaks and biological warfare (Sergeev et al., 2006; Wang et al., 2007). Since multiple genetic properties can be analyzed and their flexibility in developing arrays that are specific for certain analyses, DNA microarrays make an excellent tool in epidemiological studies and food safety control. However, improvements in microarrays are still required for them to become more economical and practical for public health laboratories and food industries (Garaizar et al., 2006; Kostrzynska and Bachand, 2006).

1.2.7.3 BIOSENSORS

According to Lazcka et al. (2007), in general, biosensor technology is the fastest growing technology compared to PCR, immunology, culture methods, and gel electrophoresis. A biosensor consists of biological material, biologically derived material, or a biomimic that is associated or integrated to a transducer. This transducer can be physiochemical or biological that converts the detected change or presence of various analyte in the analyzed sample into a measurable signal (Lazcka et al., 2007). These devices can be used to detect analytes, such as carcinogens, pollutants, drugs, pesticides, and pathogens from water, waste, soil, and foods (Arora et al., 2006). Various detection methods based on biosensors have been applied in the detection of food pathogens.

These sensors have been based on DNA, immunology, and phage display peptides (Table 1.1). Transducers are based on optical, acoustical, and electrochemical signal detection (Lazcka et al., 2007). Optical biosensors measure changes in fluorescence, luminescence, absorbance, or refractive index. Fluorescence techniques are based on direct measurement of fluorescent indicator compounds, or in the case of fluorescence resonance energy transfer biosensors (Baeumner, 2003), a donor fluorophore donates energy to an acceptor fluorophore, which then emits light (Ko and Grant, 2003). Surface plasmon resonance is based on detecting changes in refractive index caused by structural alterations of a thin film metal surface, such as gold (Cooper, 2003). Acoustic sensors measure changes in resonance frequency, due to a mass change of a bio-molecular surface, such as piezoelectric crystals, e.g., quartz (O'Sullivan and Guilbault, 1999). Another sensor type is electrochemical biosensors designed to measure changes as current and potential at the sensor/sample matrix interface (Lazcka et al., 2007). These sensors are classified with respect to what they measure: amperometric (current), potentiometric (potential), and impedimetric (impedance).

TABLE 1.1 Different Quantification Processes are Summarized with Their System Name, Approaches, Advantages, and Their Disadvantages

Systems	Approaches	Advantages	Disadvantages
Plate count	Spread, Streak	Simplicity	Time-consuming, labor-intensive
Staining	Dye exclusion, Slide gram stain, Filtergram stain	Easy and fast	Inefficient, expensive
Microscopic	Basic, Immunofluorescence, Epifluorescence, Confocal scanning laser microscope, Fluorescence	Rapid process	Instrument sensitive
DP count	-	Useful for non-swarming bacteria	Not useful for swarming bacteria
Instrumentation	-	Efficient process	Instrument sensitive
ATP, rRNA	Specific enzymatic activity	Efficient processes	Difficult

Whole cells and higher organisms (plants, algae, nematodes, and animal tissues) have been used as detectors in biosensors (Baeumner, 2003). In this chapter, the emphasis is placed on bacterial biosensors. Viable microbes produce metabolites, such as carbon dioxide, ammonia, acids, or they are bioluminescent as exemplified by *Vibrio fischeri*, which can be used to monitor viability (D'Souza, 2001). Many microbial biosensors are based on light emission from luminescent or fluorescent bacteria that are genetically engineered to express fluorescent or luminescent proteins, such as GFP or luciferase protein (D'Souza, 2001; Baeumner, 2003). So far, microbial biosensors and bioassays have been applied more prevalently in the detection of food additives and food contaminants than in direct-monitoring of food pathogens themselves.

1.2.8 OTHER METHODS

Other methods such as ATP (Venkateswaran et al., 2003) and rRNA measurements (Karner and Furhman, 1997) assays are also reported. Another assay for viability estimation is based on specific enzymatic activity prevalent in the cell. These methods are usually species or genus specific (Rompre et al., 2002). Another process to identify live and dead cells is 4′,6-diamidino–2-phenylindole using DAPI staining methodology (Auty et al., 2001). Schmidt et al. (2007) has also shown quantification of pure cultures from more than one culture by using terminal-restriction fragment length polymorphism. Therefore, problems in differentiating live and dead cells in a microbial culture persist with limitations including lack of sensitivity, expense, intensive labor and lot of consuming time. The main differences between live and dead cells are well represented in the next paragraph.

1.3 METHYLENE BLUE DYE REDUCTION TEST (MBRT)

Evaluation of live cells is one of the most challenged parameters in biological experiments. Several methods are used for live cell (metabolic cell activity) test. But all processes used to determine CFU are tedious, expensive, and higher error method (Auty et al., 2001; Rompre et al., 2002). Hence, there is required to search for a quick and less error-prone method to quantify active metabolic cells. In this chapter, one easy, quick, inexpensive, and less error-prone method is introduced to demonstrate live cells within 3 minutes. The main aim of this chapter is to describe this unique method for detection of CFU based on aerobic microorganisms, such as *Escherichia coli* and *Bacillus subtilis*. Only live cells those are metabolic active in the growth phase demonstrate methylene blue dye reduction test (MBRT). Therefore, this method also distinguishes between exponential and death phase of the microorganism by using MBRT slope with a highly accurate measurement result. In this chapter, the ability of the MBRT method and its importance for basic physiology is demonstrated. This study indicated that MBRT has the potential to be a global method for another microorganism like yeast and fungi. Therefore, this method will soon be extended to the application in industry. MBRT was developed to quantify MACC of different microorganisms. MBRT was also described to check the microbial contamination in the industrial reactor to save batch time.

In this chapter, MBRT is used to detect pure culture MACC from their mixtures and also quantifies the external effect of antibiotic and exhibition process on several microorganisms. Therefore, extended the current methodology to determine pure culture from more than one culture and it fits very well if the start cell concentrations are known (Nandy and Venkatesh, 2008, 2010).

1.4   CONCLUSION AND FUTURE PERSPECTIVES

Biotechnology and biochemical work with industrial microorganisms usually has a close connection with their industrial applications, because basic biological phenomena are generally best studied in organisms that very well characterized in certain respects. Contamination is one of the 'old problems' that can affect in growing of different industrial microorganisms under complex medium. During cultivations, metabolically active cell numbers have an important effect on the basic and advanced physiology in lab-scale or in industrial work. To solve these difficulties, several kits or methods are used for live and dead cell quantification and check microbial contamination. Mostly, MBRT used to check the contamination in dairy or cheese industry and therefore, the focus of MBRT study to solve the contamination problem and study MBRT process for microbial metabolic active cells quantification to describe some ways of dealing with a solution of many difficulties. Adding the dye methylene blue color in the milk will disappear quickly or slowly based on the microbial load presence and always tested by MBRT. Removal of the oxygen from the milk and the formation of the reducing agents because of metabolic activity of microorganism cause the blue color to disappear. Therefore, oxygen removal from the milk happens because of the active microbe's present. If the microbial load is higher in the milk, blue color of MB vanishes or fades quickly and vice versa. Many species will quickly use up all the oxygen. Simultaneously, the total metabolic reactions happening at the cell surface of the microorganism be an important significance for this incidence.

KEYWORDS

- colony forming units
- flow cytometry
- fluorescence
- green fluorescent protein
- immunofluorescence assay
- metabolic active cell count
- methylene blue dye reduction test
- mixed culture
- polymerase chain reaction
- pure culture
- staining

REFERENCES

Álvarez-Barrientos, A., Arroyo, J., Cantón, R., Nombela, C., & Sánchez-Pérez, M., (2000). Applications of flow cytometry to clinical microbiology. *Clin. Microbiol. Rev., 13,* 167–195.

Arora, K., Chand, S., & Malhotra, B. D., (2006). Recent developments in bio-molecular electronics techniques for food pathogens. *Anal. Chim. Acta., 568,* 259–274.

Auty, M. A. E., Gardiner, G. E., McBrearty, S. J., O'Sullivan, E. O., Mulvihill, D. M., Collins, J. K., et al., (2001). Direct *in situ* viability assessment of bacteria in probiotic dairy products using viability staining in conjunction with confocal scanning laser microscopy. *Appl. Env. Microbiol., 67*, 420–425.

Baeumner, A. J., (2003). Biosensors for environmental pollutants and food contaminants. *Anal. Bioanal. Chem., 377*, 434–445.

Bailey, J. E., Fazel-Madjlessi, J., McQuitty, D. N., Lee, L. Y., Allred, J. C., & Oro, J. A., (1977). Characterization of bacterial growth by means of flow micro-fluorometry. *Science, 198*, 1175–1176.

Beveridge, T. J., & Davies, J. A., (1983). Cellular responses of *Bacillus subtilis* and *Escherichia coli* to the Gram stain. *J. Bacteriol., 156*, 846–858.

Bottger, E. C., (1989). Rapid determination of bacterial ribosomal RNA sequences by direct sequencing of enzymatically amplified DNA. *FEMS Microbiol. Lett., 53*, 171–176.

Breeuwer, P., & Abee, T., (2000). Assessment of viability of microorganisms employing fluorescence techniques. *Int. J. Food Microbiol., 55*, 193–200.

Brehm-Stecher, B. F., & Johnson, E. A., (2004). Single-cell microbiology: Tools, technologies, and applications. *Microbiol. Mol. Biol. Rev., 68*, 538–559.

Caldwell, D. E., Korber, D. R., & Lawrence, J. R., (1992). Confocal laser microscopy and digital image analysis in microbial ecology. *Adv. Microbiol. Ecol., 12*, 1–67.

Call, D. R., Borucki, M. K., & Loge, F. J., (2003). Detection of bacterial pathogens in environmental samples using DNA microarrays. *J. Microbiol. Meth., 53*, 235–243.

Callaway, N. L., Riha, P. D., Bruchey, A. K., Munshi, Z., & Ganzalez-Lima, F., (2004). Methylene blue improves brain oxidative metabolism and memory retention in rats. *Pharmacol. Biochem. Behav., 77*, 175–181.

Callister, S. J., Ayala-del-Rio, H. L., & Asma, S. A., (2003). Quantification of a single population in a mixed microbial community using a laser integrated microarray scanner. *J. Env. Eng. Sci., 2*, 247–253.

Chiang, Y. C., Yang, C. Y., Li, C., Ho, Y. C., Lin, C. K., & Tsen, H. Y., (2006). Identification of *Bacillus* spp., *Escherichia coli*, *Salmonella* spp., *Staphylococcus* spp., and *Vibrio* spp. with 16S ribosomal DNA-based oligonucleotide array hybridization. *Int. J. Food Microbiol., 107*, 131–137.

Cooper, M. A., (2003). Label-free screening of bio-molecular interactions. *Anal. Bioanal. Chem., 377*, 834–842.

D'Souza, S. F., (2001). Microbial biosensors. *Biosens. Bioelectron, 16*, 337–353.

Davey, H. M., & Kell, D. B., (1996). Flow cytometry and cell sorting of heterogeneous microbial populations: The importance of single-cell analyses. *Microbiol. Rev., 60*, 641–696.

Davies, J. A., Anderson, G. K., Beveridge, T. J., & Clark, H. C., (1983). Chemical mechanism of the Gram stain and synthesis of a new electron-opaque marker for electron microscopy which replaces the iodine mordant of the stain. *J. Bacteriol., 156*, 837–845.

De Boer, E., & Beumer, R. R., (1999). Methodology for detection and typing of foodborne microorganisms. *Int. J. Food Microbiol., 50*, 119–130.

Decre, D., Shen, J., Vesey, G., Bell, P., Bissinger, P., & Veal, D., (1998). Flow cytometry and cell sorting for yeast viability assessment and cell selection. *Yeast, 14*, 147–160.

Deishing, A. K., & Thompson, M., (2004). Strategies for the detection of *Escherichia coli* O157: H7 in foods. *J. Appl. Microbiol., 96*, 419–429.

Drancourt, M., Bollet, C., Carlioz, A., Martelin, R., Gayral, J. P., & Raoult, D., (2000). 16S ribosomal DNA sequence analysis of a large collection of environmental and clinical unidentifiable bacterial isolates. *J. Clin. Microbiol., 38*, 3623–3630.

Dunne, W. M., Sheth, N. K., & Franson, T. R., (1987). Quantitative epifluorescence assay of adherence of coagulase-negative staphylococci. *J. Clin. Microbial, 25*, 741–743.

Eom, H. S., Hwang, B. H., Kim, D. H., Lee, I. B., Kim, Y. H., & Cha, H. J., (2007). Multiple detection of food-borne pathogenic bacteria using a novel 16S rDNA-based oligonucleotide signature chip. *Biosens. Bioelectron, 22*, 845–853.

Ferris, M. M., Yoshida, T. M., Marrone, B. L., & Keller, R. A., (2005). Fingerprinting of single viral genomes. *Anal. Biochem., 337*, 278–288.

Garaizer, J., Rementeria, A., & Porwollik, S., (2006). DNA microarray technology: A new tool for the epidemiological typing of bacterial pathogens? *FEMS Immunol. Med. Mic., 47*, 178–189.

Glynn, B., Lahiff, S., Wernecke, M., Barry, T., Smith, T. J., & Maher, M., (2006). Current and emerging molecular diagnostic technologies applicable to bacterial food safety. *Int. J. Dairy Technol., 59*, 126–139.

Gram, C., (1884). Uber die isolierte Farbung der Schizomycenten in Schmitt-und Trockenpraparaten. *Fortschr. Med., 2*, 185–189.

Grennan, B., O'Sullivan, N. A., Fallon, R., Carroll, C., Smith, T., Glennon, M., & Maher, M., (2001). PCR-ELISAs for the detection of *Campylobacter jejuni* and *Campylobacter coli* in poultry samples. *Biotechniques, 30*, 602–610.

Gunasekera, T. S., Attfield, P. V., & Veal, D. A., (2000). A flow cytometry method for rapid detection and enumeration of total bacteria in milk. *Appl. Env. Microbiol., 66*, 1228–1232.

Hauschka, T. S., Mitchell, J. T., & Niedergruem, D. J., (1959). A reliable frozen tissue bank: Viability and stability of 82 neoplastic and normal cell types after prolonged storage at −78°C. *Cancer Res., 19*, 643–653.

Herigstad, B., Hamilton, M., & Heersink, J., (2001). How to optimize the drop plate method for enumerating bacteria. *J. Microbiol. Meth., 44*, 121–129.

Holst-Jensen, A., Rønning, V., Løvseth, A., & Berdal, K. G., (2003). PCR technology for screening and quantification of genetically modified organisms (GMOs). *Anal. Bioanal. Chem., 375*, 985–993.

Huang, Z., Jett, J. H., & Keller, R. A., (1999). Bacteria genome fingerprinting by flow cytometry. *Cytometry, 35*, 169–175.

Huang, Z., Perry, J. T., O'Quinn, B., Longmire, J. L., Brown, N. C., Jett, J. H., & Keller, R. A., (1996). Large DNA fragment sizing by flow cytometry: Application to the characterization of P1 artificial chromosome (PAC) clones. *Nucleic Acids Res., 24*, 4202–4209.

Karner, M., & Fuhrman, J. A., (1997). Determination of active marine bacterioplankton: A comparison of universal 16S rRNA probes, autoradiography, and nucleoid staining. *Appl. Env. Microbiol., 63*, 1208–1213.

Kepner, R. J., & Pratt, J. R., (1994). Use of fluorochromes for direct enumeration of total bacteria in environmental samples: Past and present. *Microbiol. Rev., 58*, 603–615.

Kirkpatrick, C., Maurer, L. M., Oyelakin, N. E., Yoncheva, Y. N., Maurer, R., & Slonczewski, J. L., (2001). Acetate and formate stress: Opposite responses in the proteome of *Escherichia coli*. *J. Bacteriol., 183*, 6466–6477.

Ko, S., & Grant, S. A., (2003). Development of a novel FRET method for detection of *Listeria* or *Salmonella*. *Sensor Actuator B., 96*, 372–378.

Kostrzynska, M., & Bachand, A., (2006). Application of DNA microarray technology for detection, identification, and characterization of foodborne pathogens. *Can. J. Microbiol., 52*, 1–8.

Laplace-Builhé, C., Hahne, K., Hunger, W., Tirilly, Y., & Drocourt, J. L., (1993). Application of flow cytometry to rapid microbial analysis in food and drinks industries. *Biol. Cell, 78*, 123–128.

Lazcka, O., Del Campo, F. J., & Munoz, F. X., (2007). Pathogen detection: A perspective of traditional methods and biosensors. *Biosens. Bioelectron, 22*, 1205–1217.

Link, A. J., Jeong, K. J., & Georgiou, G., (2007). Beyond toothpicks: New methods for isolating mutant bacteria. *Nature, 5*, 680–688.

Longobardi, G. A., (2001). *Flow Cytometry: First Principles.* Wiley-Liss, Inc., New York, USA, (p. 296).

Malacrinò, P., Zapparoli, G., Torriani, S., & Dellaglio, F., (2001). Rapid detection of viable yeasts and bacteria in wine by flow cytometry. *J. Microbiol. Meth., 45*, 127–134.

Malorny, B., Tassios, P. T., Rådström, P., Cook, N., Wagner, M., & Hoorfar, J., (2003). Standardization of diagnostic PCR for the detection of foodborne pathogens. *Int. J. Food Microbiol., 83*, 39–48.

Mandrell, R. E., & Wachtel, M. R., (1999). Novel detection techniques for human pathogens that contaminate poultry. *Curr. Opin. Biotech., 10*, 273–278.

Monis, P. T., & Giglio, S., (2006). Nucleic acid amplification-based technique for pathogen detection and identification. *Infect. Genet. Evol., 6*, 2–12.

Nandy, S. K., & Venkatesh, K. V., (2008). Effect of carbon and nitrogen on the cannibalistic behavior of *Bacillus subtilis. Appl. Biochem. Biotech., 151*(2 & 3), 424–432.

Nandy, S. K., & Venkatesh, K. V., (2010). Application of methylene blue dye reduction test (MBRT) to determine growth and death rates of microorganisms. *Afr. J. Microbiol. Res., 4*(2), 61–70.

Nebe-von-Caron, G., Stephens, P. J., Hewitt, C. J., Powell, J. R., & Badley, R. A., (2000). Analysis of bacterial function by multi-color fluorescence flow cytometry and single cell sorting. *J. Microbiol. Meth., 42*, 97–114.

O'Connor, L., Joy, J., Kane, M., Smith, T., & Maher, M., (2000). Rapid polymerase chain reaction/DNA probe membrane-based assay for detection of *Listeria* and *Listeria monocytogenes* in food. *J. Food Protect, 63*, 337–342.

O'Sullivan, C. K., & Guilbault, G. G., (1999). Commercial quartz crystal microbalances – theory and applications. *Biosens. Bioelectron, 14*, 663–670.

O'Sullivan, N. A., Fallon, R., Carroll, C., Smith, T., & Maher, M., (2000). Detection and differentiation of *Campylobacter jejuni* and *Campylobacter coli* in broiler chicken samples using a PCR/DNA probe membrane-based colorimetric detection assay. *Mol. Cell Probes, 14*, 7–16.

Olsen, G., & Woese, C. R., (1993). Ribosomal RNA: A key to phylogeny. *FASEB J., 7*, 113–123.

Oravcova, K., Trncíkova, T., Kuchta, T., & Kaclíkova, E., (2008). Limitation in the detection of *Listeria monocytogenes* in food in the presence of competing *Listeria innocua. J. Appl. Microbiol., 104*(2), 429–437.

Paau, A. S., Cowles, J. R., & Oro, J., (1977). Flow-microfluorometric analysis of *Escherichia coli, Rhizobium meliloti*, and *Rhizobium japonicum* at different stages of the growth cycle. *Can. J. Microbiol., 23*, 1165–1169.

Pappenheimer, A. M., (1917). Experimental studies upon lymphocytes. I. The reactions of lymphocytes under various experimental conditions. *J. Exp. Med., 25*, 633–650.

Pettipher, G. L., Mansell, R., McKinnon, C. H., & Cousins, C., (1980). Rapid membrane filtration epifluorescent microscopy technique for direct enumeration of bacteria in raw milk. *Appl. Env. Microbiol., 39,* 423–429.

Rahman, S., Lipert, R. J., & Porter, M. D., (2006). Rapid screening of pathogenic bacteria using solid phase concentration and diffuse reflectance spectroscopy. *Anal. Chim. Acta., 569,* 83–90.

Robertson, B. H., & Nicholson, J. K. A., (2005). New microbiological tools for public health and their implications. *Annu. Rev. Publ. Health, 26,* 281–302.

Rodríguez-Lázaro, D., Lombard, B., Smith, H., Rzezutka, A., D'Agastino, M., Helmuth, R., et al., (2007). Trends in analytical methodology in food safety and quality: Monitoring microorganisms and genetically modified organisms. *Trends Food Sci. Tech., 18,* 306–319.

Romero, S., Schell, R. F., & Pennell, D. R., (1988). Rapid method for the differentiation of gram-positive and gram-negative bacteria on membrane filters. *J. Clin. Microbiol., 26,* 1378–1382.

Rompre, A., Servais, P., Baudart, J., De-Roubin, M., & Laurent, P., (2002). Detection and enumeration of coliforms in drinking water: Current methods and emerging approaches. *J. Microbiol. Meth., 49,* 31–54.

Saiki, R. K., Scharf, S., Faloona, F., Mullis, K. B., Horn, G. T., Erlich, H. A., & Arnheim, N., (1985). Enzymatic amplification of β-globin genomic sequences and restriction site analysis for diagnosis of sickle cell anemia. *Science, 230,* 1350–1354.

Sampson, J. J., (1924). Determination of the resistance of leukocytes. *Arch. Int. Med., 34,* 490–502.

Schmidt, J. K., Konig, B., & Reichl, U., (2007). Characterization of a three bacteria mixed culture in a Chemostat: Evaluation and application of a quantitative Terminal-Restriction Fragment Length Polymorphism (T-RFLP) analysis for absolute and species-specific cell enumeration. *Biotechnol. Bioeng., 96,* 738–756.

Sergeev, N., Distler, M., Courtney, S., Al-Khadi, S. F., Volokhov, D., Chizhikov, V., et al., (2006). Microarray analysis of *Bacillus cereus* group virulence factors. *J. Microbiol. Meth., 65,* 488–502.

Settanni, L., & Corsetti, A., (2007). The use of multiplex PCR to detect and differentiate food and beverage-associated microorganisms: A review. *J. Microbiol. Meth., 69,* 1–22.

Steen, H. B., (2000). Flow cytometry of bacteria: Glimpses from the past with a view to the future. *J. Microbiol. Meth., 42,* 65–74.

Touron, A., Berthe, T., Pawlak, B., & Petit, F., (2005). Detection of *Salmonella* in environmental water and sediment by a nested-multiplex polymerase chain reaction assay. *Res. Microbiol., 156,* 541–553.

Trousseillier, M., Courties, C., & Vaguer, A., (1993). Recent applications of flow cytometry in aquatic microbial ecology. *Biol. Cell, 78,* 111–121.

Tyagi, S., & Kramer, F. R., (1996). Molecular beacons: Probes that fluoresces upon hybridization. *Nat. Biotechnol., 14,* 303–308.

Ueckert, J., Breeuwer, P., Abee, T., Stephens, P., Nebe Von Caron, G., & Ter Steeg, P. F., (1995). Flow cytometric applications in physiological study and detection of foodborne microorganisms. *Int. J. Food Microbiol., 28,* 317–326.

Venkateswaran, K., Hattori, N., La Duc, M. T., & Kern, R., (2003). ATP as a biomarker of viable microorganisms in clean-room facilities. *J. Microbiol. Meth., 52,* 367–377.

Virta, M., Lineri, S., Kankaanpaa, P., Karp, M., Peltonen, K., Nuutila, J., & Lilius, E. M., (1998). Determination of complement-mediated killing of bacteria by viability staining and bioluminescence. *Appl. Environ. Microb., 64*, 515–519.

Wang, X. W., Zhang, L., Jin, L. Q., Jin, M., Shen, Z. Q., An, S., et al., (2007). Development and application of an oligonucleotide microarray for the detection of foodborne bacterial pathogens. *Appl. Microbiol. Biotechnol., 76*, 225–233.

Weisburg, W. G., Barns, S. M., Pellettier, D. A., & Lane, D. J., (1991). 16S ribosomal DNA amplification for phylogenetic studies. *J. Bacteriol., 173*, 697–703.

Woese, C. R., (1987). Bacterial evolution. *Microbiol. Rev., 51*, 221–271.

Yamaguchi, N., Sasada, M., Yamanaka, M., & Nasu, M., (2003). Rapid detection of respiring *Escherichia coli* O157: H7 in apple juice, milk, ground beef by flow cytometry. *Cytom. Part A., 54A*, 27–35.

Yang, H., & Irudayaraj, J., (2003). Rapid detection of foodborne microorganisms on food surface using Fourier Transform Raman Spectroscopy. *J. Mol. Struct., 646*, 35–43.

Yaron, S., & Matthews, K. R., (2002). A reverse transcriptase-polymerase chain reaction assay for detection of viable *Escherichia coli* O157: H7: Investigation of specific target genes. *J. Appl. Microbiol., 92*, 633–640.

Zhang, Y., & Shen, J., (2006). Effect of temperature and iron concentration on the growth and hydrogen production of mixed bacteria. *Int. J. Hydrogen Energ., 31*, 441–446.

CHAPTER 2

CULTIVATION STRATEGIES WITH SPECIAL REFERENCE TO BIOREACTOR DESIGN AND OPERATION FOR INDUSTRIAL PRODUCTION IN BIOTECHNOLOGY

ANUPREET KAUR and JASPREET KAUR

Department of Biotechnology, University Institute of Engineering and Technology, Panjab University, Chandigarh–160014, India

2.1 INTRODUCTION

Cultivation of microbes is a biological process of growing organisms under controlled conditions and defined media for various productions. Commercial fermentation products are controlled by operational parameters such as pH, temperature, rates of aeration and agitation and medium components (Wang et al., 2005). Fermentation requires a special vessel called bioreactor, made up of steel or glass, designed to support the growth of microorganisms such as fungi, bacteria, algae, plants, cells or tissues for various productions, via adequate mixing, contacting, mass transfer and heat transfer conditions towards controlled environment as well as protection from the contamination. The cost incurred in any fermentation is a function of raw material and the bioreactor operation itself and is based on the process, location of the plant, labor, and energy costs. Hence, it becomes essential to understand that how significant is the role of reactor engineering in improving the overall process performance.

The bioreactor can be built to cater to numerous objectives such as mass production of biomass, metabolites, enzymes, recombinant products as well as immobilized enzymes using silica gel and many other supports

(Urban et al., 2012). It has also been used for the hydrolysis of oils (Chen et al., 2010, Anuar et al., 2011; Liang et al., 2012). A bioreactor design should be capable of carrying out enzymatic reactions, product separation, and recovery of the enzyme. A fixed bed biofilm system for making clean drinking water has been constructed for scrubbing of perchlorate from water (Li et al., 2012). Design of pilot-scale membrane bioreactors to isolate hemoglobin from bovine erythrocytes is an interesting example (Stojanovic et al., 2010). Production of mesenchymal stem cells has been developed in an open, agitated, and stirred bioreactor via microcarriers which are not only costly but impractical for a small number of patients. To circumvent these problems, a closed bioreactor has been proposed on the single-use basis for the isolation and cultivation of human placental mesenchymal stem cells (Timmins et al., 2012). Stem cell cultivation in bioreactors has also been reviewed by Rodrigues et al. (2011). Reactor operation is a function of many factors including the type of product and microorganism, reaction conditions leading to the substrate or product inhibition. The current chapter explains the different types of cultivation techniques for bioprocesses. Broadly, the bioreactor operation can be classified as batch, continuous, fed-batch, and modified continuous systems with complete or partial cell-recycle. These techniques will be briefly described.

2.2   BIOREACTOR OPERATION

Bioreactor operation and its configuration depending on the producer organism, the operating conditions for product formation and scale of the production. The design also takes into account for capital investment and cost-effectiveness. It can be generalized that: (i) large volume and low-value products requires simple fermenter and do not ask for stringent aseptic condition; (ii) high value and low volume products such as antibiotics and antibodies need a stringent and aseptic operation. The size of the bioreactors can vary over several orders of dimension: the shake flask (100–1000 ml), the laboratory scale (1–100 L), pilot scale (0.3–10 m³) to production scale (2–500 m³).

Before finalizing the reactor design, one must consider the cost determining factors or cost structure of the process. This is influenced by factors such as raw materials, research, and development, bioreactor operation and downstream processing. Maximizing substrate consumption to enhance productivity in the reactor becomes crucial when the raw material is expensive which contributes significantly to the overall cost. For instance, if the running cost of bioprocess is high, the reactor operation should be optimized for maximizing the product concentration. An ideal fermenter should be complete in the following capabilities: (1) It should be capable of providing all facilities for the growth of a diverse range of organisms and also capable of producing a number of fermentation products; (2) It should provide proper agitation and aeration to fulfill the metabolic needs of the fermentation culture; (3) It should allow the feeding of nutrients and other essential supplements during process; (4) It should have pH control throughout the processing; and (5) It should be free of contamination and it should not allow any unwanted microorganisms to enter into the system.

Achieving high volumetric rates at large scale, maximum catalyst activity at the highest concentration should be achieved by the reactor operation. Product formation needs the procurement of huge quantities of enzyme.

The biotechnology productions highly depend on Continuous Stirred Tank Reactors (CSTRs) because it allows the introduction of required nutrients and sufficient oxygen into media for cells to survive and grow. CSTRs are standard bioreactors and used in most of the biological process and preferred because of their well-mixing state. Designing or scale up for such system is a function of operating variables such as external power, mixing time, mass transfer rate and impeller tip speed. It is crucial to keep these variables constant in the scaling-up process. Results from research work or practice emphasize on a lot of improvements/developments needed in reactor designing for complex systems such as cultivation of animal cells and genetically modified/engineered microorganisms. With these improvements in fermenter design, a number of industries for the production of biofuels, biopharmaceuticals, and recombinant products will be able to increase production and reduce the costs, which benefit the economy.

2.3 CLASSIFICATION OF REACTORS

Reactors can be classified into the following types as per the number of phases co-existing. (i) Homogeneous: Single phase either gas or liquid exists in the reactor. Further, the reactor type can be plug, CSTR, or batch as per its hydrodynamic flow properties. Here, CSTR and batch are well CSTRs. (ii) Heterogeneous: Reactants or catalyst coexist as two distinct phases in a reaction system and can also be called as multi-phase reactors.

This category may be classified into the following subcategories: (a) Catalytic reactors: gas or liquid phase is in contact with a catalyst which is solid, but could be another liquid phase. Examples of this category include the catalytic packed-bed reactor and the three-phase trickle-bed reactor for carrying catalytic gas-liquid reaction. (b) Non-catalytic reactors: liquid-liquid or gas-liquid reactions can be carried in different contact vessels such as the gas-liquid continuously stirred tank reactor (Figure 2.1).

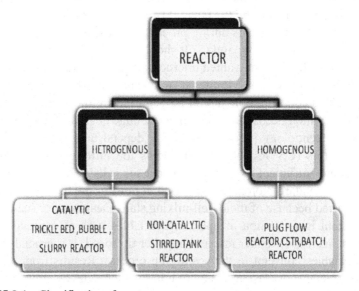

FIGURE 2.1   Classification of reactors.

2.3.1 HOMOGENOUS BIOREACTORS

An ideal homogenous bioreactor operation (stirred tank) can be classified as a batch, fed-batch, and continuous operation. Their characteristics are described as given in the following subsections.

2.3.1.1 BATCH OPERATION

Majority of the fine chemical and biotech industry productions are carried as a pure batch operation. During the batch operation, a change in substrate and product concentration with time is observed. Emptying, cleaning, filling, are the non-productive periods during batch operation. A batch operation can be described with following attributes: (1) unsteady-state conditions, (2) discontinuous production, (3) non-productive time period during cleaning and filling operations, (4) flexible, and, (5) reactors can be used for multiple purposes. This is advantageous for the fine chemical industry where multiple products are produced in one plant.

Kinetics of cell-growth and kefiran production in batch culture of *L. kefiranofaciens* has been investigated in a 16 L bioreactor achieving the 2.76 g L^{-1} of maximal cell mass with 1.91 g L^{-1} of kefiran production (Dialin et al., 2016). This has been established with the help of optimized media and cultivation conditions as has been used in shake flask cultivations. The enhanced cell mass and kefiran production are attributed to improved cultivation conditions in terms of improved aeration and agitation. Increase in cell growth and productivities have been reported for the production of several primary and secondary metabolites when scaled up from shake flask to bioreactor cultivation.

Oxygen supply being a critical factor controlling growth and product formation in aerobic bioprocesses which can also be used as a parameter for bioprocess scale up. The effects of oxygen transfer, on the production of sonorensin (a biopreservative in fruit products), was investigated and characterized in terms of volumetric oxygen transfer coefficient ($k_L a$) using optimized medium composition in a bioreactor. Production of sonorensin was performed in 5.0 L bioreactor with a working volume of 2.0 L. Experiments were performed out using optimized media at 40°C, initial pH of 7.5 and agitation rate of 200 rpm (Chopra et al., 2015).

Production of 1,3-butanediol enhanced from the metabolically engineered *E. coli* under strict controlled cultivation conditions in a bioreactor system has been investigated (Kataoka et al., 2014). The influence of oxygen supply to the fermentation medium on yield and rate of production of 1,3-butanediol in a batch fermentation system was also studied. It was confirmed that subsequent glucose feedings enable the recombinant strain to produce a higher yield of 1,3-butanediol in the fed-batch fermentation.

The influence of extracellular factors including dispersion in the broth, fluctuations in the levels of dissolved oxygen concentration, pH on lipase production by *Y. lipolytica* in a 20-L batch bioreactor has been successfully investigated.

2.3.1.2 CONTINUOUS OPERATION

Cell-growth, product formation, and substrate utilization declines or ceases beyond a particular time because of continuously changing culture conditions in a batch culture. On the contrary, growth, and product formation can be prolonged as fermentation nutrients are continually supplied and products and cells are withdrawn during continuous culture. Continuous cultivation devices primarily include chemostat and turbidostat (Plug flow reactor in some cases). In a chemostat, cell-growth is usually decided by a limiting nutrient concentration and concentrations of nutrient, product, and cell-mass are maintained constant at a steady-state, justifying its meaning, i.e., constant chemical environment. Maintenance of the constant cell-concentration in the bioreactor by monitoring the optical density and controlling the feed flow rate is the norm in the turbidostat. A continuous bioreactor operation has the following features: (1) continuous production, (2) steady-state conditions after start-up, hence minimized variations of concentrations with time, (3) consistent reaction rates, and (4) easy to determine kinetics.

Large-scale productions are mainly based on continuous reactors. A chemostat can be used to probe the influence of environmental changes on cell physiology and mutation with a special application in the selection of special organisms. Productions in chemostat are associated with a significant problem of cell-degeneration or contamination-within especially with cells containing recombinant DNA where the productivity of the most productive cell is retarded and eventually washed-out by less productive cells. In the tanks-in-series reactor configuration, the output from one reactor enters as the input for the second one and so on. Increasing the number of tanks in series lands up in plug flow reactor behavior. The improved control and homogeneity, easy adjustment, are inherent benefits of a multistage CSTR over a PFR.

Most industrial scale bioprocesses are operated as a batch reactor; continuous systems are in use to make Single-Cell Protein (SCP). Other modified forms of continuous cultures are being used in waste-water treatments, ethanol production and also for large-volume, growth-linked metabolites such as lactic acid.

Fermentation targeting secondary metabolite production requires growth and the product formation steps to be separate so as to optimize conditions for each step. Factors such as temperature, pH, and/or limiting nutrient can be varied at each stage resulting in different cell physiology and products via multistage systems. At laboratory-scale, (CSTR) as single-stage tank reactors are the most common continuous system used for anaerobic digestion (Jayalakshmi et al., 2009; Kotsupoulos, 2009) in hydrogen production. A remarkable reactor design to assess the potential of kitchen waste for the hydrogen conversion a pilot-scale, plug-flow reactor and inclined at a 20° angle to the horizontal with internal screw arrangement has been proposed (Venetsaneas et al., 2009). A two-stage system combining hydrogen fermentation and methane production is recommended for the complete degradation of substrates such as livestock waste and food waste (Koutrouli et al., 2009; Wang et al., 2009). Another two-stage system comprised of the dark and photo-fermentation described (Steyer et al., 2010) proceeds in a sequential batch reactor.

Study investigating fundamental mechanisms as to how anaerobic biomass cope with salinity using solutes such as sodium chloride, potassium for biological wastewater treatment was carried and findings were applied to a continuous Submerged Anaerobic Membrane Bioreactors (SAMBRs) (Vyrides et al., 2010).

2.3.1.3 COMPARISON OF BATCH AND CONTINUOUS OPERATION

Following characteristics (A-Advantage; D-Disadvantage) are associated with batch operation: (1) Suitable for small-scale production (A), (2) High flexibility for multiple-product formation (A), (3) Low capital cost (A), (4) Easy shut down and cleaning procedures (A), (5) Inherent downtime between batches (D), and (6) Difficult process control and inconsistent product quality (D).

Following characteristics are associated with continuous operation: (1) Suitable for indefinitely long production runs of one product or set of products (A), (2) Flexible operation (D), (3) Relatively high capital cost (D), (4) Shut down for cleaning very demanding and costly (D), (5) No downtime except for scheduled and emergency maintenance (A); (6) Relatively low operating cost (A), (7) Process control and obtaining uniformity of product easier because of steady-state operation (A), and (8) Ease of coupling with continuous down-stream operation (A).

2.3.1.4 SEMI-BATCH AND SEMI-CONTINUOUS OPERATION

In this cultivation technique, a substrate feed stream is provided inter-mittently as the reactor input to increase the selectivity with the optional product removal. Such operations are the norm in the chemical industry and in bioprocesses. For instance, a continuous supply of oxygen in aerobic fermentations in a batch mode can also be treated as a special case of fed-batch culture. Fed-batch cultures have been successful in control-ling catabolite repression. Enhancement in the production of *A. oryzae* in fed-batch culture owes to the prevention of the cells to a high glucose concentration relative to batch culture.

Possible feeding policies for reactor control are: (1) Constant feeding of the medium results in consistent reaction rate and low accumulation, (2) To maintain limiting substrate concentration constant via exponential feeding, and (3) Feeding with feedback control based on measurements such as temperature or concentration. Special features of semi-continuous operation are: (1) Suitable for small-scale production (A), (2) Capital cost usually between batch and continuous operation (A, D), (3) Ease of adjust-ment of operating parameters for increased reaction selectivity (A), (4) Suitable operation for many runaway–type reactions (A), (5) Relatively high operational costs compared to batch operation (D), (6) Inherent downtime (D), (7) Unsteady-operation demands optimal feed or often complex control strategy (D), and (8) Complicated reactor design (D). Dynamic modeling improves the situation.

Fermentation contributes significantly for the development of pharmaceuticals and health products with an objective of fermentation in research and industry to achieve the highest volumetric productivity thus high cell densities are a prerequisite for high productivities. High Cell Density (HCD) cultivation of bacteria is important for the majority of industrial processes to achieve higher productivity of the desired product. A fed-batch bioprocess using glucose as a carbon source for the HCD *Pseudomonas putida* KT2440 in the absence of oxygen has been described. Growth kinetics data from batch fermentations were based on designing of feeding strategies. Biomass as a biocatalyst was established for the production of the biodegradable polymer polyhydroxyalkanoate (PHA) (Davis et al., 2015) with nonanoic acid (NA) to the glucose-grown cells of *P. putida* KT2440, achieving the of 32% accumulation of cell dry weight as PHA in 11 h.

The repeated-batch culture is superior over the batch culture as it provides the better substrate utilization, the reuse of microbial cells for subsequent fermentation runs, higher cell densities in the culture and less time required for process operation. Moreover, the repeated-batch culture is expected to increase cell productivity ensuring a high cell growth rate (Huang et al., 2008). It has been established that the repeated-batch culture is influenced by certain operational factors such as harvesting times and harvesting volumes of the culture broth (Masuda et al., 2011). As described in a recent research lipid production by *C. bainieri* 2A1 in the batch culture and the repeated-batch culture on a nitrogen-limited medium has been compared. Enhancement of lipid production was observed in the batch culture of *C. bainieri* 2A1 grown under the nitrogen-limited medium under higher agitation rates. Increased agitation rates resulted in higher glucose consumption favoring biomass production and lipid accumulation. It was also established that the repeated-batch culture is more reliable fermentation system for lipid production compared to the batch culture.

HCD fed-batch fermentation may be promising for low-cost production of lipases. In enzyme production, HCD can be achieved through the fed-batch operation and not through cumbersome perfusion culture. The feeding strategies and oxygen transfer used in fed-batch operations for lipase productions have been reviewed (Salehmin et al., 2013).

2.3.1.5   BATCH, FED-BATCH AND CONTINUOUS CULTURES

In spite of the spectrum of advantages such as high productivity, long runs with a low proportion of downtime, steady-state conditions resulting in easier process control and consistent product quality, associated with large-scale continuous cultivations in industry, the majority of the bioprocesses are performed as the batch cultivations. Batch processing is suitable for non-growth associated products with improved genetic stability and reduced risk of contamination. Fed-batch cultivation is superior to batch cultivation as it is more suitable in the case of substrate inhibition where changing substrate concentrations influence the overall productivity of the desired product. Their use for the production of primary and secondary metabolites, proteins, and other bio-products have been well established.

2.3.2   MULTI-PHASE REACTORS

Multi-phase reactors are the reactors having a minimum of two distinct phases for carrying out various gas-liquid-solid and gas-liquid-liquid reactions and can also be named as heterogeneous reactors. The reaction between gas and liquid is facilitated with the help of a porous solid catalyst typically at a catalytic site. The transport steps include gas dissolution into the liquid, movement of dissolved gas to the catalyst surface followed by diffusion and reaction in the catalyst. Process systems with multiple phases are found in a number of diverse areas including the petroleum-based products and fuels, specialty chemicals, pharmaceuticals, herbicides, and pesticides proving that the multiphase reactor is the heart of the process. Multiphase reactors for new processes such as fluid catalytic cracking and slurry bubble column reactors for gas conversion are still evolving. Improved multiphase reactor models that can be used for existing processes and scale-up of new processes are yet not developed.

2.3.2.1   PACKED-BED REACTORS (PBRS)

Catalytic packed-bed reactors (PBRs) can be operated as multiphase reactors which can be: (1) trickle bed, having a gas phase in continuous and a liquid phase as distributed, and (2) bubble column, with a distributed gas and a continuous liquid phase, and the major resistance to mass transport is in the liquid phase. For three-phase reactions (gas/liquid with a solid catalyst), trickle- or PBRs, in which the catalyst particles are held stationary, and slurry reactors, where the catalyst particles are suspended in the liquid phase. Gas and liquid streams move co-currently downflow or gas is fed counter currently up to the flow. Commercially, the liquid phase is allowed to flow mainly through the catalyst particles in the form of films or droplets. The optimum catalyst and reactor selection are possible through a detailed knowledge of each reactor type are very important.

2.3.2.2   TRICKLE BED REACTORS

Trickle beds can be operated with or without recycling, but recycling allows higher loading and gives better flow distribution, which is even more critical than in submerged packed bed operation. Cleaning is again possible by flooding the filter and proceeding as with submerged fixed bed reactors. In the case of gas purification, the gas is cleaned in a single passage, and the liquid is there both as an absorption fluid and as nutrient supply to the biomass on the packing (usually wood shavings or bark). Excess biomass is settled out of this stream, and no other cleaning is needed. Trickle beds with the counter flow of gas and liquid are used on a large scale for vinegar production, as biofilters for gas clean-up and deodorization, for water purification and for ore leaching.

2.4 STIRRED TANK DESIGN PARAMETERS

Designing a stirred tank is a difficult task because of its high experimental approach used by researchers. The most important part of the design is matching the fermenter ability that depends on the oxygen demand of the fermentation culture. Following is the detailed diagram for any stirred tank bioreactor which can be used for numerous fermentation or bioproduction (Figure 2.2).

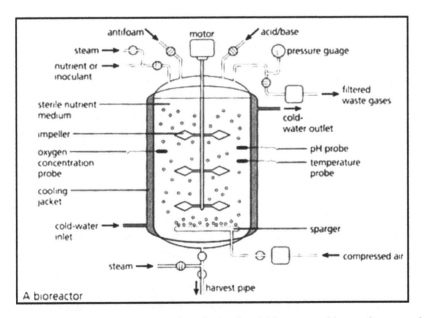

FIGURE 2.2 Schematic representation of stirred tank bioreactor with complete control (http://chrisbiology.blogspot.com/2012/02/58-interpret-and-label-diagram-of.html).

Some novel non-mechanically agitated and more energy-efficient reactors based on mild shear stress relative to CSTRs are the airlift reactors. The aseptic operation can be achieved easily as the shaft seals in airlift reactors are absent (Xu et al., 2012). The successful differentiation of cell cultures depends on the physiochemical environment, mainly pH, dissolved oxygen, carbon dioxide concentrations and temperature (Silva et al., 2012).

2.4.1 DESIGN PARAMETERS OF STRS

Some guidelines have been outlined so as to enhance the mass transfer rate in STRs, but these parameters do not provide valid information on how different aspects of the tank (like impeller and baffle construction) affect oxygen transfer rate (OTR) in stirred vessels. The availability of valid data related to these experimental methods is not well documented. It means chances of errors in the existing data are more due to probe times and unsteady state measurements. Based on the studies, the following important characteristics are being discussed.

2.4.2 HEIGHT-TO-DIAMETER RATIO OR ASPECT RATIO

The height-to-diameter (H/D) ratio of the vessel is desired between of 2 and 3; however, the taller vessels up to (H/D) = 4.0 have also been used to decrease the power requirement of the impellers. Typical tanks also employ a dish-shaped bottom to improve the mixing and prevent form dead zones.

2.4.3 AGITATION

Agitators or impellers facilitate proper mixing of nutrients. The agitator is a power transmission system. The mixing effect of the agitator gives rapid dispersion of injected fluids such as, for pH control: nutrients and acid, for temperature control: enhance the heat transfer. The agitator is made up of one or more impellers mounted on a drive shaft aligned with the vessel. The objective of mixing is to maintain a uniform environment throughout the vessel contents. Agitation speed can be measured by power consumed through agitator shaft. The tachometer can be used to control the agitation speed. Wattmeter is usually used in large-scale process and dynamometer is used in small scale. Commonly used impellers are radial flow, axial flow, and mixed flow types which throw liquid radially outwards towards vessel walls and axially towards the base. The ratio of the impeller diameter (d_i) to the diameter of the tank (d_t) should be between 0.3 and 0.5. In radial flow impellers, the approximate ratio should be 0.3. Very small impellers may not generate enough fluid dispersion whereas very large impellers require much more power are thus less efficient. Advancement in impeller design for adequate mixing has led to impeller design with low power consumption. The typical or ideal bioreactors employ standard Rushton turbines. The Rushton disc turbine is most often used for highly aerobic fermentations and is better characterized. The spacing between the two impellers should be 1.0 d_i to 2.0 d_i. The impeller is placed at 1.0 d_i from the base of the tank.

2.4.4 BAFFLES

Stirred tank bioreactors mainly employ baffles to disrupt the bulk fluid flow in the tank and also for mixing and gas dispersion in the tank. It prevents vortex and improves aeration efficiency. The agitation effect is slightly increased with an increase in width of baffles, but drops sharply with narrower baffles. Generally, four to eight baffles are incorporated.

2.4.5 AERATION

Aeration is done in order to fulfill the oxygen demand of the microbes in submerged culture fermentation. The aeration is provided with the help of sparger. A sparger is a device used for the supply of air into liquid fermenter or bioreactor. Sparger design and its location affect the aeration rate and productivity. Aeration and agitation are complementary to each other. Common types of sparger design include the nozzle sparger or open/partially closed pipe, the orifice sparger or perforated pipe, and the porous sparger. In small stirred tank bioreactor the orifice sparger is installed below the impeller in the form of crosses or rings, approximately three-quarters of the impeller diameter. In order to minimize the pressure drop and prevent smaller holes getting blocked, the sparger whole diameter should be 6 mm.

2.4.6   OXYGEN TRANSFER RATE (OTR)

In aerobic fermentations, OTR to the liquid phase in bioreactors has long been a serious concern for various bioprocesses. An increase in the OTR would overcome oxygen limitation. The volumetric oxygen coefficient of mass transfer (k_La) depends on the air pressure in a stirred lab-scale bioreactor. The cheapest and easily available source of oxygen supplied to microbial culture in the form of air. In laboratory scale, cultures should be aerated by means of the shake-flask, shaken on a specially designed platform in a controlled environment chamber. In pilot and production/industrial scale, air is provided into the cultures through specific types of fermenter known as bubble fermenter. Oxygen transfer from air bubble to cell component takes place via a sequence of steps: (a) the transfer of oxygen from the interior of an air bubble to the gas phase interface in a solution, (b) movement of air through the gas-liquid interface, (c) diffusion through the stationary liquid film surrounding the bubbles, (d) transfer of oxygen through the bulk liquid, (e) diffusion into the stationary liquid film surrounding the cells, (f) transport of air across the liquid-cell interface, (g) transport through the cell-site to the site of reaction, (h) dissolved oxygen transfer through the fermentation medium to the microbial cell, and, (i) consumption of the dissolved oxygen by the cell. It can be said that the OTR is a function of driving force and the mass-transfer area (a):

$$OTR \ \alpha \ (a)* \ (driving \ force)$$

where α is replaced and substituted by a mass transfer coefficient k_L. The OTR from the air bubble to the liquid phase is given by the following equation:

$$dC_L/dt = K_La \ (C*- C)$$

where C_L = concentration of dissolved oxygen in the fermentation broth, units moles dm^{-3}, t = time, units hours.

dC_L/dt = change in oxygen concentration over a time period, moles $O_2 \ dm^{-3} \ h^{-1}$
K_L = mass transfer coefficient, units cm h^{-1}.
a = gas/liquid interface area per unit liquid volume, $cm^2 \ cm^{-3}$.
$C*$ = saturated concentration of dissolved oxygen, mmoles dm^{-3}.

where k_L is defined as the sum of the reciprocals of the resistance to the transfer rate of oxygen from gas to liquid and ($C*-C_L$) and may be defined as the 'driving force' across the resistance. It is very difficult to measure both 'k_L' and 'a' in a fermentation separately, therefore the two terms are combined in the form of $k_L a$, known as the volumetric mass transfer coefficient and units are reciprocal of time (h^{-1}). The volumetric mass transfer coefficient ($k_L a$) is used as a measurement of aeration capacity of a bioreactor vessel during fermentation. The aeration capacity of the bioreactor system is increased if the value of $k_L a$ is high, and the value of $k_L a$ depends on the design criteria and operating conditions of the fermenter and will be affected by some variables such as aeration, agitation, and impeller design of a bioreactor. These variables affect 'k_L' by means of decreasing the resistance to transfer and 'a' affected by changing the number, size, and residence time of air bubbles. Several techniques are available for the $k_L a$ assessment:

1. In the sulfite oxidation technique, oxygen transfer to the solution gets consumed instantly by the oxidation of sulfite reaction, so that the sulfite oxidation rate of reaction is assumed to the oxygen-transfer rate (OTR).
2. Gassing-out techniques, which are further divided into two types as the static method of gassing out and the dynamic method of gassing out. In the static method of gassing out, the concentration of oxygen in the solution is reduced by passing nitrogen gas into the liquid, in this way all the oxygen will be removed from the solution. So, these techniques have the advantage instead of the sulfite oxidation method in this way; it is very fast generally taking up to 15 minutes and utilize the fermentation medium. In the dynamic method of gassing out, the supply of air to the fermentation is stopped which results in a linear fall in the dissolved oxygen concentration due to the respiration of live cells in the fermentation broth. The increase in the concentration of dissolved oxygen which is monitored by using some form of dissolved oxygen probe. This method has the advantage over the gassing-out methods to calculating the $k_L a$ during fermentation and used to determine $k_L a$ values at different stages in the process.
3. The oxygen-balance technique: In this technique, the transportation amount of oxygen from air bubble into the fermentation medium in a given period of time is measured and thus leads to the measurement of $k_L a$ in a fermenter.

2.5   WORKING OF STIRRED TANK BIOREACTORS

There are three types of sensors are used in bioreactors for control and the monitoring of a fermentation system: (a) In-line sensor: It forms an integral part of bioreactor and its directly measured value controls the process in fermenter, for example: Antifoam probe; (b) On-line sensor: It forms an integral part of bioreactor and its measured value entered into the control system to control the process in fermenter, for example, Ion specific sensors, mass spectrophotometer, and, (c) Off-line sensors: It is not an integral part of bioreactor, and its measured value must be entered into the control system for data collection. Monitoring and control of several important fermentation operating variables are described below:

1. Temperature measuring device: These are classified into three major classes: (a) Non-electrical: Bimetallic thermometer, Liquid-in-gas thermometer, Pressure thermometer (Mercury in steel, constant volume, Vapor pressure); (b) Electrical: Electrical resistance (Metallic resistance, Semiconductor resistance), Thermo electrical resistance, and (c) Radiation: Total radiation, Partial radiation, IR pyrometer.

2. Gas flow rate measurement device: Gas flow rate can be measured by simple variable area meter. (a) Rotameter: It is used to measure gas and liquid flow rates. (b) Thermal mass flowmeter: Two thermistors (T_1 and T_2) are placed in upstream and downstream. The mass flow rate Q is calculated from the below equation as: $H = Q$ $Cp (T_1-T_2)$, where, Cp is the specific flow rate of the Gas, T_1 is the temperature of gas before heat transfer and T_2 is the temperature of gas after heat transfer.

3. Liquid flow rate measurement device: Rotameter can be used to measure liquid flow rate. A syringe pump is also used as for controlling the liquid flow rate. A peristaltic pump is another useful device for liquid flow control.

4. Pressure measuring and controlling device: Bourdon tube pressure gauge is used to measure the pressure changes. Diaphragm gauge is another device for measuring the pressure in aseptic condition. The piezoelectric transducer is a solid crystal, any change in shape due to pressure produce an equal external electric charge on the opposite face of the crystal is produced.

5. Agitation measuring and controlling device: Wattmeter is used to measure power consumed by the agitator. It is used for large-scale processes. Torsion dynamometer is used for small-scale processes. The tachometer is used to control the agitation speed.

6. Foam sensing: (a) Additional metering of antifoam is based on the information given by the foam sensor. This additional control is essential to avoid the possible overdose of antifoam to prevent the decrease in mass transfer efficiency. (b) Mechanical metering of foam is a special disk type foam-breaking mixer that is installed in the bioreactor's upper cover.

7. pH monitoring devices: Glass electrode and combined electrode are used to control pH. Any pH electrode requires both a sensing electrode and a reference electrode. The voltage difference between the two electrodes is used to calculate the pH of the unknown solution by using the Nernst equation.

2.6 RECENT DEVELOPMENTS IN BIOPROCESSES

Production of bioproducts at a commercial scale requires knowledge of subjects such as microbiology, molecular biology, and bioprocess engineering. Molecular biology helps expression of genes from plants, microorganisms or from an animal origin that can be used for industrial productions (Mitra et al., 2012; Stoger, 2012) while bio-process engineering enables us to develop large-scale operation for growing and subsequent purification and formation of products (Roosta et al., 2012). The scale-up of the reactor was thought as a matter of using larger volumes in earlier times. Such an approach results not only in product variability, both in terms of product and quality, but also higher operating costs. Hence, a detailed study of process engineering principles is a requisite for scaling up and operation of biotechnological processes for manufacturing various products. To cope-up with regulations on product quality, stability, and consistency of a bioprocess, an exhaustive knowledge of fermenter dynamics and stability is required, which helps in optimizing and controlling of a process. In this regard, several exhaustive studies have been reported recently (Krzeminski et al., 2012; Upadhyaya et al., 2012).

In this context, conventional stirred bioreactors can contribute during the scaling-up procedure but are probably not the best choice (Achilli et al., 2011). Cell cultures are well-established platforms for the production of new products with novel applications. The establishment of new bioprocesses based on cell culture relies on the inputs the field of genetic engineering to express the desired compound, medium definition, culture conditions, and optimal culture strategy needs to be precisely defined and optimized to replicate such bioprocesses to the production scale (Delvigne et al., 2014).

Like novel bioreactor, perfusion bioreactors can be used to provide culture medium through solid porous 3D scaffolds pores; hereby proving that this set-up enhances growth and differentiation of chondrocyte as well as deposition of the mineralized matrix by bone cells (Salehi-Nik et al., 2013).

Another interesting application wherein a bioreactor has been used for use both in conventional and organic agriculture and soil-restoration programs, i.e., bioconversion of olive oil industry waste by low-cost biostabilization using the earthworm *Eisenia andrei* (Melgar et al., 2009).

High Throughput Technology (HTP) is widely used for manufacturing of a variety of proteins during process development phase (Retallack et al., 2012; Royle et al., 2013). It is mainly focusing around reduction in culture volume, real-time monitoring, and control at micro-scale and full automation of the systems (Kusterer et al., 2008; Rohe et al., 2012). Bioreactors miniaturization is a controlling factor in translating HTP cultivation practically and economically feasible. There is a huge demand for HTP cultivation driven by the emerging areas of biotechnology towards the establishment of the mutant library such as soluble protein expression, metabolic engineering and synthetic biology (Long et al., 2014). Interest in continuous bioprocessing in biologics manufacturing is growing because of the obvious advantages of consistent product quality and higher productivities (Croughan et al., 2015).

2.7 CONCLUSION

Process development has been the final breakthrough for achieving commercially important microbial products cost-effectively. Ongoing research and development in the field of bioprocessing shall definitely prove a driving force for the production of enzymes and drugs, especially for generic biopharmaceuticals through enhanced competitiveness and quality.

KEYWORDS

- baffles
- batch operation
- biofuels
- biopharmaceuticals
- bioreactor
- catalytic reactors
- continuous operation
- esterification
- fed-batch fermentation
- fermentation
- foam sensing
- homogenous bioreactors
- oxygen transfer rate
- packed-bed reactors
- rotameter
- semi-batch operation
- semi-continuous operation
- single-cell protein
- trickle bed reactors

REFERENCES

Achilli, A., & Marchand, E. A. C. A., (2011). A performance evaluation of three membrane bioreactor systems: Aerobic, anaerobic, and attached-growth. *Water Sci. Technol., 63*(12), 2999–3005.

Anuar, S. T., Villegas, C., & Mugo, S. M. C. J., (2011). The development of flow-through bio-catalyst microreactors from silica microstructured fibers for lipid transformations. *Lipids, 46*(6), 545–555.

Chen, J. P. L. G., (2010). Optimization of biodiesel production catalyzed by fungus cells immobilized in fibrous supports. *Appl. Biochem. Biotechnol., 161*(1–8), 181–194.

Chopra, L., Singh, G., Jena, K. K., Verma, H., & Sahoo, D. K., (2015). Bioresource technology bioprocess development for the production of sonorensin by *Bacillus sonorensis* MT93 and its application as a food preservative. *Bioresource Technology, 175*, 358–366.

Croughan, M. S., Konstantin, B., & Konstantinov, C. C., (2015). The future of industrial bioprocessing: Batch or continuous? *Biotechnology and Bioengineering, 112*(4), 648–651.

Dailin, D. J., Elsayed, E. A., Othman, N. Z., Malek, R., Phin, H. S., Aziz, R., & Wadaan, M. E. E. H., (2016). Bioprocess development for kefiran production by *Lactobacillus kefiranofaciens* in semi-industrial scale bioreactor. *Saudi J. Biol. Sci., 23*(4), 495–502.

Davis, R., Duane, G., Kenny, S. T., Cerrone, F., Guzik, M. W., Babu, R. P., et al., (2015). High cell density cultivation of *Pseudomonas putida* KT2440 using glucose without the need for oxygen enriched air supply. *Biotechnology and Bioengineering, 112*(4), 725–733.

Delvigne, F., Zune, Q., Lara, A. R., Al-Soud, W., & Sørensen, S. J., (2014). Metabolic variability in bioprocessing: Implications of microbial phenotypic heterogeneity. *Trends in Biotechnology, 32*(12), 608–616.

Huang, W. C., Chen, S. J., & Chen, T. L., (2008). Production of hyaluronic acid by repeated batch fermentation. *Biochemical Engineering Journal, 40*(3), 460–464.

Jayalakshmi, S., Joseph, K., & Sukumaran, V., (2009). Biohydrogen generation from kitchen waste in an inclined plug flow reactor. *International Journal of Hydrogen Energy, 34*(21), 8854–8858.

Kataoka, N., Vangnai, A. S., Ueda, H., Tajima, T., & Kato, J., (2014). Enhancement of (R)-1,3-butanediol production by engineered *Escherichia coli* using a bioreactor system with strict regulation of overall oxygen transfer coefficient and pH. *Bioscience, Biotechnology, and Biochemistry, 78*(4), 695–700.

Kotsopoulos, T. A., Fotidis, I. A., Tsolakis, N., & Martzopoulos, G. G., (2009). Biohydrogen production from pig slurry in a CSTR reactor system with mixed cultures under hyper-thermophilic temperature (70°C). *Biomass and Bioenergy, 33*(9), 1168–1174.

Koutrouli, E. C., Kalfas, H., Gavala, H. N., Skiadas, I. V., Stamatelatou, K., & Lyberatos, G., (2009). Bioresource technology hydrogen and methane production through two-stage mesophilic anaerobic digestion of olive pulp. *Bioresource Technology, 100*(15), 3718–3723.

Krzeminski, P., Van der Graaf, J. H., & Van Lier, J. B., (2012). Specific energy consumption of membrane bioreactor (MBR) for sewage treatment. *Water Science and Technology, 65*(2), 380–392.

Kusterer, A., Krause, C., Kaufmann, K., Arnold, M., & Weuster-Botz, D., (2008). Fully automated single-use stirred-tank bioreactors for parallel microbial cultivations. *Bioprocess and Biosystems Engineering, 31*(3), 207–215.

Li, X., Yuen, W., & Morgenroth, E, R. L., (2012). Backwash intensity and frequency impact the microbial community structure and function in a fixed-bed biofilm reactor. *Appl. Microbiol. Biotechnol., 96*(3), 815–827.

Liang, Z. X., Li, L., Li, S., Cai, Y. H., Yang, S. T., & Wang, J. F., (2012). Enhanced propionic acid production from Jerusalem artichoke hydrolysate by immobilized *Propionibacterium acidipropionici* in a fibrous-bed bioreactor. *Bioprocess Biosyst. Eng., 35*(6), 915–921.

Long, Q., Liu, X., Yang, Y., Li, L., Harvey, L., McNeil, B., & Bai, Z., (2014). The development and application of high throughput cultivation technology in bioprocess development. *Journal of Biotechnology, 192*, 323–338.

Masuda, M., Das, S. K., Fujihara, S., Hatashita, M., & Sakurai, A., (2011). Production of cordycepin by a repeated batch culture of a *Cordyceps militaris* mutant obtained by proton beam irradiation. *Journal of Bioscience and Bioengineering, 111*(1), 55–60.

Melgar, R., Benitez, E., & Nogales, R., (2009). Bioconversion of wastes from olive oil industries by vermicomposting process using the epigeic earthworm *Eisenia andrei*. *Journal of Environmental Science and Health, Part B, Pesticides, Food Contaminants, and Agricultural Wastes, 44*(5), 488–495.

Mitra, D., Rasmussen, M. L., Chand, P., Reddy, V., Yao, L., Grewell, D., et al., (2012). Bioresource technology value-added oil and animal feed production from corn-ethanol stillage using the oleaginous fungus *Mucor circinelloides*. *Bioresource Technology, 107*, 368–375.

Retallack, D. M., Jin, H., & Chew, L., (2012). Reliable protein production in a *Pseudomonas fluorescens* expression system. *Protein Expression and Purification, 81*(2), 157–165.

Rodrigues, C. A., Fernandes, T. G., Diogo, M. M., & Da Silva, C. L., (2011). Stem cell cultivation in bioreactors. *Biotechnol. Adv., 29*(6), 815–829.

Rohe, P., Venkanna, D., Kleine, B., Freudl, R., & Oldiges, M., (2012). An automated workflow for enhancing microbial bioprocess optimization on a novel microbioreactor platform. *Microbial Cell Factories, 11*(1), 144, doi: 10.1186/1475-2859-11-144.

Roosta, A., Jahanmiri, A., Mowla, D., Niazi, A., & Sotoodeh, H., (2012). Optimization of biological sulfide removal in a CSTR bioreactor. *Bioprocess and Biosystems Engineering, 35*(6), 1005–1010.

Royle, K. E., Jimenez, I., & Kontoravdi, C., (2013). Integration of models and experimentation to optimize the production of potential biotherapeutics. *Drug Discovery Today, 18*(23-24), 1250–1255.

Salehi-Nik, N., Amoabediny, G., Pouran, B., Tabesh, H., Shokrgozar, M. A., Haghighipour, N., et al., (2013). Engineering parameters in bioreactor's design: a critical aspect in tissue engineering. *Biomed. Res. Int., 762132*, doi: 10.1155/2013/762132.

Salehmin, M. N. I., Annuar, M. S. M., & Chisti, Y., (2013). High cell density fed-batch fermentations for lipase production: Feeding strategies and oxygen transfer. *Bioprocess Biosyst. Eng., 36*, 1527–1543.

Silva, C. R., Zangirolami, T. C., Rodrigues, J. P., Matugi, K., Giordano, R. C., & Giordano, R. L. C., (2012). Enzyme and microbial technology an innovative biocatalyst for production of ethanol from xylose in a continuous bioreactor. *Enzyme and Microbial Technology, 50*(1), 35–42.

Steyer, J., Guo, X. M., Trably, E., & Latrille, E., (2010). Hydrogen production from agricultural waste by dark fermentation: A review. *International J. Hydr. Ener., 35*, 10660–10673.

Stoger, E., (2012). Plant bioreactors – the taste of sweet success. *Biotechnol. J., 7*, 475–476.

Stojanović, R., Ilić, V., Manojlović, V., Bugarski, D., & Dević, M. B. B., (2012). Isolation of hemoglobin from bovine erythrocytes by controlled hemolysis in the membrane bioreactor. *Appl. Biochem. Biotechnol., 166*(6), 1491–506.

Timmins, N. E., Kiel, M., Günther, M., Heazlewood, C., Doran, M. R., & Brooke, G. A. K., (2012). Closed system isolation and scalable expansion of human placental mesenchymal stem cells. *Biotechnol. Bioeng., 109*(7), 1817–1826.

Upadhyaya, G., Clancy, T. M., Snyder, K. V., Brown, J., Hayes, K. F., & Raskin, L., (2012). Effect of air-assisted backwashing on the performance of an anaerobic fixed-bed bioreactor that simultaneously removes nitrate and arsenic from drinking water sources. *Water Research, 46*(4), 1309–1317.

Urban, J., Svec, F., & Fréchet, J. M., (2012). A monolithic lipase reactor for biodiesel production by transesterification of triacylglycerides into fatty acid methyl esters. *Biotechnology and Bioengineering, 109*(2), 371–380.

Venetsaneas, N., Antonopoulou, G., Stamatelatou, K., & Kornaros, M., (2009). Bioresource technology using cheese whey for hydrogen and methane generation in a two-stage continuous process with alternative pH controlling approaches. *Bioresource Technology, 100*(15), 3713–3717.

Vyrides, I., Santos, H., Mingote, A., Ray, M. J., & Stuckey, D. C., (2010). Are compatible solutes compatible with biological treatment of saline wastewater? Batch and continuous studies using submerged anaerobic membrane bioreactors (SAMBRs). *Environ. Sci. Technol., 44*(19), 7437–7442.

Wang, L., Ridgway, D., Gu, T., & Moo-Young, M., (2005). Bioprocessing strategies to improve heterologous protein production in filamentous fungal fermentations. *Biotechnology Advances, 23*(2), 115–129.

Wang, X., & Zhao, Y., (2009). A bench scale study of fermentative hydrogen and methane production from food waste in integrated two-stage process. *International Journal of Hydrogen Energy, 34*(1), 245–254.

Xu, Z., Li, S., Fu, F., Li, G., Feng, X., Xu, H., & Ouyang, P., (2012). Production of D-tagatose, a functional sweetener, utilizing alginate immobilized *Lactobacillus fermentum* CGMCC2921 cells. *Applied Biochemistry and Biotechnology, 166*(4), 961–973.

CHAPTER 3

DIVERSITY AND APPLICATION OF ACETIC ACID BACTERIA IN ASIA

SOMBOON TANASUPAWAT[1], TAWEESAK MALIMAS[1], NITTAYA PITIWITTAYAKUL[2], HUONG THI LAN VU[3], PATTARAPORN YUKPHAN[4], and YUZO YAMADA[5]

[1]*Department of Biochemistry and Microbiology, Faculty of Pharmaceutical Sciences, Chulalongkorn University, Bangkok 10330, Thailand*

[2]*Department of Agricultural Technology and Environment, Faculty of Sciences and Liberal Arts, Rajamangala University of Technology Isan, Nakhon Ratchasima Campus, Nakhon Ratchasima 30000, Thailand*

[3]*Faculty of Biology and Biotechnology, University of Science, Vietnam National University-HCM City, Ho Chi Minh City, Vietnam*

[4]*BIOTEC Culture Collection (BCC), National Center for Genetic Engineering and Biotechnology (BIOTEC), National Science and Technology Development Agency (NSTDA), 113 Thailand Science Park, Pathumthani 12120, Thailand*

[5]*Department of Applied Biological Chemistry, Faculty of Agriculture, Shizuoka University, Suruga-Ku, Shizuoka 422-8529, Japan*

3.1 INTRODUCTION

Acetic acid bacteria (AAB) are ubiquitous organisms that are distributed widely in sugary, acidic, and alcoholic environments (Komagata et al., 2014; Yamada, 2016). They are important bacteria in the food industry because of their ability to oxidize many types of sugars and alcohols to the corresponding organic acids as the end products. AAB commonly produced acetic acid by ethanol oxidization. However, they can involve not only a positive role in the production of vinegar and kombucha beverage, but also they can spoil other foods and beverages, such as wine, beer, soft drinks, and fruits. These group of bacteria are also used in cellulose and sorbose production (Raspor and Goranovič, 2008; Sengun and Karabiyikli, 2011).

In Southeast Asia, Indonesia, Thailand, Philippines, and Vietnam, AAB were distributed in traditional fermented foods, fruits, flowers, and other sources such as ragi, tofu, and palm sugar. *Acetobacter* strains were found in fermented foods, in tropical fruits and flowers while most of the *Gluconobacter* strains inhabited in fruits and flowers. *Gluconacetobacter* and *Komagataeibacter* strains were mostly found in nata de coco. A large number of *Asaia* strains were isolated from flowers while *Kozakia baliensis* strains were found in ragi and brown palm sugar so far (Navarro and Komagata, 1999; Lisdiyanti et al., 2000, 2001, 2002, 2003b, 2006; Yukphan et al., 2008; Tanasupawat et al., 2011a). The pseudacetic acid strains of *Frateuria aurantia* have been found in fruits and flowers in Indonesia (Lisdiyanti et al., 2003b; Yamada and Yukphan, 2008). In addition, the strains of *Neoasaia chiangmaiensis* were isolated from a flower of red ginger (Yukphan et al., 2005). *Tanticharoenia sakaeratensis* strains were from soil (Yukphan et al., 2008), *K. baliensis* from fruit and *Swaminathania salitolerans* from seed of *Ixora* (Kommanee et al., 2008–2009), *Ameyamaea chiangmaiensis* strains from the flowers of red ginger (Yukphan et al., 2009), the strains of *Neokomagataea thailandica* and *Neokomagataea tanensis* were from flowers of lantana and candle bush, respectively (Yukphan et al., 2011), while *Swingsia samuiensis* strains were from the flowers in Thailand (Malimas et al., 2013). In Vietnam, *Nguyenibacter vanlangensis* strains were isolated from the rhizosphere of Asian rice (Vu et al., 2013) and *Tanticharoenia aidae* strains were from the stems of sugar cane (Vu et al., 2016a). In this chapter, we describe the isolation, cultivation, taxonomy, diversity, oxidative products and applications of AAB in Asia and related countries.

3.2   TAXONOMY AND METABOLISM

AAB are gram-negative, obligately aerobic, ellipsoids to rods, measuring 0.4 to 1.0 by 1.0 to 4.0 μm. They are generally catalase positive and oxidase negative. They grow at pH 3–4 and the optimum pH for growth are 5–6.5 (Kersters et al., 2006; Komagata et al., 2014). They oxidize ethanol to acetic acid except for *Asaia*, *Nguyenibacter*, and *Bombella*. In addition,

Acetobacter, Gluconacetobacter, and *Acidomonas* strains can oxidize the resulting acetic acid to CO_2 and H_2O, named the overoxidation of ethanol. In contrast, *Gluconobacter* is absent, and *Asaia* and *Kozakia* are weak in the oxidative activity. They grow in the presence of 0.35% acetic acid except for *Asaia, Saccharibacter, Neokomagataea,* and *Swingsia* strains, and they produce gluconic acid from glucose. *Acetobacter* strains contain ubiquinone Q–9, where as all other AAB possess Q–10 (Malimas et al., 2017).

All members of AAB are obligately aerobic, and their metabolism is strictly respiratory with oxygen as the terminal electron acceptor. Eighteen genera of AAB accommodated in the family *Acetobacteraceae,* the *Alphaproteobacteria* are *Acetobacter, Gluconobacter, Acidomonas, Gluconacetobacter, Asaia, Kozakia, Swaminathania, Saccharibacter, Neoasaia, Granulibacter, Tanticharoenia, Ameyamaea, Neokomagataea, Komagataeibacter, Endobacter, Nguyenibacter, Swingsia, and Bombella* (Kersters et al., 2006; Komagata et al., 2014). *Acetobacter* species are easily differentiated from *Gluconobacter, Gluconacetobacter, Acidomonas, Asaia,* and *Kozakia* species by their ability to oxidize acetate and lactate to CO_2 and H_2O and by containing Q–9 as the major ubiquinone (Yamada et al., 1969; Kersters et al., 2006; Komagata et al., 2014; Malimas et al., 2017).

AAB produce acetic acid from ethanol by two enzymes, membrane-bound quinohemoprotein alcohol dehydrogenase (ADH) and membrane-bound quinohemoprotein aldehyde dehydrogenase (ALDH) (Matsushita et al., 1994). Vinegar production, L-sorbose production, and D-gluconate production are catalyzed by PQQ-dependent dehydrogenases (quino-proteins), while 2-keto-D-gluconic acid (2KGA), 2,5-diketo-D-gluconic acid (2,5DKGA) and 5-keto-D-fructose production involve flavoprotein dehydrogenases (Adachi et al., 2003). *Acetobacter* and *Gluconobacter* strains contain the pentose phosphate cycle instead of Embden/Meyerhof/Parnas pathway (Cheldelin, 1961; De Ley, 1961; Arai et al., 2016; Bringer and Bott, 2016). In addition, *Gluconobacter* strains lack the tricarboxylic acid (TCA) cycle, differing from those of *Acetobacter* strains, and the pentose phosphate cycle acts as a terminal oxidation system. Some AAB may include a TCA cycle function in the metabolism, which enables them to completely transform acetic acid to CO_2 and water (De Ley et al., 1984). AAB have the direct oxidation system for alcohols, sugars, and sugar alcohols and produce a large amount of the oxidation products, i.e., acetic acid from ethanol, gluconic acid, 2-ketogluconic acid, 5-ketogluconic acid, and 2,5-diketogluconic acid from glucose, fructose from mannitol, L-sorbose from sorbitol, and 5-ketofructose from fructose or sorbitol (Cheldelin 1961; De Ley, 1961; Komagata et al., 2014). This partial or incomplete oxidation is traditionally called "oxidative fermentation" and carried out by the membrane-bound dehydrogenases that are linked to the energy-yielding or the non-energy-yielding respiratory chain (Matsushita et al., 2004; Komagata et al., 2014; Adachi and Yakushi, 2016; Matsushita and Matsutani, 2016). *Frateuria* strains belonging to *Gammaproteobacteria* can grow at pH 3.5, produce acetic acid from ethanol, and oxidize D-glucose to D-gluconate, 2-keto-D-gluconate and 2,5-diketo-D-gluconate similar to AAB, but their growth is inhibited by acetic acid (Lisdiyanti et al., 2003b; Yamada and Yukphan, 2008).

AAB can oxidize alcohols into sugars; mannitol into fructose; sorbitol into sorbose or erythritol into erythrulose (González et al., 2005). *Gluconobacter* is an industrially important genus for the production of L-sorbose from D-sorbitol; D-gluconic acid, 2-keto- and/or 5-keto-D-gluconic acid (2KGA and/or 5KGA) from D-glucose; and dihydroxyacetone (DHA) from glycerol (Gupta et al., 2001). However, the strains produced 2KGA and/or 5KGA such as of *Gluconobacter oxydans* strains were based on D-gluconate dehydrogenase (GADH) activity. Therefore, the best strain for 2KGA and/or 5KGA production should be selected (Sainz et al., 2016). In addition, *Gluconacetobacter liquefaciens* strains were reported to produce 4-keto-D-arabonate (4KAB) (Adachi et al., 2010). AAB are involved in some important industrial process (Raspor and Goranovič, 2008; De Vero et al., 2010). They can produce high concentrations of acetic acid, which makes them important to the vinegar industry. The other known application of AAB is to produce sorbose and cellulose (González et al., 2005). On the other hand, AAB are sometimes involved the changes of aroma and flavor in foods and beverages in detrimental way, such as in wine (Bartowsky and Henschke, 2008).

3.2.1   *ISOLATION AND CULTIVATION*

AAB are widely distributed in various sources, and they were isolated by enrichment procedure using different media, the medium I, II or III for *Acetobacter* strains, medium I for *Gluconobacter*, *Gluconacetobacter*, and *Kozakia* strains, medium II and IV for *Asaia*, adjusted the pH to 3.5 (Yamada et al., 2000b; Lisdiyanti et al., 2003a) as shown in Table 3.1. They were cultivated at 30°C and incubated for 2–5 days. The culture in the medium composed of 2.0% glucose, 0.5% ethanol (99.8%) v/v, 0.3% peptone, 0.3% yeast extract, 0.7% calcium carbonate (precipitated), and 1.5% agar (Yamada et al., 1976, 1999) was streaked onto agar plates. Colonies showing halo formation on the agar plates are picked up and transferred onto the agar slants and are preserved for the further study. *Acetobacter* strains were cultivated on the medium composed of 30 g glucose, 5 g yeast extract, 3 g peptone, 20 g agar and 1000 ml of distilled water (pH 6.8) supplemented with 3% of ethanol and 10 g $CaCO_3$ and incubated at 30°C while *Gluconobacter* strains were cultivated on mannitol agar medium composed of 25 g mannitol, 5 g yeast extract, 3 g peptone, 20 g agar and 1000 ml of distilled water (pH 6.0) (Asai et al., 1964). A nitrogen-fixing bacteria, *Gluconacetobacter diazotrophicus* was isolated by using LGI medium (Cavalcante and Dobereiner, 1988; Nishijima et al., 2013; Vu et al., 2016b) while *Saccharibacter* was isolated by YUG medium (Jojima et al., 2004) (Table 3.1).

TABLE 3.1   Isolation Media of AAB

Genera	Media	References
Acetobacter	I (1.0% D-glucose, 0.5% ethanol, 1.5% peptone, 0.8% yeast extract, 0.3% acetic acid, 0.01% cycloheximide, pH 3.5)	Yamada et al., 2000b; Lisdyanti et al., 2003a
	II (2.0% D-sorbitol, 0.5% peptone, 0.3% yeast extract, 0.01% cycloheximide, pH 3.5)	Lisdyanti et al., 2003a
	III (2.0% D-mannitol, 0.5% peptone, 0.3% yeast extract, 0.2% acetic acid, 0.01% cycloheximide, pH 3.5)	Lisdyanti et al., 2003a
Gluconobacter	I, II	Lisdyanti et al., 2003a
Gluconacetobacter	I	Yamada et al., 2000a; Lisdyanti et al., 2003a
Asaia	II	Lisdyanti et al., 2003a
	IV (2.0% dulcitol, 0.5% peptone, 0.3% yeast extract, 0.01% cycloheximide, pH 3.5)	Lisdyanti et al., 2003a
Kozakia	I	Yamada et al., 2000b; Lisdyanti et al., 2003a
Asaia, Frateuria	V (0.15% glucose, 2.0% methanol, 0.5% peptone, 0.3% yeast extract, 0.01% cycloheximide, pH 3.5)	Lisdyanti et al., 2003a, 2003b

TABLE 3.1 *(Continued)*

Genera	Media	References
Saccharibacter	YUG medium (1.0% yeast extract, 0.1% urea, 20% glucose)	Jojima et al., 2004
Ga. diazotrophicus, Tanticharoenia, Nguyenibacter	LGI medium (10% sucrose, 0.06% KH_2PO_4, 0.02% $MgSO_4$, 0.002% $CaCl_2$, 0.001% $FeCl_3$, 0.0002% Na_2MoO_4, pH 6.0)	Cavalcante and Döbereiner, 1988; Vu et al., 2013
Ga. tumulisoli, Ga. takamatsuzukensis, Ga. aggeris, Ga. asukensis	Glucose (10 g); ethanol (5 ml); yeast extract (5 g); peptone (3 g); $CaCO_3$ (5 g); distilled water (900 ml); 10% (v/v) potato extract per litre (pH 4.2–4.5)	Nishijima et al., 2013
Ga. tumulicola, Ga. asukensis	10 g glucose, 5 ml ethanol, 5 g yeast extract, 3 g peptone, 5 g $CaCO_3$ and 15 g agar in 900 ml distilled water, 10% (v/v) potato extract (pH unadjusted)	Tazato et al., 2012
Ga. (Acetobacter) diazotrophicus	LGI semi-solid medium and LGI medium	Cavalcante and Döbereiner, 1988; Li and MacRae, 1991; Muthukumarasamy et al., 1999a
Ga. diazotrophicus, A. peroxydans	N-free semisolid LGI medium with an initial pH 6.0 with 1, 10 or 30% cane sugar (w/v) and 0.005% yeast extract (w/v)	Poly et al., 2001; Muthukumarasamy et al., 2005
Ga. diazotrophicus	LGI with 10% sucrose for *Gluconacetobacter* and *Acetobacter* species	Muthukumarasamy et al., 2007; Cavalcante and Döbereiner, 1988
K. xylinus, K. intermedius	$CuSO_4$ solution in modified Watanabe and Yamanaka medium	Neera Ramana and Batra, 2015
K. kakiaceti	I GYP GYP (1.0% D-glucose, 1.0% glycerol, 1.0% peptone, 0.5% yeast extract, 0.7% $CaCO_3$, 1.5% agar)	Lisdyanti et al., 2003a; Suzuki et al., 2010; Iino et al., 2012b
Ameyamaea	2.0% D-sorbitol, 0.3% peptone, 0.3% yeast extract, pH 3.5	Yukphan et al., 2009
Neokomagataea, Swingsia	Glucose/ethanol/peptone/yeast extract medium (2.0% D-glucose, 0.5% ethanol, 0.5% peptone, 0.3% yeast extract, pH 3.5)	Yukphan et al., 2011

3.2.2 IDENTIFICATION

On the identification of AAB in genus-level using phenotypic characteristics and chemotaxonomic characteristics is relatively easy when compared with species level identification (Yukphan et al., 2006a; Vu et al., 2007a; Kommanee et al., 2012). The molecular techniques have been developed to determine the disadvantages of phenotyping identification of AAB. Recently, the principles of identification techniques were used for *Acetobacter*, *Gluconobacter*, and *Asaia* strains isolated from fruits, flowers, and fermented products in Thailand based on 16S–23S rRNA gene internal transcribed spacer restriction and 16S rRNA gene sequence analysis (Tanasupawat et al., 2009, 2011c). *Acetobacter* strains from Bioresource Collection and Research Center (BCRC), Taiwan were identified using phenotypic and 16S rDNA sequence analysis including the *hsp60* gene sequences and resolved the species of *A. aceti* using novel species-specific PCR combined with the mini-sequencing technology (Huang et al., 2014). *Acetobacter pasteurianus* from rice vinegar in Japan (Nanda et al., 2001) and *A. pasteurianus* strains from cereal vinegars in China were identified (Li et al., 2014) based on enterobacterial repetitive intergenic consensus (ERIC)-PCR. *Acetobacter sicerae* isolated from cider and kefir, and the species of the genus *Acetobacter* were identified by *dnaK*, *groEL*, and *rpoB* sequence analysis (Li et al., 2014). *Acetobacter* strains from fruits, flowers, mushrooms, and fermented products were examined based on 16S rRNA gene and *groEL* gene sequences (Pitiwittayakul et al., 2015b). *Gluconobacter frateurii* strains were reidentified at the species level on the basis of restriction analysis of 16S–23S rDNA ITS regions by digestion with six restriction endonucleases: *Bsp*1286I, *Mbo*II, *Ava*II, *Taq*I, *Bso*BI, and *Bst*NI (Malimas et al., 2006). *Asaia* strains were identified based on 16S rDNA restriction analysis (Vu et al., 2007a). Phylogenetic relationships between the genera *Swaminathania* and *Asaia*, with reference to the genera *Kozakia* and *Neoasaia*, were determined based on 16S rDNA, 16S–23S rDNA ITS, and 23S rDNA sequences (Yukphan et al., 2006b). Repetitive sequences based on polymerase chain reaction (rep-PCR fingerprinting) using the $(GTG)_5$ primer $(GTG)_5$-PCR fingerprinting), was used to differentiate *Gluconobacter uchimurae* strains and related species (Tanasupawat et al., 2011b) while AFLP DNA fingerprinting was used to differentiate *Gluconacetobacter kombuchae*, a later heterotypic synonym of *Gluconacetobacter hansenii* (Cleenwerck et al., 2009).

Recently, MALDI-TOF analyses of *Acetobacter, Gluconacetobacter,* and *Gluconobacter* strains from vinegar by the analysis of MS spectra obtained from single colonies were cross-checked by comparative sequence analysis of 16S rRNA gene fragments allowed them to be identified, and it was possible to differentiate them from mixed cultures and non-AAB. The results showed that MALDI-TOF MS analysis was a rapid and reliable method for the clustering and identification of AAB species (Andrés-Barrao et al., 2013). The molecular biological methods for identification of AAB, with a focus on the 16S–23S rRNA gene ITS regions and alternative method for identification, MALDI-TOF MS were recently described (Trček and Barja, 2015).

The identification and proposal of a new acetic acid bacterial species was based on the phenotypic characterization and molecular techniques (sequencing of the 16S rRNA gene, DNA G+C content determinations, and DNA-DNA hybridization). Fluorometric and colorimetric determination of the DNA-DNA relatedness among species using microplate reader were useful (Ezaki et al., 1989) while the membrane method, spectrophotometer initial renaturation method have been used (Cleenwerck and DeVos, 2008). Cellular fatty acid profiles and ubiquinone components are additional characteristics. Therefore, a polyphasic study is the recommended technique for an accurate identification of AAB strains (Cleenwerck and De Vos, 2008).

3.3 DIVERSITY OF AAB IN ASIA

In Asia, Indonesia, Philippines, Thailand, India, Bangladesh, Pakistan, Sri Lanka, Iran, Turkey, China, Japan, and Korea, AAB are widely distributed in fruits, flowers, fermented foods and related materials (Tables 3.2–3.5). The strains of *A. aceti, A. orleanensis, A. cibinongensis, A. pasteurianus, A. pomorum, A. peroxydans, A. nitrogenifigens, A. lovaniensis, A. indonesiensis, A. tropicalis, A. syzygii, A. orientalis, A. okinawensis, A. papayae, A. ghanensis, A. persici, A. fabarum, A. thailandicus* and *A. suratthanensis* were distributed in fruits, flowers, mushroom, root interior of wheat, wine, vinegar, sugarcane, Kombucha tea, Nata de coco, moromi soya, curd of tofu, fermented rice flour and fermented foods (Table 3.2), while *G. albidus, G. frateurii, G. japonicus, G. kanchanaburiensis, G. kondonii, G. morbifer, G. nephelii, G. oxydans, G. roseus, G. sphaericus, G. thailandicus, G. uchimurae, G. wancherniae, Gluconobacter* strains were

TABLE 3.2 Diversity of *Acetobacter* species

Country	Species	Sources	References
Bangladesh	*Acetobacter* sp.	Grapes, mangoes, oranges, pineapples, safeda	Diba et al., 2015
	A. aceti	Rotten papaya for vinegar production, sugarcane bagasse	Kowser et al., 2015
	A. orleansis	Sugarcane bagasse	Arifuzzaman et al., 2014
	A. cibinongensis	Sugarcane juice, red grape	Arifuzzaman et al., 2014
	A. pasteurianus	Red grape	Arifuzzaman et al., 2014
China	*A. pasteurianus*, *Acetobacter* sp.	Solid-state fermentation of cereal vinegars, Zhenjiang vinegar grains	Wu et al., 2010; Wang et al., 2015; Chen et al., 2016; Zhao et al., 2016
	A. aceti	Zhenjiang vinegar grains	Wang et al., 2015
	A. pomorum	Solid-state fermentation of cereal vinegars	Zhao et al., 2016
	Acetobacter sp.	Solid-state fermentation of cereal vinegars	Zhao et al., 2016
India	*A. peroxydans*	Root and stem of wetland rice	Muthukumarasamy et al., 2005
	A. nitrogenifigens	Kombucha tea	Dutta and Gachhui, 2006
Indonesia	*A. pasteurianus*	Rice wine, palm vinegar, palm wine, pickle, water of nata, fermented rice, tape (fermented cassava)	Lisdiyanti et al., 2001, 2003a
	A. orleanensis	Flowers, rotten starfruit, rice wine, nata de coco, guava, sapodilla plum	Lisdiyanti et al., 2001, 2003a
	A. lovaniensis	Nata de coco, moromi soya, mangosteen, mango, sapodilla plum, fruit, javagrape, palm wine, tape cassva, pickle, coconut, rotten strafruit, markisa	Lisdiyanti et al., 2001, 2003a
	A. indonesiensis	Rotten starfruit, durian, cendol, palm wine, banana, papaya, zirzak, mango, coconut	Lisdiyanti et al., 2001, 2003a

TABLE 3.2 *(Continued)*

Country	Species	Sources	References
	A. tropicalis	Guava, orange, palm wine, rice wine, lime, coconut, coconut juice	Lisdiyanti et al., 2001, 2003a
	A. syzygii	Flowers, rotten starfruit, vinegar, malay rose apple	Lisdiyanti et al., 2001, 2003a
	A. cibinongensis	Curd of tofu, fruit	Lisdiyanti et al., 2001, 2003a
	A. orientalis	Canna flower, curd of tofu, rotten starfruit, tempe, coconut	Lisdiyanti et al., 2001, 2003a
Iran	*Acetobacter* sp.	Iranian white-red cherry vinegar, peach, pear, flame fruit	Beheshti Maal and Shafiee, 2010; Sharafi et al., 2010
	A. pasteurianus	Apple, fig, grape, apricot, vinegar	Beheshti Maal and Shafiee, 2010; Sharafi et al., 2010
Japan	*A. aceti*	Rice vinegar mash, sake vinegar mash	Ohmori et al., 1980
	A. pasteurianus	Surface of vinegar mash, soils in vinegar factories, rice vinegar, unpolished-rice vinegar	Ohmori et al., 1980; Nanda et al., 2001
	Acetobacter sp.	Rice vinegar, unpolished-rice vinegar, coconut wine	Nanda et al., 2001; Kadere et al., 2008
	A. orientalis	Fermented milk	Ishida et al., 2005
	A. okinawensis	Sugarcane, grape, plum, oriental melon, broad-leaf vetch	Iino et al., 2012a
	A. papayae	Papaya, peach, pomegranate	Iino et al., 2012a
	A. persicus	Traditional rice vinegar	Iino et al., 2012a
Korea	*Acetobacter* sp.	Vinegar fermented	Chun and Kim, 1993
	A. pomorum	Korean traditional vinegar	Baek et al., 2014
	A. aceti	Vinegar fermented, Korean black raspberry vinegar, alcohol vinegar	Lee et al., 2015; Song et al., 2016; Zahoor et al., 2006

TABLE 3.2 *(Continued)*

Country	Species	Sources	References
Pakistan	*A. pasteurianus*	Root interior of wheat	Majeed et al., 2015
Philippines	*A. pasteurianus*	Fermented foods	Lisdiyanti et al., 2001
	A. orleanensis	Fermented foods, fruit, flower	Lisdiyanti et al., 2001
	A. lovaniensis	Fermented foods, fruits	Lisdiyanti et al., 2001
	A. indonesiensis	Fermented foods, fruits, flowers	Lisdiyanti et al., 2001
	A. syzygii	Vinegar	Lisdiyanti et al., 2001
Sri Lanka	*A. pasteurianus*	Coconut water vinegar	Perumpuli et al., 2014
Thailand	*A. rancens or A. pasteurianus*	Longan, mango, sapodilla, banana, mushroom, flowers, fermented foods	Saeki et al., 1997b; Lisdiyanti et al., 2001; Seearunruangchai et al., 2004; Tanasupawat et al., 2009; Kanchanarach et al., 2010; Tanasupawat et al., 2011c; Kommanee et al., 2008a, 2012; Pitiwittayakul et al., 2015b
	A. lovaniensis	Fruits, fermented rice flour, jujube, mango, rambutan, makrut, longan, pineapple, tamarind, golden fig flower	Saeki et al., 1997b; Tanasupawat et al., 2011c; Kommanee et al., 2008a, 2012
	A. aceti	Fruit	Saeki et al., 1997b
	A. farinalis	Fermented rice flour	Tanasupawat et al., 2011a
	A. orleanensis	Fermented foods, fruits, strawberry, star fruit flower	Lisdiyanti et al., 2001; Kommanee et al., 2012
	A. tropicalis	Palm wine, rice wine, fruits, guava, musk-melon, Loog-pang khao mak, kaffir, flower, mushroom, banana, rambeh	Lisdiyanti et al., 2001; Muramatsu et al., 2009; Tanasupawat et al., 2009; Pitiwittayakul et al., 2015b

TABLE 3.2 *(Continued)*

Country	Species	Sources	References
	Acetobacter sp.	Fruits, flower, fermented rice flour, kaffir lime, Indian gooseberry, pineapple, star fruit, cantaloupe, dragon fruit	Hanmoungjai et al., 2007; Tanasupawat et al., 2011c; Klawpiyapamornkun et al., 2015
	A. cerevisiae	Sapodilla	Muramatsu et al., 2009
	A. orientalis	Fruits, flowers, flower of canna, lime, musk melon, fermented rice flour, peach, red grape, sala	Seearunruangchai et al., 2004; Muramatsu et al., 2009; Tanasupawat et al., 2009; Kommanee et al., 2008a, 2012; Pitiwittayakul et al., 2015b
	A. indonesiensis	Flowers, fruit of *Aglaia* sp., guava, hog plum, jujube, musk-melon, mushroom, fermented rice flour	Muramatsu et al., 2009; Tanasupawat et al., 2011c; Kommanee et al., 2012; Pitiwittayakul et al., 2015b
	A. peroxydans	Guava	Muramatsu et al., 2009
	A. syzygii	Flowers, fermented rice flour	Muramatsu et al., 2009; Tanasupawat et al., 2009
	A. ghanensis	Makrut, jujube, pineapple, rose apple, palmyra palm fruit pulp, banana, orange	Kommanee et al., 2012; Pitiwittayakul et al., 2015b; Artnarong et al., 2016
	A. persici	Fruits, fermented rice	Pitiwittayakul et al., 2015b
	A. cibinongensis	Fermented rice	Pitiwittayakul et al., 2015b
	A. papayae	Fruits	Pitiwittayakul et al., 2015b
	A. fabarum	Mushroom	Pitiwittayakul et al., 2015b
	A. okinawensis	Flower	Pitiwittayakul et al., 2015b
	A. thailandicus	Flower of the blue trumpet vine (*Thunbergia laurifolia*)	Pitiwittayakul et al., 2015a
	A. suratthanensis	Fruits	Pitiwittayakul et al., 2016

TABLE 3.2 *(Continued)*

Country	Species	Sources	References
Turkey	*A. lovaniensis*	Vinegar	Çoban and Biyik, 2011
Vietnam	*A. persici, A. papayae, A. indonesiensis, A. pasteurianus, A. syzygii, A. ghanensis, A. fabarum, A. lovaniensis*	Fruits	Vu et al., 2016b
	A. ghanensis	Flower	Vu et al., 2016b
	A. tropicalis	Stem of *Oryza sativa*	Vu et al., 2016b

found in coconut wine, fermented foods, fruits, flowers, gut of *Drosophila melanogaster* and the dwarf honey bee (*Apis florea*), Asian honey bee (*A. cerena*) and giant honey bee (*A. dorsata*) (Kappeng and Pathom-aree, 2009) (Table 3.3). *As. astilbes, As. bogorensis, As. krungthepensis, As. lannensis, As. platicodi, As. prunellae, As. siamensis* and *As. spathodeae* strains were isolated from fruit, seeds, mushroom, and flowers (Table 3.4). *Ga. aggeris, Ga. asukensis, Ga. diazotrophicus, Ga. persimmonis, Ga. takamatsuzukensis, Ga. tumulicola* and *Ga. tumulisoli* strains were found in plant root and stem, fermentation starter of vinegar, soil, and related meterials (Table 3.5). *K. hansenii, K. intermedius, K. kombuchae, K. kakiaceti, K. nataicola, K. oboediens, K. rhaeticus, K. swingsii, K. sucrofermentans, K. xylinus*, and *Komagataeibacter* sp. strains were isolated from vinegar, fruits, nata de coco and flowers (Table 3.5). In addition, the novel genera, *Kozakia baliensis* from ragi and palm brown sugar (Lisdiyanti et al., 2002), *Neoasaia chiangmaiensis* from a flower of red ginger collected in Thailand (Yukphan et al., 2005) and from banana in India (Patil and Shinde, 2014), *Tanticharoenia sakaeratensis* strains from soil (Yukphan et al., 2008), *Ameyamaea chiangmaiensis* strains from the flowers of red ginger collected in Chiang Mai, Thailand (Yukphan et al., 2009), *Neokomagataea thailandica* and *Neokomagataea tanensis*, the osmotolerant AAB isolated from flowers of lantana and candle bush, respectively, collected in Thailand were isolated (Yukphan et al., 2011) whereas *Nguyenibacter vanlangensis* strains from the rhizosphere of Asian rice and the root of Asian rice (Vu et al., 2013) and *Tanticharoenia aidae* strains from the stems of sugar cane were found in Vietnam (Vu et al., 2016a). A diversity of AAB and a lot of suspected novel species from various sources in Vietnam was reported (Vu et al., 2016b).

3.4 OXIDATIVE PRODUCTS AND APPLICATIONS

In Asian countries, China, and Japan, vinegars have long been used as traditional seasonings. *A. pasteurianus* strains spontaneously involved almost pure cultures during a century of vinegar fermentation, in China and Japan. The polished rice (komesu) and unpolished rice (kurosu) vinegars are produced by the same process including rice saccharification, alcohol fermentation and ethanol oxidation (Nanda et al., 2001). *Acetobacter* strains have been used in industrial vinegar production by submerged acetic acid fermentation at about 30°C. However, *Acetobacter*

TABLE 3.3 Diversity of *Gluconobacter* species

Country	Species	Sources	References
Bangladesh	*Gluconobacter* sp.	Sugarcane bagasse, sugarcane juice, apple, red grape, white grape	Arifuzzaman et al., 2014
China	*G. oxydans*	Red tea fungi	Jia et al., 2004
India	*Gluconobacter* sp.	Fermented sliced carrots, carbonated soft drink, oil-spilled seawater, soil, rotten apple	Malik and Garg, 2011; Dhail and Jasuja, 2012; Shanker et al., 2012; Sivasakaran et al., 2014
Indonesia	*G. oxydans*	Flower of pea, waste, nata de coco, brown sugar, sawokecik, papaya, coconut juice, ragi	Yamada et al., 1999; Lisdiyanti et al., 2003a
	G. frateurii	Rotten fruit, nata de coco, mash of soybean paste, sawokecik	Yamada et al., 1999; Lisdiyanti et al., 2003a
Iran	*Gluconobacter* sp.	Nectarine, sloe, peach, pear, flame fruit	Sharafi et al., 2010
Japan	*G. cerinus*	Fruits	Asai, 1935; Yamada and Akita, 1984; Katsura et al., 2002
	G. frateurii	Persimmon, strawberry, cherry	Malimas et al., 2006
	G. albidus	Flower of dahlia	Yukphan et al., 2004b
	G. thailandicus	Strawberry, cherry	Malimas et al., 2006
	G. japonicus	Strawberry	Malimas et al., 2009b
	G. kondonii	Strawberry	Malimas et al., 2008b
	G. roseus	Persimmon	Asai, 1935; Malimas et al., 2008b
	G. sphaericus	Grape	Ameyama, 1975; Malimas et al., 2008c
	Gluconobacter sp.	Coconut wine	Kadere et al., 2008
Korea	*G. morbifer*	Gut of *Drosophila melanogaster*	Roh et al., 2008
Philippines	*G. frateurii*	Fermented foods, fruit, flower	Lisdiyanti et al., 2003a

TABLE 3.3 *(Continued)*

Country	Species	Sources	References
Sri Lanka	*G. frateurii*	Coconut vinegar	Perumpuli et al., 2014
Thailand	*G. frateurii*	Fruit, apple, guava, sugar apple, longan, pum melon, flower, mushroom	Moonmangmee et al., 2000; Vu et al., 2007b; Muramatsu et al., 2009; Kommanee et al., 2008b; Tanasupawat et al., 2009
	G. thailandicus	Flowers, jack fruit, langsat, mango, mangosteen	Tanasupawat et al., 2004; Kommanee et al., 2008b
	Gluconobacter sp.	Fruits, flower, water of pond	Hanmoungjai, 2007; Vu et al., 2007b; Klawpiyapamornkun et al., 2015
	G. albidus	Flowers, guava	Muramatsu et al., 2009
	G. oxydans	Pineapple, grape, jack fruit, kaffir, lime, papaya, petunia, sapodilla, longan, musk-melon, mangosteen, strawberry	Kommanee et al., 2008b; Muramatsu et al., 2009
	G. kondonii	Flower	Muramatsu et al., 2009
	G. japonicus	Flower, manila, tamarind, plum, mango, logan	Tanasupawat et al., 2009
	G. kanchanaburiensis	Spoiled fruit	Malimas et al., 2009a
	G. uchimurae	Rakam fruit, litchi fruit, logan fruit, jujube fruit	Tanasupawat et al., 2011b
	G. wancherniae	Fruit, seed	Yukphan et al., 2010
	G. nephelii	Rambutan	Kommanee et al., 2011
Vietnam	*Gluconobacter* species	Fruits, flowers, root of *Oryza sativa*, stem of *Saccharum* sp.	Vu et al., 2016b

TABLE 3.4 Diversity of *Asaia* species

Country	Species	Sources	References
India	As. bogorensis	Inner tissues of *Michalia champaca*, abdominal sap of *Anopheles mosquito* and petiole portion of ant *Tetraponera rufonigra*	Samaddar et al., 2011
	As. platicodii	Tropical flower *Michalia champaca*	Samaddar et al., 2011
Indonesia	As. bogorensis	Flowers of the orchid tree (*Bauhinia purpurea*) and of plumbago (*Plumbago auriculata*) and from fermented glutinous rice	Yamada et al., 2000b; Lisdyanti et al., 2003a
	As. siamensis	Flower	Lisdyanti et al., 2003a
Japan	As. astilbes	Astibel, many spiny knotweed, Asian dayflower	Suzuki et al., 2010
	As. platicodii	Balloon flower, unidentified flower	Suzuki et al., 2010
	As. prunellae	Self-heal (*Prunella vulgaris*)	Suzuki et al., 2010
Philippines	As. bogorensis	Flower	Lisdyanti et al., 2003a
	As. siamensis	Flower	Lisdyanti et al., 2003a
Thailand	As. siamensis	Tropical flowers, fruit, seeds, mushroom	Katsura et al., 2001; Vu et al., 2007a; Muramatsu et al., 2009
	As. bogorensis	Flower, fruit, seeds, mushroom	Vu et al., 2007a; Muramatsu et al., 2009
	As. krungthepensis	Heliconia flower, flower, fruit, seeds, mushroom	Yukphan et al., 2004a; Vu et al., 2007a; Muramatsu et al., 2009
	As. lannensis	Flower of spider lily, flower, fruit, seeds, mushroom	Malimas et al., 2008a; Muramatsu et al., 2009
	As. spathodeae	Flower of African tulip	Kommanee et al., 2010
Vietnam	As. bogorensis	Stem of *Oryza sativa*	Vu et al., 2016b
	As. siamensis	Flowers	Vu et al., 2016b

TABLE 3.5 Diversity of *Gluconacetobacter* and *Komagataeibacter* species

Country	Species	Sources	References
China	*Gluconacetobacter* sp.	Fermentation starter of vinegar	Du et al., 2011
India	*Ga. diazotrophicus*	Root and stem of *Eleusine coracana* L., sugarcane, root of sugarcane, tissues of surface sterilized roots, stems, and leaves of crop plant, root, and stem of wetland rice (*Oryza sativa*)	Loganathan et al., 1999; Muthukumarasamy et al., 1999a,b; Suman et al., 2001; Madhaiyan et al., 2004; Muthukumarasamy et al., 2005
Japan	*Ga. tumulicola*	A black viscous substance in a plaster hole at the center of the ceiling in the stone chamber	Tazato et al., 2012
	Ga. asukensis	A brown viscous gel on the north-east area of the ceiling in the stone chamber	Tazato et al., 2012
	Ga. tumulisoli	Clay soil	Nishijima et al., 2013
	Ga. takamatsuzukensis	Burial mound soil	Nishijima et al., 2013
	Ga. aggeris	Burial mound soil	Nishijima et al., 2013
Korea	*Ga. persimmonis*	Persimmon vinegar	Yeo et al., 2004
Thailand	*Ga.liquefaciens*	Sugar cane juice, palm juice, coconut juice	Seearunruangchai et al., 2004
Vietnam	*Ga. liquefaciens*	Fruit, stem of *Saccharum* sp.	Vu et al., 2016b
China	*K. intermedius*	Rotten mandarin fruit	Yang et al., 2013
India	*K. kombuchae*	Kombucha tea	Dutta and Gachhui, 2007; Yamada et al., 2012a,b
	K. hansenii	Contaminated grape wine	Usha Rani and Anu Appaiah, 2013
	K. sucrofermentans		Cleenwerck et al., 2010
	K. xylinus	Fruit samples and Kombucha tea	Neera Ramana and Batra, 2015
Indonesia	*K. xylinus*	Nata de coco, vinegar	Lisdyanti et al., 2003a; Ahmed et al., 2014; Sarkono et al., 2014; Srikanth et al., 2015

TABLE 3.5 *(Continued)*

Country	Species	Sources	References
Israel	*K. hansenii*	Local vinegar	Gosselé et al., 1983; Lisdiyanti et al., 2006; Yamada et al., 1998; Yamada et al., 2012a,b
	K. xylinus		Shigematsu et al., 2005; Bae and Shoda, 2005
Japan	*K. sucrofermentans*	Cherry	Toyosaki et al. 1995; Cleenwerck et al., 2010; Yamada et al., 2012a,b
	K. kakiaceti	Traditional kaki vinegar	Iino et al., 2012b; Yamada et al., 2014
	Komagataeibacter sp.	Persimmon vinegar	Wee et al., 2011
Philippines	*K. xylinus*	Nata de coco, vinegar	Navarro and Komagata, 1999; Lisdyanti et al., 2003a
	K. hansenii	Nata de coco, pineapple	Lisdyanti et al., 2003a
	K. nataicola	Nata de coco	Lisdyanti et al., 2006
Thailand	*K. xylinus*	Fruits	Saeki et al., 1997b
	K. hansenii	Beleric myrobalan, Fetid passion flower, Lady's finger, banana, lychee, star fruit, wild lemon	Suwanposri et al., 2013
	K. sucrofermentans	Fetid passion flower, lychee	Suwanposri et al., 2013
	K. oboediens	Governor's plum, mamao, sapodilla, water melon	Suwanposri et al., 2013
	K. rhaeticus	Rambutan	Suwanposri et al., 2013
	K. swingsii	Sapodilla, water melon	Suwanposri et al., 2013
	Komagataeibacter sp.	Grape, java plum, mamao, mangosteen, papaya, rambutan, sugar apple	Suwanposri et al., 2013, 2014; Valera et al., 2015

aceti 1023, could produce acetic acid in continuous submerged culture at
35°C (Ohmori et al., 1980). In Thailand, thermotolerant AAB produced
acetic acid at 38 to 40°C. They oxidized higher ethanol up to 9% without
a lag time compared to the mesophilic strains under the same conditions.
Vinegar fermentation at higher temperatures was successful in submerged
culture as well as static culture (Saeki et al., 1997a; Saeki et al., 1997b).
In addition, *Acetobacter* strain SKU 1111 oxidized much acetate, which is
not desired in vinegar manufacturing (Saeki et al., 1999). *Acetobacter* sp.
I14–2, from spoiled banana in Taiwan, produced acetic acid two and three
times higher than *A. aceti* IFO 3283 and *Acetobacter* sp. CCRC 12326,
respectively (Lu et al., 1999). *A. aceti, A. indonesiensis, A. cibinongensi,*
and *A. syzygii* were isolated from traditional Iranian dairy products. Only
A. cibinongensis 34L strain, which was isolated from curd, could be introduced as novel candidate probiotics (Haghshenas et al., 2015).

Gluconobacter strains exhibited their ability to incompletely oxidize
a wide range of carbohydrates and alcohols to provide various products,
such as L-sorbose (vitamin C synthesis) from D-sorbitol and 6-amino-
L-sorbose (synthesis of the antidiabetic drug miglitol) (Deppenmeier
et al., 2002). *Gluconobacter* strains can be used industrially to produce
D-gluconic acid, 5-keto- and 2-ketogluconic acids from D-glucose; and
DHA from glycerol. Its possible application in biosensor technology has
also been worked out (Gupta et al., 2001). Recently, *Gluconobacter* strains
involved in the production of L-ribose and miglitol by the biotransformations, both of the productions are very promising pharmaceutical lead
molecules. Among them are dextran dextrinase, capable of transgluco-
sylating substrate molecules, and intracellular NAD-dependent polyol
dehydrogenases, of interest for co-enzyme regeneration (De Muynck et
al., 2007). *G. frateurii* strains from fruits and flowers were able to produce
L-sorbose and D-fructose at 37°C. The effect of increased initial sorbitol
concentrations was investigated in the sorbitol to sorbose bioconversion
process, and the accumulation of sorbose with an increase in productivity
was found in fed-batch culture (Giridhar and Srivastava, 2002). The strains
CHM16 and CHM54 were rapidly oxidized D-mannitol and D-sorbitol to
D-fructose and L-sorbose and gave high fermentation efficiency within 24
h at 30°C (Moonmangmee et al., 2000). *G. frateurii* CHM43 was adapted
to keep GLDH as holoenzyme with the increased PQQ production, and
therefore it produced more L-sorbose and grew better under higher
temperature (Hattori et al., 2012). The mutant strains produced 5KGA

exclusively, and the final yields were over 90% at 30°C and 50% at 37°C (Saichana et al., 2009). *G. frateurii* CHM 43 produced L-erythrulose from meso-erythritol at higher temperatures (Moonmangmee et al., 2002). *G. frateurii* strain transformed glycerol into D-glyceric acid (D-GA) and provided a yield of more than 80 g/l D-GA (Habe et al., 2009a). *Gluconobacter* sp. NBRC 3259 could also convert glycerol to glyceric acid (GA). Using a raw glycerol sample with activated charcoal pretreatment, 45.9 g/l of GA and 28.2 g/l of DHA were produced from 174 g/l of glycerol (Habe et al., 2009b). *G. oxydans* IFO 3244 posed NADP-dependent shikimate dehydrogenae (SKDH, EC 1.1.1.25) that proved to be a useful catalyst for shikimate production from 3-dehydroshikimate (Adachi et al., 2006).

An electrochemical system consisted of *G. oxydans* as a bacterium and 2-hydroxy–1,4-naphthoquinone (HNQ) as a mediator has been set up to determine the effect of initial carbon sources on the glucose detection (Lee et al., 2002). High-yield production (87%) of 5-keto-D-gluconate (5KG) from D-glucose by *Gluconobacter* strain was achieved by controlling the medium pH strictly in a range of pH 3.5–4.0 (Ano et al., 2011). A mutant strain of *G. japonicus* NBRC 3271 expressed the *kgdSLC* genes and engineered to produce 2KG efficiently from a mixture of D-glucose and D-gluconate. This mutant strain consumed D-glucose and D-gluconate to produce 2,5-dike to-D-gluconate (2,5DKG) efficiently and homogeneously. This compound was the intermediate for D-tartrate and also vitamin C production (Kataoka et al., 2015). *G. oxydans* NBRC 3292 possed 4-keto-D-arabonate synthase (4KAS), which converted 2,5DKG to 4-keto-D-arabonate (4KA) in D-glucose oxidative fermentation (Tazoe et al., 2016). *Gluconobacter* strain was reported to use the raw glycerol derived from biodiesel fuel production and the effect of methanol supplementation on GA production (Sato et al., 2013). AAB produced lactobionic acid (β-D-galactopyranosyl-(1→4)-D-gluconic acid, LacA) that has health benefits, such as increased in the intestinal calcium absorption (Brommage et al., 1993). *A. orientalis* KYG22 produced LacA from lactose. *A. orientalis* isolated from Caucasian fermented milk produced lactobionic acid that was suitable for applications in food industries (Kiryu et al., 2009, 2012). *G. frateurii* NBRC3285 produced lactobionic acid and also cellobionic acid (Kiryu et al., 2015). Cellobionic acid is expected to be used for functional foods. D-Hexosaminate production by oxidative fermentation was carried out mainly with *G. frateurii* IFO 3264 (Moonmangmee et al., 2004).

A. aceti and *G. roseus* strains (A-MFC and G-MFC) were used for current generation in batch-type microbial biofuel cells (Karthikeyan et al., 2009). The material, chitosan (chit) was used in batch type microbial fuel cells operated with a mixed consortium of *A. aceti* and *G. roseus* as the biocatalysts and bad wine as a feedstock showed the best performance and have potentials for further development (Karthikeyan et al., 2016). *G. oxydans* ZJB09112 could increase 1,3-dihydroxyacetone production when five times of glycerol feeding were made (Hu et al., 2010). The stable mutant *G. oxydans* ZJB11001 exhibited high DHA productivity after 72 h of fed-batch fermentation at 30°C (Hu et al., 2011). *G. frateurii* CGMCC 5397 could convert crude glycerol to DHA with high yield and productivity when fed-batch fermentation was carried out (Liu et al., 2013). DHA production with immobilized *G. oxydans* cells was done by using modified Haldane substrate-inhibition model (Dikshit and Moholkar, 2016). *G. frateurii* CGMCC 5397 produced high DHA (131.16g/L) in submerged bioreactors (Zheng et al., 2016). *G. oxydans* produced xylonic acid (102.3±3.2 g/L) and 3-dihydroxyacetone (40.6±1.8 g/L) using the system of compressed oxygen supply-sealed and stirred tank reactor system (COS-SSTR) (Zhou et al., 2016a). *G. oxydans* NL71 gave the highest level of glycerol bioconversion into DHA (Zhou et al., 2016b). *G. oxydans* strain DSM 2003 could transform 2,3-butanediol into acetoin, ahigh-valuee feedstock that can be used in dairy and cosmetic products, and chemical synthesis (Wang et al., 2013). A novel water-soluble dextran was synthesized from maltodextrin by cell-free extract of *G. oxydans* DSM 2003 (Wang et al., 2011).

Ga. persimmonis strain KJ 145[T] isolated from the Korean traditional fermented persimmon vinegar was found to form a thick cellulose pellicle (Yeo et al., 2004). *A. pasteurianus* strains IFO 3283, SKU1108 and MSU10 exhibited pellicle formation directly related to acetic acid resistance ability (Kanchanarach et al., 2010). The isolate BPR 2001, a potent cellulose producer in agitation cultures, was examined to determine its taxonomic characteristics and was proposed as a new subspecies, *A. xylinum* subsp. *sucrofermentans* (Toyosaki et al., 1995). Bacterial cellulose (BC) production by BPR2001 using lower molasses concentration was essential for efficient BC production in jar fermentor (Bae and Shoda, 2005). *Ga. xylinus* strain BPR 2001 isolated as a high BC producer when using fructose as the carbon source (Shigematsu et al., 2005). *Gluconacetobacter* strains produced cellulose when cultivated in Hestrin-Schramm (HS) medium statically and isolate PAP1 gave the highest yield (1.15 g/L) of BC. However, the BC yield increased threefold (3.5 g/L) when D-glucose in HS medium was replaced by D-mannitol (Suwanposri et al., 2013). The biocellulose (BC)-producing *Gluconacetobacter* strains isolated from rotten tropical fruits collected in Thailand identified as *Ga. xylinus* group (*Ga. oboediens*, subgroup I, five isolates), *Ga. rhaeticus*, subgroup II, 1 isolate), *Ga. hansenii*, subgroup III, 7 isolates), *Ga. swingsii*, subgroup IV, 2 isolates) and *Ga. sucrofermentans*, subgroup V, 2 isolates) were reported (Suwanposri et al., 2013).

Komagataeibacter sp. PAP1 produced BC using a low-cost medium, soya bean whey (SBW) (Suwanposri et al., 2014). *Ga. xylinus* ANG–29 was reported to produce BC in static fermentation method, shaking at 50 rpm, 100 rpm and 150 rpm, respectively (Sarkono et al., 2014). *Ga. xylinus* strain DFBT produced soft cellulose (yield 5.6 g/L) in non conventional media such as pineapple juice medium and hydrolyzed corn starch medium (Neera Ramana and Batra, 2015). *Acetobacter xylinum* NCIM 2526 produced levan by batch fermentation insucrose-richh medium was characterized using various physicochemical techniques (Ahmed et al., 2014; Srikanth et al., 2015). Levan, a fructose homopolymer offered the general properties such as biocompatibility, biodegradability, renewability, flexibility, and eco-friendliness, and also biomedical properties such as anti-oxidant, anti-inflammatory, anti-carcinogenic, anti-AIDS, and hyperglycaemic inhibitor.

Ga. diazotrophicus strains were used as biofertilizers supported sugar cane yields equivalent to or greater than yields supported by N fertilizers (Muthukumarasamy et al., 1999a,b). These strains exhibited antagonistic potential against *Colletotrichum falcatum* Went, a causal organism of red-rot of sugarcane (Muthukumarasamy, 2000). Some strains from root tissues of carrot (*Daucus carota* L.), raddish (*Raphanus sativus* L.), beet root (*Beta vulgaris* L.) and coffee (*Coffea arabica* L.) produced IAA in the presence of tryptopha, and the isolates efficiently solubilized phosphorus and zinc (Madhaiyan et al., 2004). *Ga. diazotrophicus* isolate IS100 was efficient in promoting plant growth and nutrient uptake in sugarcane (Suman et al., 2005). *Ga. diazotrophicus* and *A. peroxydans*, nitrogen-fixing bacteria were isolated from four different wetland rice varieties cultivated in the state of Tamil Nadu, India. *A. peroxydans* strain was found to present *nifH* genes which were the association of N_2-fixing (Muthukumarasamy et al., 2005). Cell-free culture culture supernatant of *Ga. diazotrophicus* strains showed antagonistic efficacy against root-knot nematode (*Meloidogyne incognita*) infecting cotton *in vitro* and *in planta* including reduced egg hatching by more than 95% and exhibited a paralyzing effect on the infective juveniles (Bansal et al., 2005). Some AAB species can fix atmospheric nitrogen. *Ga. diazotrophicus, Ga. azotocaptans* and *Ga. johannae* are known as nitrogen-fixing species while *S. salitolerans, A. peroxydan,* and *A. nitrogenifigens* were also identified asnitrogen-fixingg bacteria (Dutta and Gachhui, 2006). *Ga. diazotrophicus,* the diazotrophs was isolated from Korean wetland rice varieties grown with long-term application of N and compost and their short-term inoculation effect on rice plants where sugarcane is not cultivated (Muthukumarasamy et al., 2007). *Ga. diazotrophicus* showed variations in their solubilization potential with the strains used and ZnO were effectively solubilized Zn compounds tested (Saravanan et al., 2007). The plant growth, chlorophyll content and plant zinc content were found to be enhanced due to inoculation of *Ga. diazotrophicus* with insoluble zinc compounds (Sarathambal et al., 2009). Application of ZnO with *Ga. diazotrophicus* showed better uptake of the nutrient by the plant (Sarathambal et al., 2010).

3.5 CONCLUSION

AAB are obligate aerobic Gram-negative or Gram-variable, ellipsoidal to rod-shaped cells which are widespread in nature on various plants and fermented foods. They are important bacteria in food industry because of their ability to oxidize many types of sugars and alcohols to organic acids as end products during fermentation. In Asian countries, they were isolated by using various media. These phenotypic characteristics for the identification have now been complemented or replaced by molecular techniques, which are DNA and RNA based techniques. The vinegar production is the best known industrial application of AAB. The unique metabolism of the AAB, governed by their specialized enzymes, enables their use in food production, pharmaceutics, medicinal biotechnology, biosensorics, biotransformations, fine chemicals production, and alternative energy source production. They are also used in cellulose and sorbose production. On the other hand, the oxidizing ability of AAB could have spoilage effect in some products such as in wine. Thus, this chapter covered the diversity and application of AAB in Asia.

KEYWORDS

- acetic acid bacteria
- *Acetobacter*
- *Ameyamaea*
- *Asaia*
- bacterial cellulose
- biofertilizers
- dihydroxyacetone
- *Frateuria*
- *Gluconacetobacter*
- *Gluconobacter*
- keto-D-gluconate
- *Komagataeibacter*

- *Kozakia*
- lactobionic acid
- *Neoasaia*
- *Neokomagataea*
- *Nguyenibacter*
- nitrogen-fixing bacteria
- oxidative products
- *Saccharibacter*
- *Swaminathania*
- *Swingsia*
- *Tanticharoenia*
- vinegar fermentation

REFERENCES

Adachi, O., & Yakushi, T., (2016). Membrane-bound dehydrogenases of acetic acid bacteria. In: Matsushita, K., Toyama, H., Tonouchi, N., & Kainuma, A., (eds.), *Acetic Acid Bacteria: Ecology and Physiology* (pp. 273–298). Springer-Verlag, New York.

Adachi, O., Ano, Y., Toyama, H., & Matsushita, K., (2006). Purification and properties of NADP-dependent shikimate dehydrogenase from *Gluconobacter oxydans* IFO 3244 and its application to enzymatic shikimate production. *Biosci. Biotechnol. Biochem., 70,* 2786–2789.

Adachi, O., Hours, R. A., Akakabe, Y., Tanasupawat, S., Yukphan, P., Shinagawa, E., et al., (2010). Production of 4-keto-D-arabonate by oxidative fermentation with newly isolated *Gluconacetobacter liquefaciens. Biosci. Biotechnol. Biochem., 74,* 2555–2558.

Adachi, O., Moonmangmee, D., Toyama, H., Yamada, M., Shinagawa, E., & Matsushita, K., (2003). New developments in oxidative fermentation. *Appl. Microbiol. Biotechnol., 60,* 643–653.

Ahmed, K. B., Kalla, D., Uppuluri, K. B., & Anbazhagan, V., (2014). Green synthesis of silver and gold nanoparticles employing levan, a biopolymer from *Acetobacter xylinum* NCIM 2526, as a reducing agent and capping agent. *Carbohydr. Polym., 112,* 539–545.

Ameyama, M., (1975). *Gluconobacter oxydans* subsp. *sphaericus*, new subspecies isolated from grapes. *Int. J. Syst. Bacteriol., 25*(4), 365–370.

Andrés-Barrao, C., Benagli, C., Chappuis, M., Ortega, P. R., Tonolla, M., & Barja, F., (2013). Rapid identification of acetic acid bacteria using MALDI-TOF mass spectrometry fingerprinting. *Syst. Appl. Microbiol., 36,* 75–81.

Ano, Y., Shinagawa, E., Adachi, O., Toyama, H., Yakushi, T., & Matsushita, K., (2011). Selective, high conversion of D-glucose to 5-keto-D-gluoconate by *Gluconobacter suboxydans. Biosci. Biotechnol. Biochem., 75,* 586–589.

Arai, H., Sakurai, K., & Ishii, M., (2016). Metabolic features of *Acetobacter aceti.* In: Matsushita, K., Toyama, H., Tonouchi, N., & Kainuma, A., (eds.), *Acetic Acid Bacteria: Ecology and Physiology* (pp. 252–272). Springer-Verlag, New York.

Arifuzzaman, M., Zahid, H. M., Badier, R. S. M., & Kamruzzaman, P. M., (2014). Isolation and characterization of *Acetobacter* and *Gluconobacter* spp. from sugarcane and rotten fruits. *Res. Rev. Biosci., 8*(9), 359–365.

Artnarong, S., Masniyom, P., & Maneesri, J., (2016). Isolation of yeast and acetic acid bacteria from palmyra palm fruit pulp (*Borassus flabellifer* Linn.). *Int. Food Res. J., 23*(3), 1308–1314.

Asai, T., (1935). Taxonomic study of acetic acid bacteria and allied oxidative bacteria in fruits and a new classification of oxidative bacteria, *Nippon Nogeikagaku Kaishi, 11,* 674–708.

Asai, T., Iizuka, H., & Komagata, K., (1964). The flagellation and taxonomy of genera *Gluconobacter* and *Acetobacter* with reference to the existence of intermediate strains. *J. Gen. Appl. Microbiol., 10,* 95–126.

Bae, S. O., & Shoda, M., (2005). Production of bacterial cellulose by *Acetobacter xylinum* BPR 2001 using molasses medium in a jar fermentor. *Appl. Microbiol. Biotechnol., 67,* 45–51.

Baek, C. H., Park, E. H., Baek, S. Y., Jeong, S. T., Kim, M. D., Kwon, J. H., et al., (2014). Characterization of *Acetobacter pomorum* KJY8 isolated from Korean traditional vinegar. *J. Microbiol. Biotechnol., 24*(12), 1679–1684.

Bansal, R. K., Dahiya, R. S., Narula, N., & Jain, R. K., (2005). Management of *Meloidogyne incognita* in cotton using strains of the bacterium *Gluconacetobacter diazotrophicus. Nematol. Mediter., 33,* 101–105.

Bartowsky, E. J., & Henschke, P. A., (2008). Acetic acid bacteria spoilage of bottled red wine – a review. *Int. J. Food Microbiol., 125*(1), 60–70.

Beheshti, M. K., & Shafiee, R., (2010). Isolation and characterization of *Acetobacter* strain from Iranian white-red cherry as a potential strain for cherry vinegar production in microbial biotechnology. *Asian J. Biotechnol., 2*(1), 53–59.

Bringer, S., & Bott, M., (2016). Central carbon metabolism and respiration in *Gluconobacter oxydans.* In: Matsushita, K., Toyama, H., Tonouchi, N., & Kainuma, A., (eds.), *Acetic Acid Bacteria: Ecology and Physiology* (pp. 235–254). Springer-Verlag, New York.

Brommage, R., Binacua, C., Antille, S., & Carrié, A. L., (1993). Intestinal calcium absorption in rats is stimulated by dietary lactulose and other resistant sugars. *J. Nutr., 123,* 2186–2194.

Cavalcante, V. A., & Dobereiner, J., (1988). A new acid-tolerant nitrogen-fixing bacterium associated with sugarcane. *J. Plant Soil, 108,* 23–31.

Cheldelin, V. H., (1961). *Metabolic Pathways of Microorganisms.* Willey, New York, (pp. 1–29).

Chen, H., Chen, T., Giudici, P., & Chen, F., (2016). Vinegar functions on health: Constituents sources and formation mechanisms. *Compr. Rev. Food Sci. Food Saf., 15*(6), 1124–1138.

Chun, H. S., & Kim, S. J., (1993). Characterization of a new acidophilic *Acetobacter* sp. strain HA isolated from Korean traditional fermented vinegar. *J. Microbiol. Biotechnol., 3*(2), 108–114.

Cleenwerck, I., & De Vos, P., (2008). Polyphasic taxonomy of acetic acid bacteria: An overview of the currently applied methodology. *Int. J. Food Microbiol., 125*, 2–14.

Cleenwerck, I., De Vos, P., & De Vuyst, L., (2010). Phylogeny and differentiation of species of the genus *Gluconacetobacter* and related taxa based on multilocus sequence analyses of housekeeping genes and reclassification of *Acetobacter xylinus* subsp. *sucrofermentans* as *Gluconacetobacter sucrofermentans* (Toyosaki et al. 1996) sp. nov., comb. nov. *Int. J. Syst. Evol. Microbiol., 60*, 2277–2283.

Cleenwerck, I., De Wachter, M., Gonzalez, A., De Vuyst, L., & De Vos, P., (2009). Differentiation of species of the family *Acetobacteraceae* by AFLP DNA fingerprinting: *Gluconacetobacter kombuchae* is a later heterotypic synonym of *Gluconacetobacter hansenii. Int. J. Syst. Evol. Microbiol., 59*, 1771–1786.

Çoban, E. P., & Biyik, H., (2011). Effect of various carbon and nitrogen sources on cellulose synthesis by *Acetobacter lovaniensis* HBB5. *Afr. J. Biotechnol., 10*(27), 5346–5354.

De Ley, J., (1961). Comparative carbohydrate metabolism and a proposal for a phylogenetic relationship of the acetic acid bacteria. *J. Gen. Microbiol., 24*, 31–50.

De Ley, J., Gossele, F., & Swings, J., (1984). Genus I *Acetobacter*. In: Williams and Wilkens, (eds.), *Bergey's Manual of Systematic Bacteriology* (Vol. 1, pp. 268–274). Baltimore, USA.

De Muynck, C., Pereira, C. S., Naessens, M., Parmentier, S., Soetaert, W., & Vandamme, E. J., (2007). The Genus *Gluconobacter oxydans*: Comprehensive overview of biochemistry and biotechnological applications. *Crit. Rev. Biotechnol., 27*, 147–171.

De Vero, L., Gullo, M., & Giudici, P., (2010). Acetic acid bacteria, biotechnological applications. In: Wiley, (ed.), *Encyclopedia of Industrial Biotechnology: Bioprocess Bioseparation and Cell Technology* (pp. 9–25). New York.

Deppenmeier, U., Hoffmeister, M., & Prust, C., (2002). Biochemistry and biotechnological applications of *Gluconobacter* strains. *Appl. Microbiol. Biotechnol., 60*, 233–242.

Dhail, S., & Jasuja, N. D., (2012). Isolation of biosurfactant-producing marine bacteria. *Afr. J. Environ. Sci. Technol., 6*(6), 263–266.

Diba, F., Alam, F., & Talukder, A. Z., (2015). Screening of acetic acid producing microorganisms from decomposed fruits for vinegar production. *Adv. Microbiol., 5*, 291–297.

Dikshit, P. K., & Moholkar, V. S., (2016). Kinetic analysis of dihydroxyacetone production from crude glycerol by immobilized cells of *Gluconobacter oxydans* MTCC 904. *Bioresour. Technol., 216*, 948–957.

Du, X. J., Jia, S. R., Yang, Y., & Wang, S., (2011). Genome sequence of *Gluconacetobacter* sp. strain SXCC–1, isolated from Chinese vinegar fermentation starter. *J. Bacteriol., 193*(13), 3395–3396.

Dutta, D., & Gachhui, R., (2006). Novel nitrogen-fixing *Acetobacter nitrogenifigens* sp. nov., isolated from kombucha tea. *Int. J. Syst. Evol. Microbiol., 56*, 1899–1903.

Dutta, D., & Gachhui, R., (2007). Nitrogen-fixing and cellulose producing *Gluconaceto-bacter kombuchae* sp. nov., isolated from Kombucha tea. *Int. J. Syst. Evol. Microbiol., 57*, 353–357.

Ezaki, T., Hashimoto, Y., & Yabuuchi, E., (1989). Fluorometric deoxyribonucleic acid-deoxyribonucleic acid hybridization in microdilution wells as an alternative to membrane filter hybridization in which radioisotopes are used to determine genetic relatedness among bacterial strains. *Int. J. Syst. Bacteriol., 39*, 224–229.

Giridhar, R. N., & Srivastava, A. K., (2002). Productivity improvement in l-sorbose biosynthesis by fed batch cultivation of *Gluconobacter oxydans*. *J. Biosci. Bioeng., 94*, 34–38.

González, Á., Hierro, N., Poblet, M., Mas, A., & Guillamón, J. M., (2005). Application of molecular methods to demonstrate species and strain evolution of acetic acid bacteria population during wine production. *Int. J. Food Microbiol., 102*, 295–304.

Gosselé, J., Swings, J., Kersters, K., & De Ley, J., (1983). Numerical analysis of phenotypic features and protein gel electrophoregrams of *Gluconobacter* Asai 1935, 689 emend. mut. char. Asai, Iizuka and Komagata 1964. *Int. J. Sys. Bacteriol., 33*, 65–81.

Gupta, A., Singh, V. K., Qazi, G. N., & Kumar, A., (2001). *Gluconobacter oxydans*: Its biotechnological applications. *J. Mol. Microbiol. Biotechnol., 3*, 445–456.

Habe, H., Fukuoka, T., Kitamoto, D., & Sakaki, K., (2009a). Biotechnological production of D-glyceric acid and its application. *Appl. Microbiol. Biotechnol., 84*, 445–452.

Habe, H., Shimada, Y., Fukuoka, T., Kitamoto, D., Itagaki, M., Watanabe, K., et al., (2009b). Production of glyceric acid by *Gluconobacter* sp. NBRC 3259 using raw glycerol. *Biosci. Biotechnol. Biochem., 73*, 1799–1805.

Haghshenas, B., Nami, Y., Abdullah, N., Radiah, D., Rosli, R., Abolfazl, B. A., & Khosroushahi, A. Y., (2015). Potentially probiotic acetic acid bacteria isolation and identification from traditional dairies microbiota. *Int. J. Food Sci. Technol., 50*(4), 1056–1064.

Hanmoungjai, W., Chukeatirote, E., Pathom-aree, W., Yamada, Y., & Lumyoung, S., (2007). Identification of acidotolerant acetic acid bacteria isolated from Thailand sources. *Res. J. Microbiol., 2*(2), 194–197.

Hattori, H., Yakushi, T., Matsutani, M., Moonmangmee, D., Toyama, H., Adachi, O., & Matsushita, K., (2012). High-temperature sorbose fermentation with thermotolerant *Gluconobacter frateurii* CHM43 and its mutant strain adapted to higher temperature. *Appl. Microbiol. Biotechnol., 5*, 1531–1540.

Hu, Z. C., & Zheng, Y. G., (2011). Enhancement of 1,3-dihydroxyacetone production by a UV-induced mutant of *Gluconobacter oxydans* with DO control strategy. *Appl. Biochem. Biotechnol., 165*, 1152–1160.

Hu, Z. C., Liu, Z. Q., Zheng, Y. G., & Shen, Y. C., (2010). Production of 1,3-dihydroxyacetone from glycerol by *Gluconobacter oxydans* ZJB09112. *J. Microbiol. Biotechnol., 20*, 340–345.

Huang, C. H., Chang, M. T., Huang, L., & Chua, W. S., (2014). Molecular discrimination and identification of *Acetobacter* genus based on the partial heat shock protein 60 gene (*hsp60*) sequences. *J. Sci. Food Agric., 94*, 213–218.

Iino, T., Suzuki, R., Kosako, Y., Ohkuma, M., Komagata, K., & Uchimura, T., (2012a). *Acetobacter okinawensis* sp. nov., *Acetobacter papayae* sp. nov., and *Acetobacter persicus* sp. nov., novel acetic acid bacteria isolated from stems of sugarcane, fruits and a flower in Japan. *J. Gen. Appl. Microbiol., 58*, 235–243.

Iino, T., Suzuki, R., Tanaka, N., Kosaka, Y., Ohkuma, M., Komagata, K., & Uchimura, T., (2012b). *Gluconacetobacter kakiaceti* sp. nov., an acetic acid bacterium isolated from a traditional Japanese fruit vinegar. *Int. J. Syst. Evol. Microbiol., 62*, 1465–1469.

Ishida, T., Yokota, A., Umezawa, Y., Toda, T., & Yamada, K., (2005). Identification and characterization of Lactococcal and *Acetobacter* strains isolated from traditional Caucasusian fermented milk. *J. Nutr. Sci. Vitaminol., 51*(3), 187–193.

Jia, S., Ou, H., Chen, G., Choi, D., Cho, K., Okabe, M., & Cha, W. S., (2004). Cellulose production from *Gluconobacter oxydans* TQ-B2. *Biotechnol. Bioprocess Eng., 9*(3), 166–170.

Jojima, Y., Mihara, Y., Suzuki, S., Yokozeki, K., Yamanaka, S., & Fudou, R., (2004). *Saccharibacter floricola* gen. nov., sp. nov., a novel osmophilic acetic acid bacterium isolated from pollen. *Int. J. Syst. Evol. Microbiol., 54*, 2263–2267.

Kadere, T. T., Miyamoto, T., Oniango, R. K., Kutima, P. M., & Njoroge, S. M., (2008). Isolation and identification of the genera *Acetobacter* and *Gluconobacter* in coconut toddy (mnazi). *Afr. J. Biotechnol., 7*(16), 2963–2971.

Kanchanarach, W., Theeragool, G., Inoue, T., Yakushi, T., Adachi, O., & Matsushita, K., (2010). Acetic acid fermentation of *Acetobacter pasteurianus*: Relationship between acetic acid resistance and pellicle polysaccharide formation. *Biosci. Biotechnol. Biochem., 74*(8), 1591–1597.

Kappeng, K., & Pathom-aree, W., (2009). Isolation of acetic acid bacteria from honey. *Maejo. Int. J. Sci. Technol., 3*(1), 71–76.

Karthikeyan, R., Krishnaraj, N., Selvam, A., Wong, J. W., Lee, P. K., Leung, M. K., & Berchmans, S., (2016). Effect of composites based nickel foam anode in microbial fuel cell using *Acetobacter aceti* and *Gluconobacter roseus* as a biocatalysts. *Bioresour. Technol., 217*, 113–120.

Karthikeyan, R., Sathish, K. K., Murugesan, M., Berchmans, S., & Yegnaraman, V., (2009). Bioelectrocatalysis of *Acetobacter aceti* and *Gluconobacter roseus* for current generation. *Environ. Sci. Technol., 43*, 8684–8689.

Kataoka, N., Matsutani, M., Yakushi, T., & Matsushita, K., (2015). Efficient production of 2,5-diketo-D-gluconate via heterologous expression of 2-ketogluconate dehydrogenase in *Gluconobacter japonicas*. *Appl. Environ. Microbiol., 81*, 3552–3560.

Katsura, K., Kawasaki, H., Potacharoen, W., Saono, S., Seki, T., Yamada, Y., et al., (2001). *Asaia siamensis* sp. nov., an acetic acid bacterium in the α-*Proteobacteria*. *Int. J. Syst. Evol. Microbiol., 51*, 559–563.

Katsura, K., Yamada, Y., Uchimura, T., & Komagata, K., (2002). *Gluconobacter asaii* Mason and Claus 1989 is a junior subjective synonym of *Gluconobacter cerinus* Yamada and Akita 1984. *Int. J. Syst. Evol. Microbiol., 52*, 1635–1640.

Kersters, K., Lisdiyanti, P., Komagata, K., & Swings, J., (2006). The family *Acetobacteraceae*: The genera *Acetobacter, Acidomonas, Asaia, Gluconacetobacter, Gluconobacte*, and *Kozakia*. In: Dworkin, M., Falkow, S., Rosenberg, E., Schleifer, K H., & Stackebrandt, E., (eds.), *The Prokaryotes* (Vol. 5, pp. 163–200). Springer, New York.

Kiryu, T., Kiso, T., Nakano, H., & Murakami, H., (2015). Lactobionic and cellobionic acid production profiles of the resting cells of acetic acid bacteria. *Biosci. Biotechnol. Biochem., 79*, 1712–1718.

Kiryu, T., Kiso, T., Nakano, H., Ooe, K., Kimura, T., & Murakami, H., (2009). Involvement of *Acetobacter orientalis* in the production of lactobionic acid in Caucasian yogurt ("Caspian Sea Yogurt") in Japan. *J. Dairy Sci., 92*, 25–34.

Kiryu, T., Yamauchi, K., Masuyama, A., Ooe, K., Kimura, T., Kiso, T., et al., (2012). Optimization of lactobionic acid production by *Acetobacter orientalis* isolated from Caucasian fermented milk, "Caspian Sea Yogurt." *Biosci. Biotechnol. Biochem., 76*, 361–363.

Klawpiyapamornkun, T., Bovonsombut, S., & Bovonsombut, S., (2015). Isolation and characterization of acetic acid bacteria from fruits and fermented fruit juices for vinegar production. *Food Appl. Biosci. J., 3*(1), 30–38.

Komagata, K., Iino, T., & Yamada, Y., (2014). The family *Acetobacteraceae*. In: Rosenberg, E., De Long, E. F., Lory, S., Stackebrandt, E., & Thompson, F., (eds.), *The Prokaryotes, Alphaproteobacteria and Betaproteobacteria* (pp. 3–78). Springer, New York.

Kommanee, J., Akaracharunya, A., Tanasupawat, S., Malimas, T., Yukphan, P., Nakagawa, Y., & Yamada, Y., (2008a). Identification of *Acetobacter* strains isolated in Thailand based on 16S–23S rRNA gene ITS restriction and 16S rRNA gene sequence analyses. *Ann. Microbiol., 58*, 319–324.

Kommanee, J., Akaracharunya, A., Tanasupawat, S., Malimas, T., Yukphan, P., Nakagawa, Y., & Yamada, Y., (2008b). Identification of *Gluconobacter* strains isolated in Thailand based on 16S–23S rRNA gene ITS restriction and 16S rRNA gene sequence analyses. *Ann. Microbiol., 58*, 741–747.

Kommanee, J., Tanasupawat, S., Akaracharunya, A., Malimas, T., Yukphan, P., Muramatsu, Y., et al., (2008-2009). Identification of strains isolated in Thailand and assigned to the genera *Kozakia* and *Swaminathania*. *J. Culture Collections, 6*, 61–68.

Kommanee, J., Tanasupawat, S., Yukphan, P., Malimas, T., Muramatsu, Y., Nakagawa, Y., & Yamada, Y., (2010). *Asaia spathodeae* sp. nov., an acetic acid bacterium in the α-*Proteobacteria*. *J. Gen. Appl. Microbiol., 56*, 81–87.

Kommanee, J., Tanasupawat, S., Yukphan, P., Malimas, T., Muramatsu, Y., Nakagawa, Y., & Yamada, Y., (2011). *Gluconobacter nephelii* sp. nov., an acetic acid bacterium in the class *Alphaproteobacteria*. *Int. J. Syst. Evol. Microbiol., 61*, 2117–2122.

Kommanee, J., Tanasupawat, S., Yukphan, P., Thongchul, N., Moonmangmee, D., & Yamada, Y., (2012). Identification of *Acetobacter* strains isolated in Thailand based on the phenotypic, chemotaxonomic, and molecular characterizations. *Science Asia, 38*, 44–53.

Kowser, J., Aziz, M. G., & Uddin, M. B., (2015). Isolation and characterization of *Acetobacter aceti* from rotten papaya. *J. Bangladesh Agril. Univ., 13*(2), 299–306.

Lee, K. W., Shim, J. M., Kim, G. M., Shin, J. H., & Kim, J. H., (2015). Isolation and characterization of *Acetobacter* species from a traditionally prepared vinegar. *Microbiol. Biotechnol. Lett., 43*(3), 219–226.

Lee, S. A., Choi, Y., Jung, S., & Kim, S., (2002). Effect of initial carbon sources on the electrochemical detection of glucose by *Gluconobacter oxydans*. *Bioelectrochem., 57*, 173–178.

Li, L., Wieme, A., Spitaels, F., Balzarini, T., Nunes, O. C., Manaia, C. M., et al., (2014). *Acetobacter sicerae* sp. nov., isolated from cider and kefir, and identification of species of the genus *Acetobacter* by *dnaK, groEL* and *rpoB* sequence analysis. *Int. J. Syst. Evol. Microbiol., 64*, 2407–2415.

Li, R. P., & Macrae, I. C., (1991). Specific association of diazotrophic acetobacters with sugarcane. *Soil Biol. Biochem., 23*, 999–1002.

Lisdiyanti, P., Katsura, K., Potacharoen, W., Navarro, R. R., Yamada, Y., Uchimura, T., & Komagata, K., (2003a). Diversity of acetic acid bacteria in Indonesia, Thailand, and the Philippines. *Microbiol. Cult. Coll., 19*, 91–99.

Lisdiyanti, P., Katsura, K., Seki, T., Yamada, Y., Uchimura, T., & Komagata, K., (2001). Identification of *Acetobacter* strains isolated from Indonesian sources, and proposals of *Acetobacter syzygii* sp. nov., *Acetobacter cibinongensis* sp. nov., and *Acetobacter orientalis* sp. nov. *J. Gen. Appl. Microbiol., 47*, 119–131.

Lisdiyanti, P., Kawasaki, H., Seki, T., Yamada, Y., Uchimura, T., & Komagata, K., (2000). Systematic study of the genus *Acetobacter* with descriptions of *Acetobacter indonesiensis* sp. nov., *Acetobacter tropicalis* sp. nov., *Acetobacter orleanensis* (Henneberg 1950) comb. nov., *Acetobacter lovaniensis* (Frateur 1950) comb. nov., and *Acetobacter estunensis* (Carr 1958) comb. nov. *J. Gen. Appl. Microbiol., 46*, 147–165.

Lisdiyanti, P., Kawasaki, H., Widyastuti, Y., Saono, S., Seki, T., Yamada, Y., et al., (2002). *Kozakia baliensis* gen. nov., sp. nov., a novel acetic acid bacterium in the α-*Proteobacteria. Int. J. Syst. Evol. Microbiol., 52*, 813–818.

Lisdiyanti, P., Navarro, R. R., Uchimura, T., & Komagata, K., (2006). Reclassification of *Gluconacetobacter hansenii* strains and proposals of *Gluconacetobacter saccharivorans* sp. nov., and *Gluconacetobacter nataicola* sp. nov. *Int. J. Syst. Evol. Microbiol., 56*, 2101–2111.

Lisdiyanti, P., Yamada, Y., Uchimura, T., & Komagata, K., (2003b). Identification of *Frateuria aurantia* strains isolated from Indonesian sources. *Microbiol. Cult. Coll., 19*, 81–90.

Liu, Y. P., Sun, Y., Tan, C., Li, H., Zheng, X. J., Jin, K. Q., & Wang, G., (2013). Efficient production of dihydroxyacetone from biodiesel-derived crude glycerol by newly isolated *Gluconobacter frateurii. Bioresour. Technol., 142*, 384–389.

Loganathan, P., Sunita, R., Parida, A. K., & Nair, S., (1999). Isolation and characterization of two genetically distant groups of *Acetobacter diazotrophicus* from new host plant *Eleusine coracana* L. *J. Appl. Microbiol., 87*, 167–172.

Lu, S. F., Lee, F. L., & Chen, H. K., (1999). A thermotolerant and high acetic acid-producing bacterium. *Acetobacter* sp. 114–2. *J. Appl. Microbiol., 86*, 55–62.

Madhaiyan, M., Saravanan, V. S., Jovi, D. B., Lee, H., Thenmozhi, R., Hari, K., & Sa, T., (2004). Occurrence of *Gluconacetobacter diazotrophicus* in tropical and subtropical plants of Western Ghats, India. *Microbiol. Res., 159*, 233–243.

Majeed, A., Abbasi, M. K., Hameed, S., Imran, A., & Rahim, N., (2015). Isolation and characterization of plant growth-promoting rhizobacteria from wheat rhizosphere and their effect on plant growth promotion. *Front. Microbiol., 6*, 198.

Malik, K., & Garg, F. C., (2011). Microbial succession in naturally fermented sliced carrots. *As. J. Food Ag-Ind., 4*(6), 359–364.

Malimas, T., Chaipitakchonlatarn, W., Vu, H. T. L., Yukphan, P., Muramatsu, Y., Tanasupawat, S., et al., (2013). *Swingsia samuiensis* gen. nov., sp. nov., an osmotolerant acetic acid bacterium in the α-*Proteobacteria. J. Gen. Appl. Microbiol., 59*, 375–384.

Malimas, T., Vu, T. H. L., Muramatsu, Y., Yukphan, P., Tanasupawat, S., & Yamada, Y., (2017). Systematics of acetic acid bacteria. In: Sengun, I. Y., (ed.), *Acetic Acid Bacteria: Fundamentals and Food Applications* (pp. 3–43). CRC Press, Florida.

Malimas, T., Yukphan, P., Lundaa, T., Muramatsu, Y., Takahashi, M., Kaneyasu, M., et al., (2009a). *Gluconobacter kanchanaburiensis* sp. nov., a brown pigment-producing acetic acid bacterium for Thai isolates in the *Alphaproteobacteria. J. Gen. Appl. Microbiol., 55*, 247–254.

Malimas, T., Yukphan, P., Takahashi, M., Kaneyasu, M., Potacharoen, W., Tanasupawat, S., et al., (2008a). *Asaia lannaensis* sp. nov., a new acetic acid bacterium in the *Alphaproteobacteria. Biosci. Biotechnol. Biochem., 72*, 666–671.

Malimas, T., Yukphan, P., Takahashi, M., Muramatsu, Y., Kaneyasu, M., Potacharoen, W., et al., (2008b). *Gluconobacter roseus* (ex Asai 1935) sp. nov., nom. rev., a pink-colored acetic acid bacterium in the *Alphaproteobacteria. J. Gen. Appl. Microbiol., 54*, 119–125.

Malimas, T., Yukphan, P., Takahashi, M., Muramatsu, Y., Kaneyasu, M., Potacharoen, W., et al., (2008c). *Gluconobacter sphaericus* (Ameyama 1975) comb. nov., a brown pigment-producing acetic acid bacterium in the *Alphaproteobacteria. J. Gen. Appl. Microbiol., 54*, 211–220.

Malimas, T., Yukphan, P., Takahashi, M., Muramatsu, Y., Kaneyasu, M., Potacharoen, W., et al., (2009b). *Gluconobacter japonicus* sp. nov., an acetic acid bacterium in the *Alphaproteobacteria. Int. J. Syst. Evol. Microbiol., 59*, 466–471.

Malimas, T., Yukphan, P., Takahashi, M., Potacharoen, W., Tanasupawat, S., Nakagawa, Y., et al., (2006). Heterogeneity of strains assigned to *Gluconobacter frateurii* Mason and Claus 1989 based on restriction analysis of 16S–23S rDNA internal transcribed spacer regions. *Biosci. Biotechnol. Biochem., 70*, 684–690.

Matsushita, K., & Matsutani, M., (2016). Distribution, evolution, and physiology of oxidative fermentation. In: Matsushita, K., Toyama, H., Tonouchi, N., & Kainuma, A., (eds.), *Acetic Acid Bacteria: Ecology and Physiology* (pp. 159–178). Springer-Verlag, New York.

Matsushita, K., Toyama, H., & Adachi, O., (1994). Respiratory chain and bioenergetics of acetic acid bacteria. In: Rose, A. H., & Tempest, D. W., (eds.), *Advances in Microbial Physiology* (Vol. 36, pp. 247–301). Academic Press, London.

Matsushita, K., Toyama, H., & Adachi, O., (2004). Respiratory chain in acetic acid bacteria: Membrane-bound periplasmic sugar and alcohol respirations. In: Zannoni, D., (ed.), *Respiration of Archaea and Bacteria* (pp. 81–99). Springer, Dordrecht.

Moonmangmee, D., Adachi, O., Ano, Y., Shinagawa, E., Toyama, H., Theeragool, G., Lotong, N., & Matsushita, K., (2000). Isolation and characterization of thermotolerant *Gluconobacter* strains catalyzing oxidative fermentation at higher temperatures. *Biosci. Biotechnol. Biochem., 64*, 2306–2315.

Moonmangmee, D., Adachi, O., Shinagawa, E., Toyama, H., Theeragool, G., Lotong, N., & Matsushita, K., (2002). L-Erythrulose production by oxidative fermentation is catalyzed by PQQ-containing membrane-bound dehydrogenase. *Biosci. Biotechnol. Biochem., 66*, 307–318.

Moonmangmee, D., Adachi, O., Toyama, H., & Matsushita, K., (2004). D-Hexosaminate production by oxidative fermentation. *Appl. Microbiol. Biotechnol., 66*, 253–258.

Muramatsu, Y., Yukphan, P., Takahashi, M., Kaneyasu, M., Malimas, T., Potacharoen, W., et al., (2009). 16S rRNA gene sequence analysis of acetic acid bacteria isolated from Thailand. *Microbiol. Cult. Coll., 25*(1), 13–20.

Muthukumarasamy, R., Cleenwerck, I., Revathi, G., Vadivelu, M., Janssens, D., Hoste, B., et al., (2005). Natural association of *Gluconacetobacter diazotrophicus* and diazotrophic *Acetobacter peroxydans* with wetland rice. *Syst. Appl. Microbiol., 28*, 277–286.

Muthukumarasamy, R., Kang, U. G., Park, K. D., Jeon, W. T., Park, C. Y., Cho, Y. S., et al., (2007). Enumeration, isolatio, and identification of diazotrophs from Korean wetland rice varieties grown with long-term application of N and compost and their short-term inoculation effect on rice plants. *J. Appl. Microbiol., 102*, 981–991.

Muthukumarasamy, R., Revathi, G., & Lakshminarasimhan, C., (1999a). Diazotrophic associations in sugar cane cultivation in South India. *Trop. Agric., (Trinidad), 76*, 171–178.

Muthukumarasamy, R., Revathi, G., & Lakshminarasimhan, C., (1999b). Influence of N fertilization on the isolation of *Acetobacter diazotrophicus* and *Herbaspirillum* spp. from Indian sugarcane varieties. *Biol. Fertil. Soils, 29*, 157–164.

Muthukumarasamy, R., Revathi, G., & Vadivelu, M., (2000). Antagonistic potential of N_2 fixing *Acetobacter diazotrophicus* against *Colletotrichum falcatum* Went, a causal organism of red-rot of sugarcane. *Curr. Sci., 78*, 1063–1065.

Nanda, K., Taniguchi, M., Ujike, S., Ishihara, N., Mori, H., Ono, H., & Murooka, Y., (2001). Characterization of acetic acid bacteria in traditional acetic acid fermentation of rice vinegar (komesu) and unpolished rice vinegar (kurosu) produced in Japan. *Appl. Environ. Microbiol., 67*, 986–990.

Navarro, R. R., & Komagata, K., (1999). Differentiation of *Gluconacetobacter liquefaciens* and *Gluconacetobacter xylinus* on the basis of DNA base composition, DNA relatedness and oxidation products from glucose. *J. Gen. Appl. Microbiol., 45*, 7–15.

Neera, R. K. V., & Batra, H. V., (2015). Occurrence of cellulose-producing *Gluconacetobacter* spp. in fruit samples and Kombucha tea, and production of the biopolymer. *Appl. Biochem. Biotechnol., 176*, 1162–1173.

Nishijima, M., Tazato, N., Handa, Y., Tomita, J., Kigawa, R., Sano, C., & Sugiyama, J., (2013). *Gluconacetobacter tumulisoli* sp. nov., *Gluconacetobacter takamatsuzukensis* sp. nov. and *Gluconacetobacter aggeris* sp. nov., isolated from Takamatsuzuka Tumulus samples before and during the dismantling work in 2007. *Int. J. Syst. Evol. Microbiol., 63*, 3981–3988.

Ohmori, S., Masai, H., Arima, K., & Beppu, T., (1980). Isolation and identification of acetic acid bacteria for submerged acetic acid fermentation at high temperature. *Agric. Biol. Chem., 44*(12), 2901–2906.

Patil, N. B., & Shinde, S. R., (2014). Isolation, molecular characterization and plant growth promoting traits of *Neoasaia chiangmaiensis* (KD) from banana. *Int. J. Environ. Sci., 5*, 309–319.

Perumpuli, P. A., Watanabe, T., & Toyama, H., (2014). Identification and characterization of thermotolerant acetic acid bacteria strains isolated from coconut water vinegar in Sri Lanka. *Biosci. Biotechnol. Biochem., 78*(3), 533–541.

Pitiwittayakul, N., Theeragool, G., Yukphan, P., Chaipitakchonlatarn, W., Taweesak, M., Muramatsu, Y., et al., (2016). *Acetobacter suratthanensis* sp. nov., an acetic acid bacterium isolated in Thailand. *Ann. Microbiol., 66*, 1157–1166.

Pitiwittayakul, N., Yukphan, P., Chaipitakchonlatarn, W., Yamada, Y., & Theeragool, G., (2015a). *Acetobacter thailandicus* sp. nov., for a strain isolated in Thailand. *Ann. Microbiol., 65*, 1855–1863.

Pitiwittayakul, N., Yukphan, P., Sintuprapa, W., Yamada, Y., & Theeragool, G., (2015b). Identification of acetic acid bacteria isolated in Thailand and assigned to the genus *Acetobacter* by *groEL* gene sequence analysis. *Ann. Microbiol., 65*, 1557–1564.

Poly, F. L., Monrozier, J., & Bally, R., (2001). Improvement in the RFLP procedure for studying the diversity of *nifH* genes in communities of nitrogen fixers in soil. *Res. Microbiol., 152*, 95–103.

Raspor, P., & Goranovič, D., (2008). Biotechnological applications of acetic acid bacteria. *Crit. Rev. Biotechnol., 28*, 101–124.

Roh, S. W., Nam, Y. D., Chang, H. W., Kim, K. H., Kim, M. S., Ryu, J. H., et al., (2008). Phylogenetic characterization of two novel commensal bacteria involved with innate immune homeostasis in *Drosophila melanogaster. Appl. Environ. Microbiol., 74*(20), 6171–6177.

Saeki, A., Matsushita, K., Takeno, S., Taniguchi, M., Toyama, H., Theeragool, G., et al., (1999). Enzymes responsible for acetate oxidation by acetic acid bacteria. *Biosci. Biotechnol. Biochem., 63*, 2102–2109.

Saeki, A., Taniguchi, M., Matsushita, K., Toyama, H., Theeragool, G., Lotong, N., & Adachi, O., (1997a). Microbiological aspects of acetate oxidation by acetic acid bacteria, unfavorable phenomena in vinegar fermentation. *Biosci. Biotechnol. Biochem., 61*, 317–323.

Saeki, A., Theeragool, G., Matsushita, K., Toyama, H., Lotong, N., & Adachi, O., (1997b). Development of thermotolerant acetic acid bacteria useful for vinegar fermentation at higher temperatures. *Biosci. Biotech. Biochem., 61*, 138–145.

Saichana, N., Moonmangmee, D., Adachi, O., Matsushita, K., & Toyama, H., (2009). Screening of thermotolerant *Gluconobacter* strains for production of 5-keto-D-gluconic acid and disruption of flavin adenine dinucleotide-containing D-gluconate dehydrogenase. *Appl. Environ. Microbiol., 75*, 4240–4247.

Sainz, F., Torija, M. J., Matsutani, M., Kataoka, N., Yakushi, T., Matsushita, K., & Mas, A., (2016). Determination of dehydrogenase activities involved in D-glucose oxidation in *Gluconobacter* and *Acetobacter* strains. *Front. Microbiol., 7*, 1358.

Samaddar, N., Paul, A., Chakravorty, S., Chakravorty, W., Mukherjee, J., Chowdhuri, D., & Gachhui, R., (2011). Nitrogen fixation in *Asaia* sp. (Family *Acetobacteraceae*). *Curr. Microbiol., 63*, 226–231.

Sarathambal, C., Thangaraju, M., & Gomathy, M., (2009). Solubilization of insoluble zinc compounds by *Gluconacetobacter diazotrophicus* and its influence on maize. *Asian J. Bio. Sci., 4*, 110–112.

Sarathambal, C., Thangaraju, M., Paulraj, C., & Gomathy, M., (2010). Assessing the Zinc solubilization ability of *Gluconacetobacter diazotrophicus* in maize rhizosphere usinglabeledd ^{65}Zn compounds. *Indian J. Microbiol., 50*, 103–109.

Saravanan, V. S., Madhaiyan, M., & Thangaraju, M., (2007). Solubilization of zinc compounds by the diazotrophic, plant growth promoting bacterium *Gluconacetobacter diazotrophicus*. *Chemosphere, 66*, 1794–1798.

Sarkono, S., Moeljopawiro, S., Setiaji, B., & Sembiring, L., (2014). Physico-chemical properties of bacterial cellulose produced by newly strain *Gluconacetobacter xylinus* ANG–29 in static and shaking fermentations. *Biosci. Biotech. Res. Asia, 11*, 1259–1265.

Sato, S., Morita, N., Kitamoto, D., Yakushi, T., Matsushita, K., & Habe, H., (2013). Change in product selectivity during the production of glyceric acid from glycerol by *Gluconobacter* strains in the presence of methanol. *AMB Express, 3*(1), 20, doi: 10.1186/2191–0855–3–20.

Seearunruangchai, A., Tanasupawat, S., Keeratipibul, S., Thawai, C., Itoh, T., & Yamada, Y., (2004). Identification of acetic acid bacteria isolated from fruits and related materials collected in Thailand. *J. Gen. Appl. Microbiol., 50*, 47–53.

Sengun, I. Y., & Karabiyikli, S., (2011). Importance of acetic acid bacteria in food industry. *Food Control, 22*, 647–656.

Shanker, A. S., Kodaparthi, A., & Pindi, P. K., (2012). Microbial diversity in soft drinks. *J. Pharm. Sci. Innovation, 1*(3), 23–26.

Sharafi, S. M., Rasooli, I., & Beheshti-Maal, K., (2010). Isolation, characterizatio, and optimization of indigenous acetic acid bacteria and evaluation of their preservation methods. *Iran J. Microbiol., 2*(1), 38–45.

Shigematsu, T., Takamine, K., Kitazato, M., Morita, T., Naritomi, T., Morimura, S., & Kida, K., (2005). Cellulose production from glucose using a glucose dehydrogenase gene (*gdh*)-deficient mutant of *Gluconacetobacter xylinus* and its use for bioconversion of sweet potato pulp. *J. Biosci. Bioeng., 99*, 415–422.

Sivasakaran, C., Ramanujam, P. K., Xavier, V. A. J., Arokiasamy, W. J., & Mani, J., (2014). Bioconversion of crude glycerol into glyceric acid: A value added product. *Int. J. Chem. Tech. Res., 6*(12), 5053–5057.

Song, N. E., Cho, H. S., & Baik, S. H., (2016). Bacteria isolated from Korean black raspberry vinegar with low biogenic amine production in wine. *Braz. J. Microbiol., 47*(2), 452–460.

Srikanth, R., Siddartha, G., Sundhar, R. C. H., Harish, B. S., Janaki, R. M., & Uppuluri, K. B., (2015). Antioxidant and anti-inflammatory levan produced from *Acetobacter xylinum* NCIM 2526 and its statistical optimization. *Carbohydr. Polym., 123*, 8–16.

Suman, A., Gaur, A., Shrivastava, A. K., & Yadav, R. L., (2005). Improving sugarcane growth and nutrient uptake by inoculating *Gluconacetobacter diazotrophicus*. *Plant Growth Regul., 47*, 155–162.

Suman, A., Shasany, A. K., Singh, M., Shahi, H. N., Gaur, A., & Khanuja, S. P. S., (2001). Molecular assessment of diversity among endophytic diazotrophs isolated from subtropical Indian sugarcane. *World J. Microbiol. Biotechnol., 17*, 39–45.

Suwanposri, A., Yukphan, P., Yamada, Y., & Ochaikul, D., (2013). Identification and biocellulose production of *Gluconacetobacter* strains isolated from tropical fruits in Thailand. *Maejo Int. J. Sci. Technol., 7*, 70–82.

Suwanposri, A., Yukphan, P., Yamada, Y., & Ochaikul, D., (2014). Statistical optimization of culture conditions for biocellulose production by *Komagataeibacter* sp. PAP1 using soya bean whey. *Maejo Int. J. Sci. Technol., 8*, 1–14.

Suzuki, R., Zhang, Y., Iino, T., Kosako, Y., Komagata, K., & Uchimura, T., (2010). *Asaia astilbes* sp. nov., *Asaia platycodi* sp. nov., and *Asaia prunellae* sp. nov., novel acetic acid bacteria isolated from flowers in Japan. *J. Gen. Appl. Microbiol., 56*, 339–346.

Tanasupawat, S., Kommanee, J., Malimas, T., Yukphan, P., Nakagawa, Y., & Yamada, Y., (2009). Identification of *Acetobacter*, *Gluconobacter*, and *Asaia* strains isolated in Thailand based on 16S–23S rRNA gene internal transcribed spacer restriction and 16S rRNAgene sequence analysis. *Microbes Environ., 24*, 135–143.

Tanasupawat, S., Kommanee, J., Yukphan, P., Maramatsu, Y., Nalagawa, Y., & Yamada, Y., (2011a). *Acetobacter farinalis* sp. nov., an acetic acid bacterium in the α-*Proteobacteria*. *J. Gen. Appl. Microbiol., 57*, 159–167.

Tanasupawat, S., Kommanee, J., Yukphan, P., Moonmangmee, D., Muramatsu, Y., Nakagawa, Y., & Yamada, Y., (2011b). *Gluconobacter uchimurae* sp. nov., an acetic acid bacterium in the α-*Proteobacteria*. *J. Gen. Appl. Microbiol., 57*, 293–301.

Tanasupawat, S., Kommanee, J., Yukphan, P., Nakagawa, Y., & Yamada, Y., (2011c). Identification of *Acetobacter* strains from Thai fermented rice products based on the 16S rRNA gene sequence and 16S–23S rRNA gene internal transcribed spacer restriction analyses. *J. Sci. Food Agric., 91*, 2652–2659.

Tanasupawat, S., Thawai, C., Yukphan, P., Moonmangmee, D., Itoh, T., Adachi, O., & Yamada, Y., (2004). *Gluconobacter thailandicus* sp. nov., an acetic acid bacteriumin the α-*Proteobacteria*. *J. Gen. Appl. Microbiol., 50*, 159–167.

Tazato, N., Nishijima, M., Handa, Y., Kigawa, R., Sano, C., & Sugiyama, J., (2012). *Gluconacetobacter tumulicola* sp. nov. and *Gluconacetobacter asukensis* sp. nov., isolated from the stone chamber interior of the Kitora Tumulus. *Int. J. Syst. Evol. Microbiol., 62*, 2032–2038.

Tazoe, M., Oishi, H., Kobayashi, S., & Hoshino, T., (2016). A single membrane-bound enzyme catalyzes the conversion of 2,5-diketo-D-gluconate to 4-keto-D-arabonate in D-glucose oxidative fermentation by *Gluconobacter oxydans* NBRC 3292. *Biosci. Biotechnol. Biochem., 80*, 1505–1512.

Toyosaki, H., Kojima, Y., Tsuchida, T., Hoshino, K., Yamada, Y., & Yoshinaga, F., (1995). The characterization of an acetic acid bacterium useful for producing bacterial cellulose in agitation cultures: The proposal of *Acetobacter xylinum* subsp. *sucrofermentans* subsp. nov. *J. Gen. Appl. Microbiol., 41*, 307–314.

Trček, J., & Barja, F., (2015). Updates on quick identification of acetic acid bacteria with a focus on the 16S–23S rRNA gene internal transcribed spacer and the analysis of cell proteins by MALDI-TOF mass spectrometry. *Int. J. Food Microbiol., 196*, 137–144.

Usha, R. M., & Anu, A. K. A., (2013). Production of bacterial cellulose by *Gluconacetobacter hansenii* UAC09 using coffee cherry husk. *J. Food Sci. Technol., 50*(4), 755–762.

Valera, M. J., Poehlein, A., Torija, M. J., Haack, F. S., Daniel, R., Streit, W. R., et al., (2015). Draft genome sequence of *Komagataeibacter europaeus* CECT 8546, a cellulose-producing strain of vinegar elaborated by the traditional method. *Genome Announc., 3*(5), e01231–15, doi: 10.1128/genomeA.01231–15.

Vu, H. T. L., Malimas, T., Chaipitakchonlatarn, W., Bui, V. T. T., Yukphan, P., Bui, U. T. T., et al., (2016a). *Tanticharoenia aidae* sp. nov., for acetic acid bacteria isolated in Vietnam. *Ann. Microbiol., 66*, 417–423.

Vu, H. T. L., Malimas, T., Yukphan, P., Potacharoen, W., Tanasupawat, S., Loan, L. T., et al., (2007a). Identification of Thai isolates assigned to the genus *Asaia* based on 16S rDNA restriction analysis. *J. Gen. Appl. Microbiol., 53*, 259–264.

Vu, H. T. L., Malimas, T., Yukphan, P., Potacharoen, W., Tanasupawat, S., Loan, L. T., et al., (2007b). Identification of Thai isolates assigned to the genus *Gluconobacter* based on 16S–23S rDNA ITS restriction analysis. *J. Gen. Appl. Microbiol., 53*(2), 133–142.

Vu, H. T. L., Yukphan, P., Chaipitakchonlatarn, W., Malimas, T., Muramatsu, Y., Bui, U. T. T., et al., (2013). *Nguyenibacter vanlangensis* gen. nov., sp. nov., an unusual acetic acid bacterium in the α-*Proteobacteria*. *J. Gen. Appl. Microbiol., 59*, 153–166.

Vu, H. T. L., Yukphan, P., Muramatsu, Y., Dang, T. P. T., Tanaka, N., Pham, T. H., & Yamada, Y., (2016b). The microbial diversity of acetic acid bacteria in the south of Vietnam. *J. Biotechnol., 14*(1A), 397–408.

Wang, C. Y., Zhang, J., & Gui, Z. Z., (2015). *Acetobacter bacteria* are found in Zhenjiang vinegar grains. *Genet. Mol. Res., 14*(2), 5054–5064.

Wang, S., Mao, X., Wang, H., Lin, J., Li, F., & Wei, D., (2011). Characterization of a novel dextran produced by *Gluconobacter oxydans* DSM 2003. *Appl. Microbiol. Biotechnol., 91*, 287–294.

Wang, X., Lv, M., Zhang, L., Li, K., Gao, C., Ma, C., & Xu, P., (2013). Efficient bioconversion of 2,3-butanediol into acetoin using *Gluconobacter oxydans* DSM 2003. *Biotechnol. Biofuels, 6*, 155, doi: 10.1186/1754–6834–6–155.

Wee, Y. J., Kim, S. Y., Yoon, S. D., & Rye, H. W., (2011). Isolation and characterization of a bacterial cellulose-producing bacterium derived from the persimmon vinegar. *Afr. J. Biotechnol., 10*(72), 16267–16276.

Wu, J., Gullo, M., Chen, F., & Giudici, P., (2010). Diversity of *Acetobacter pasteurianus* strains isolated from solid-state fermentation of cereal vinegars. *Curr. Microbiol., 60*, 280–286.

Yamada, Y., & Akita, M., (1984). *Gluconobacter cerinus*. In validation of the publication of new names and new combinations previously effectively published outside the IJSB, List no. 16. *Int. J. Syst. Bacteriol., 34*, 503–504.

Yamada, Y., & Yukphan, P., (2008). Genera and species in acetic acid bacteria. *Int. J. Food. Microbiol., 125*, 15–24.

Yamada, Y., (2000a). Transfer of *Acetobacter oboediens* Sokollek et al. 1998 and *Acetobacter intermedius* Boesch et al. 1998 to the genus *Gluconacetobacter* as *Gluconacetobacter oboediens* comb. nov. and *Gluconacetobacter intermedius* comb. nov. *Int. J. Syst. Evol. Microbiol., 50*, 2225–2227.

Yamada, Y., (2014). Transfer of *Gluconacetobacter kakiaceti*, *Gluconacetobacter medellinensi*, and *Gluconacetobacter maltaceti* to the genus *Komagataeibacter* as *Komagataeibacter kakiaceti* comb. nov., *Komagataeibacter medellinensis* comb. nov. and *Komagataeibacter maltaceti* comb. nov. *Int. J. Syst. Evol. Microbiol., 64*, 1670–1672.

Yamada, Y., (2016). Systematics of acetic acid bacteria. In: Matsushita, K., Toyama, H., Tonouchi, N., & Kainuma, A., (eds.), *Acetic Acid Bacteria: Ecology and Physiology* (pp. 1–50). Springer-Verlag, New York.

Yamada, Y., Aida, K., & Uemura, T., (1969). Enzymatic studies on the oxidation of sugar and sugar alcohol. V. ubiquinone of acetic acid bacteria and its relation to classification of genera *Gluconobacter* and *Acetobacter*, especially of the so-called intermediate strains. *J. Gen. Appl. Microbiol., 15*, 186–196.

Yamada, Y., Hoshino, K., & Ishikawa, T., (1998). *Gluconacetobacter* corrig. (*Gluconoacetobacter* [sic]). In validation of the publication of new names and new combinations previously effectively published outside the IJSB, List no. 64. *Int. J. Syst. Bacteriol., 48*, 48327–48328.

Yamada, Y., Hosono, R., Lisdiyanti, P., Widyastuti, Y., Saono, S., Uchimura, T., & Komagata, K., (1999). Identification of acetic acid bacteria isolated from Indonesian sources, especially of isolates classified in the genus *Gluconobacter*. *J. Gen. Appl. Microbiol., 45*, 23–28.

Yamada, Y., Katsura, K., Kawasaki, H., Widyastuti, Y., Saono, S., Seki, T., et al., (2000b). *Asaia bogorensis* gen. nov., sp. nov., an unusual acetic acid bacterium in the α-*Proteobacteria*. *Int. J. Syst. Evol. Microbiol., 50*, 823–829.

Yamada, Y., Okada, Y., & Kondo, K., (1976). Isolation and characterization of polarly flagellated intermediate strains in acetic acid bacteria. *J. Gen. Appl. Microbiol., 22,* 237–245.

Yamada, Y., Yukphan, P., Vu, H. T. L., Muramatsu, Y., Ochaikul, D., & Nakagawa, Y., (2012a). Subdivision of the genus *Gluconacetobacter* Yamada, Hoshino and Ishikawa 1998: The proposal of *Komagatabacter* gen. nov., for strains accommodated to the *Gluconacetobacter xylinus* group in the α-*Proteobacteria. Ann. Microbiol., 62,* 849–859.

Yamada, Y., Yukphan, P., Vu, H. T. L., Muramatsu, Y., Ochaikul, D., Tanasupawat, S., & Nakagawa, Y., (2012b). Description of *Komagataeibacter* gen. nov., with proposals of new combinations (*Acetobacteraceae*). *J. Gen. Appl. Microbiol., 58,* 397–404.

Yang, Y., Jia, J., Chen, J., & Lu, S., (2013). Isolation and characteristics analysis of a novel high bacterial cellulose producing strain *Gluconacetobacter intermedius* Cls26. *Carbohydr. Polym., 92*(2), 2012–2017.

Yeo, S. H., Lee, O. S., Lee, I. S., Kim, H. S., Yu, T. S., & Jeong, Y. J., (2004). *Gluconacetobacter persimmonis* sp. nov., isolated from Korean traditional persimmon vinegar. *J. Microbiol. Biotechnol., 14,* 276–283.

Yukphan, P., Malimas, T., Lundaa, T., Muramatsu, Y., Takahashi, M., Kaneyasu, M., et al., (2010). *Gluconobacter wancherniae* sp. nov., an acetic acid bacterium from Thai isolates in the α-*Proteobacteria. J. Gen. Appl. Microbiol., 56,* 67–73.

Yukphan, P., Malimas, T., Muramatsu, Y., Potacharoen, W., Tanasupawat, S., Nakagawa, Y., et al., (2011). *Neokomagataea* gen. nov., with descriptions of *Neokomagataea thailandica* sp. nov. and *Neokomagataea tanensis* sp. nov., osmotolerant acetic acid bacteria of the α-*Proteobacteria. Biosci. Biotechnol. Biochem., 75,* 419–426.

Yukphan, P., Malimas, T., Muramatsu, Y., Takahashi, M., Kaneyasu, M., Potacharoen, W., et al., (2009). *Ameyamaea chiangmaiensis* gen. nov., sp. nov., an acetic acid bacterium in the α-*Proteobacteria. Biosci. Biotechnol. Biochem., 73,* 2156–2162.

Yukphan, P., Malimas, T., Muramatsu, Y., Takahashi, M., Kaneyasu, M., Tanasupawat, S., et al., (2008). *Tanticharoenia sakaeratensis* gen. nov., sp. nov., a new osmotolerant acetic acid bacterium in the α-*Proteobacteria. Biosci. Biotechnol. Biochem., 72,* 672–676.

Yukphan, P., Malimas, T., Potacharoen, W., Tanasupawat, S., Tanticharoen, M., & Yamada, Y., (2005). *Neoasaia chiangmaiensis* gen. nov., sp. nov., a novel osmotolerant acetic acid bacterium in the α-*Proteobacteria. J. Gen. Appl. Microbiol., 51,* 301–311.

Yukphan, P., Malimas, T., Takahashi, M., Kaneyasu, M., Potacharoen, W., Tanasupawat, S., et al., (2006a). Identification of strains assigned to the genus *Asaia* Yamada et al. (2000)-based on 16S rDNA restriction analysis. *J. Gen. Appl. Microbiol., 52,* 241–247.

Yukphan, P., Malimas, T., Takahashi, M., Kaneyasu, M., Potacharoen, W., Tanasupawat, S., et al., (2006b). Phylogenetic relationships between the genera *Swaminathania* and *Asaia*, with reference to the genera *Kozakia* and *Neoasaia*, based on 16S rDNA, 16S–23S rDNA ITS, and 23S rDNA sequences. *J. Gen. Appl. Microbiol., 52,* 289–294.

Yukphan, P., Potacharoen, W., Tanasupawat, S., Tanticharoen, M., & Yamada, Y., (2004a). *Asaia krungthepensis* sp. nov., an acetic acid bacterium in the α-*Proteobacteria. Int. J. Syst. Evol. Microbiol., 54,* 313–316.

Yukphan, P., Takahashi, M., Potacharoen, W., Tanasupawat, S., Nakagawa, Y., Tanticharoen, M., Yamada, Y., (2004b). *Gluconobacter albidus* (ex Kondo and Ameyama 1958) sp. nov., nom. rev., an acetic acid bacterium in the α-*Proteobacteria*. *J. Gen. Appl. Microbiol., 50*, 235–242.

Zahoor, T., Siddique, F., & Farooq, U., (2006). Isolation and characterization of vinegar culture (*Acetobacter aceti*) from indigenous sources. *Brit. Food J., 108*(6), 429–439.

Zhao, H., & Yun, J., (2016). Isolation, identification and fermentation conditions of highly acetoin-producing acetic acid bacterium from Liangzhou fumigated vinegar in China. *Ann. Microbiol., 66*, 279–288.

Zheng, X. J., Jin, K. Q., Zhang, L., Wang, G., & Liu, Y. P., (2016). Effects of oxygen transfer coefficient on dihydroxyacetone production from crude glycerol. *Braz. J. Microbiol., 47*, 129–135.

Zhou, X., Xu, Y., & Yu, S., (2016a). Simultaneous bioconversion of xylose and glycerol to xylonic acid and 1,3-dihydroxyacetone from the mixture of pre-hydrolysates and ethanol-fermented waste liquid by *Gluconobacter oxydans*. *Appl. Biochem. Biotechnol., 178*, 1–8.

Zhou, X., Zhou, X., Xu, Y., & Yu, S., (2016b). Improving the production yield and productivity of 1,3-dihydroxyacetone from glycerol fermentation using *Gluconobacter oxydans* NL71 in compressed oxygen supply-sealed and stirred tank reactor (COS-SSTR). *Bioprocess Biosyst. Eng., 39*, 1315–1318.

CHAPTER 4

SECONDARY METABOLITES AND BIOLOGICAL ACTIVITY OF ACTINOMYCETES

KHOMSAN SUPONG[1] and SOMBOON TANASUPAWAT[2]

[1]*Department of Applied Science and Biotechnology, Faculty of Agro-Industrial Technology, Rajamangala University of Technology, Twan-ok Chantaburi Campus, Chantaburi 22210, Thailand*

[2]*Department of Biochemistry and Microbiology, Faculty of Pharmaceutical Sciences, Chulalongkorn University, Bangkok 10330, Thailand*

4.1 INTRODUCTION

Research to identify new putative antibiotics has to be pursued and intensified. Natural products, especially microbe-derived metabolites, proved themselves as a good source for antibiotics and other biologically active secondary metabolites. Besides the well-known proliferative producer organisms, like the plants and fungi, currently, in the group of bacteria, actinomycetes have moved into focus (Bull et al., 2005). About 50,000 natural products produced by microorganisms, more than 10,000 of which are biological compounds, and more than 1000 metabolites are in use as antibiotics such as antitumor and agrochemical substances (Berdy, 2005). Recently, most of the antibiotics are produced by actinomycetes (Berdy, 2005; Bull et al., 2005). They produce structurally diverse secondary metabolites, distinct from the classes known so far from fungal producers. Actinomycetes are known for the production of an enormous variety of biologically active secondary metabolites, including antibiotics, immunosuppressants, and anticancer agents. In addition, these metabolites such as erythromycin and streptomycin that used clinically (Figure 4.1). Recently, the anticancer drug, salinosporamide A has been found in the marine obligate actinomycete *Salinispora tropica*. In addition, abyssomicin C was isolated from marine-derived *Verrucosispora* sp. (Donadio et al., 2010; Abdelmohsen et al., 2014). This chapter focuses on the diversity of chemical structure and biological activity of secondary metabolites from actinomycetes.

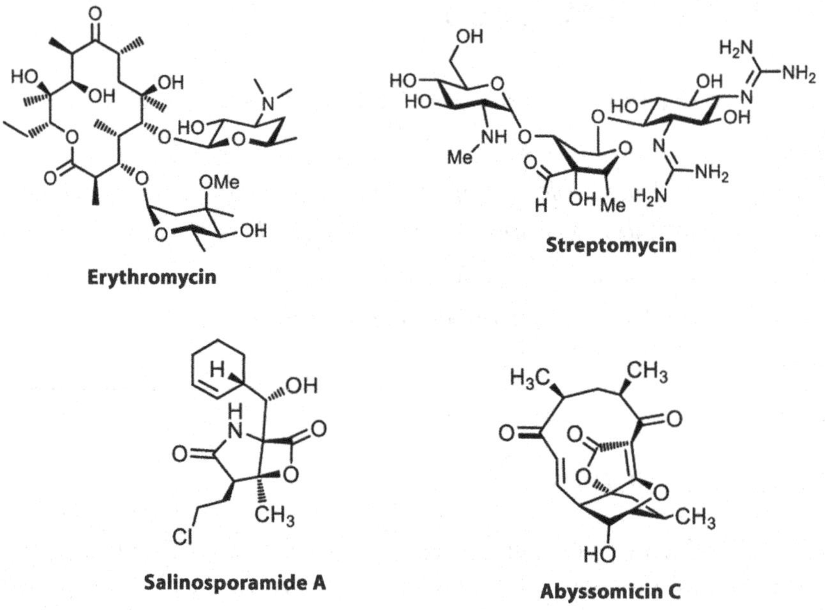

FIGURE 4.1   Examples of chemical structures of antibiotics from actinomycetes.

4.2   POLYKETIDES

In the world of secondary metabolism, polyketides are a significant biosynthetic group and account for the majority of all reported secondary metabolites. These metabolites are assembled by the incorporation of small acid units such as acetate, propionate, and to a lesser extent butyrate and others, though other small organic metabolites such as glycerol and glycolic acid can also be utilized. Even further variety can be added by the incorporation of amino acids along the chain to generate hybrid polyketides. Polyketides are found as metabolites of bacteria, fungi, phytoplankton, plants, and invertebrates. These bioactive polyketides are produced by actinomycetes, belonging to the genus *Streptomyces*. Their high capacity of secondary metabolite biosynthesis is largely different based on the diverse polyketide synthase (PKS), which gives rise to more molecular complex. The polyketide metabolites are major natural products that found in *Streptomycetes*. Lactonemacrolide bafilomycin derivatives were firstly isolated from *Streptomyces* strain. *Streptomyces* sp. YIM26 produced bafilomycins F−J (Carr et al., 2010) was isolated from marine sediment at the Queen Charlotte Islands, British Columbia. In addition, new bafilomycin derivatives such as 19-methoxybafilomycin C1 amide, 21-deoxybafilomycin A1 and 21-deoxybafilomycin A2 were produced by *Streptomyces albolongus* isolated from *Elephas maximus* feces. The bafilomycins thus block lysosomal acidification, and the activity of the acid hydrolases is responsible for autophagic degradation. The observation that bafilomycins A1, B1, D, and F−J block autophagosomal degradation is consistent with their known action. However, bafilomycin I was appeared to stimulate autophagy instead of inhibiting it was unexpected. Bafilomycins A1, B1, and D are potent and selective inhibitors of the vacuolar type H^+-ATPase responsible for maintaining the acidic pH of lysosomes. In addition, these compounds showed cytotoxicity against cancer cells. The macrolide series, micromonospolides A−C could be found in the rare actinomycetes, *Micromnospora* spp. The micromonospolide core structure is similarly formed with bafilomycin group, but it presented the long side chain from the cyclic ring. Micromonospolides A-C inhibited gastrulation of starfish embryos with MIC 0.011 and 1.6 µg/ml, respectively (Ohta et al., 2001). Ansamycin antibiotics comprise a diverse group of bioactive macrolides, including the HSP90 inhibitor geldanamycin derivatives, rifamycin, and the anticancer maytansinoid and ansamitocin P–3. Geldanamycins produced by *Streptomyces hygroscopicus* is now best known as the inhibitor of Hsp90 (DeBoer et al., 1970). This compound displayed comparable binding affinities to the Hsp90α (Wang et al., 2013). In the biosynthesis pathway, geldanamycin was derived from 17-*O*-demethylgeldanamycin by *gdmMT* gene that encoded the missing C–17 *O*-methyltransferase enzyme (Yin et al., 2011). Geldanamycin analogs such as reblastatin derivative and tetracyclic thiazinogeldanamycin were produced by plant-associated *Streptomyces* sp. and *Streptomyces hygroscopicus*, respectively (Stead et al., 2000; Wu et al., 2012). 17-Demethoxyreblastatin was an inhibitor of Hsp90 ATPase (Wu et al., 2012), and it exhibited strong cytotoxicity against NCI-H187 cell line (IC_{50} 0.28 µg/ml). The polyketides of ansamycin class, hygrocins were isolated from the gdmAI-disrupted *Streptomyces* sp. LZ35 which was isolated from intertidal soil collected at Jimei, Xiamen, China. Hygrocins C and D exhibited toxicity to human breast cancer MDA-MB–431 cells (IC_{50} 0.5 and 3.0 µM, respectively). In addition, these compounds showed cytotoxicity with prostate cancer PC3 cells (IC_{50} 1.9, and 5.0 µM, respectively), but hygrocins B and G were showed non-cytotoxicity (Lu et al., 2013). Naphthomycins (A, B, and C) were classified in the ansamycin group, previously were isolated by *Streptomyces galbus* subsp. *griseosporeus* (Keller-Schierlein et al., 1983). In addition, naphthomycin K produced from *Streptomyces* sp. CS was isolated from the medicinal plant *Maytenus hookeri*. Naphthomycin K showed evident cytotoxicity against P388 and A–549 cell lines, but showed no activities against Gram-positive bacteria, *Staphylococcus aureus* and *Mycobacterium tuberculosis* (Lu et al., 2007). In 2012, naphthomycins A, D, E, L, M, and N were produced by *Streptomyces* sp. CS. These compounds were assayed for antifungal activity, but only naphthomycins A and L displayed any evident inhibition of phytopathogenic fungi, with MICs 300 and 100 µg/ml, respectively, against *Fusarium moniliforme* and 600 and 400 µg/ml, respectively, against several other phytopathogenic fungi (*Fusarium oxysporum, Fusarium moniliforme, Gaeumannomyces graminis, Verticillium cinnabarium,* and *Phyricularia oryzae*). Naphthomycins D, E, M, and N showed no activity against phytopathogenic fungi at 800 µg/ml (Yang et al., 2012). Rare actinomycete, *Saccharothrix xinjiangensis* NRRL B–24321 produced 16-membered macrolide tianchimycins A and B (Wang et al., 2013). These compounds showed weak antibacterial activity against *Enterococcus faecalis* ATCC 29212.

The macrolide in the group of macrodiolide series such as elaiophylin also known as azalomycin B and gopalamycin, form a glycosylated macrodiolide antibiotic that produced by different *Streptomyces* species. Elaiophylin derivatives were produced from *Streptomyces* isolated from various sources, such as terrestrial *Streptomyces* (Sheng et al., 2015), marine-derived strain (Wu et al., 2013) and endophytic actinomycete (Supong et al., 2016). Previously, elaiophylin derivatives showed anti-Gram-positive bacteria against *Bacillus subtilis, S. aureus, Enterococcus faecium* and *Mycobacterium* sp. (Ritzau et al., 1998). In addition, macrodiolide efomycins G and M were obtained from the endophytic *Streptomyces* BCC 72023 that isolated from Thai rice (*Oryza sativa* L.).

These compounds exhibited antimicrobial activity against *Plasmodium falciparum* and Gram-positive bacteria *Bacillus cereus*, and *M. tuberculosis*. Efomycins G and M showed cytotoxicity against cancer cells (KB, NCF–7, NCI-H187) with IC_{50} values ranging from 1.53 to 23.50 µg/ml (Supong et al., 2016). The chemical structures are shown in Figure 4.2.

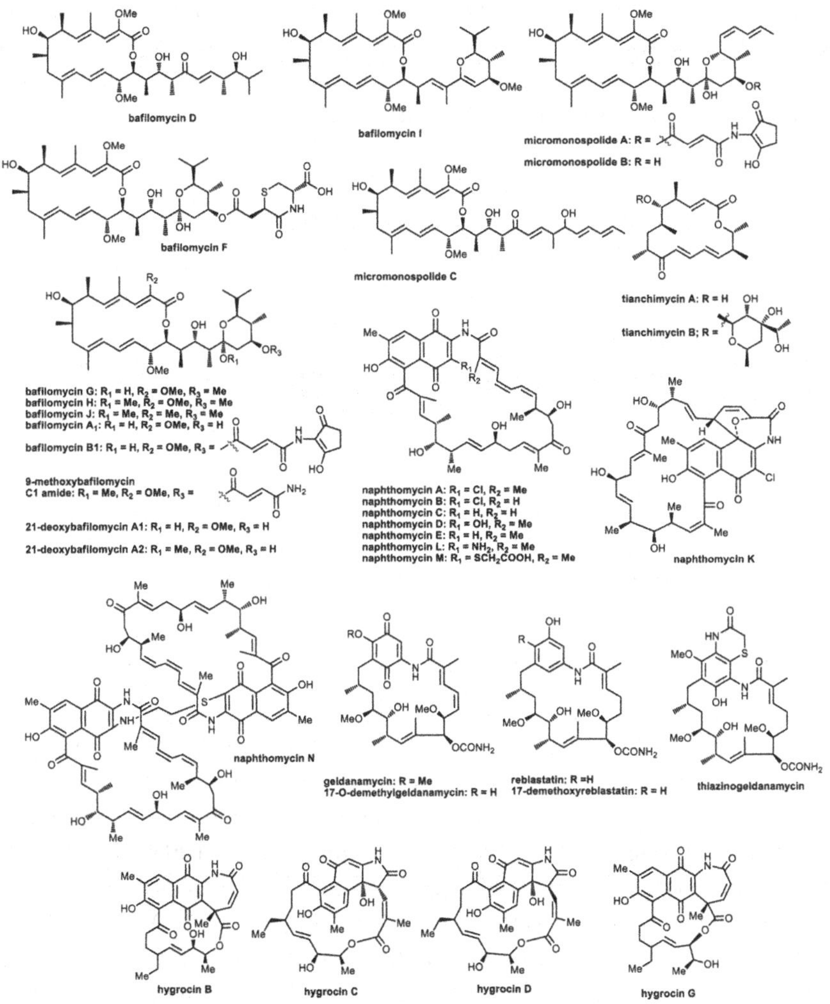

FIGURE 4.2 Polyketide metabolites from actinomycetes.

Samroiyotmycins A and B produced by *Streptomyces* sp. BCC 33756 isolated from soil collected at Khao Sam Roi Yot National Park, Prachuap Khiri Khan Province, Thailand (Dramae et al., 2013). Samroiyotmycins A and B showed antimalarial activity with IC_{50} values 3.65 and 3.16 µg/ml, respectively and exhibited weak cytotoxic to cancer cells. The marine-derived *Streptomyces* sp. MDG–04–17–069 produced macrodiolide tartrolon D which displayed strong cytotoxic activity to human tumor cell lines including lung (A549), colon (HT29), and breast (MDA-MB–231) with GI_{50} values ranging from 0.16–0.79 µM (Pérez et al., 2009). The chemical structures are shown in Figure 4.3.

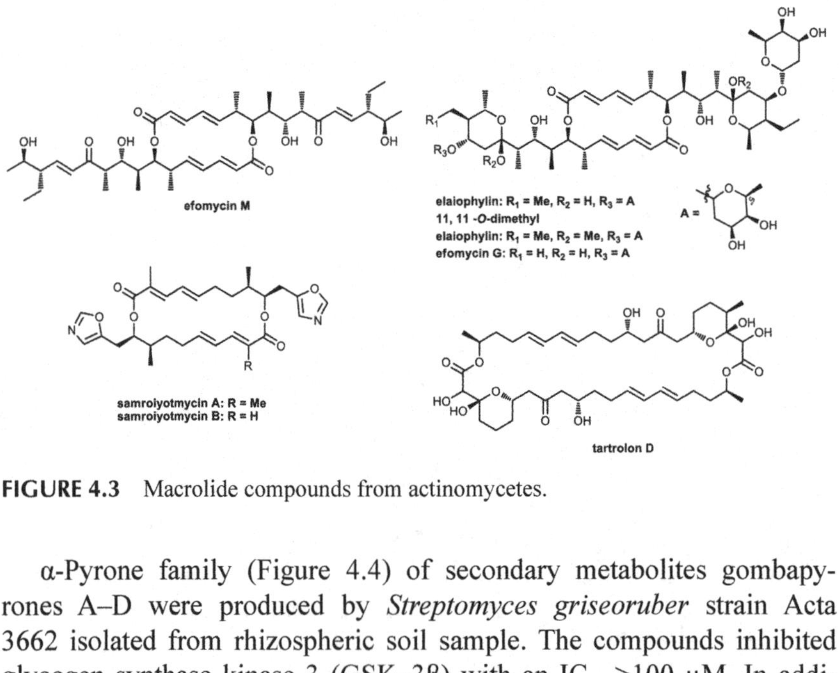

FIGURE 4.3   Macrolide compounds from actinomycetes.

α-Pyrone family (Figure 4.4) of secondary metabolites gombapy-rones A–D were produced by *Streptomyces griseoruber* strain Acta 3662 isolated from rhizospheric soil sample. The compounds inhibited glycogen synthase kinase 3 (GSK–3β) with an IC_{50} >100 µM. In addition, gombapyrones A, B, and D also exhibited weak activity to human recombinant protein tyrosine phosphatase 1B (PTPN1). Albidopyrone, α-pyrone-containing metabolite produced by marine *Streptomyces* sp. NTK 227. Albidopyrone did not show any growth inhibitory activity either against the Gram-negative bacteria, *Escherichia coli* and *Pseudomonas fluorescens*, and Gram-positive bacteria *B. subtilis* and *Staphylococcus lentus*, but did not inhibit *Candida glabrata*. Albidopyrone inhibited the

PTP1B with an IC_{50} of 128 mg/ml (Hohmann et al., 2009). Violapyrones A–G, α-pyrone derivatives were obtained from *Streptomyces violascens* that isolated from *Hylobates hoolock* feces that collected from Yunnan Wild Animal Park, Kunming, Yunnan Province, P. R. China (Zhang et al., 2013). These compounds showed inactive against the five cell lines used to evaluate cytotoxicity (IC_{50} > 200 μM). Violapyrones A–C exhibited moderate antibacterial activities against *B. subtilis* and *S. aureus* (4 to 32 μg/ml of MIC values) while violapyrones D–G displayed weak activity or were no activity to *B. subtilis* and *S. aureus*, and none of them were effective against other tested microorganisms.

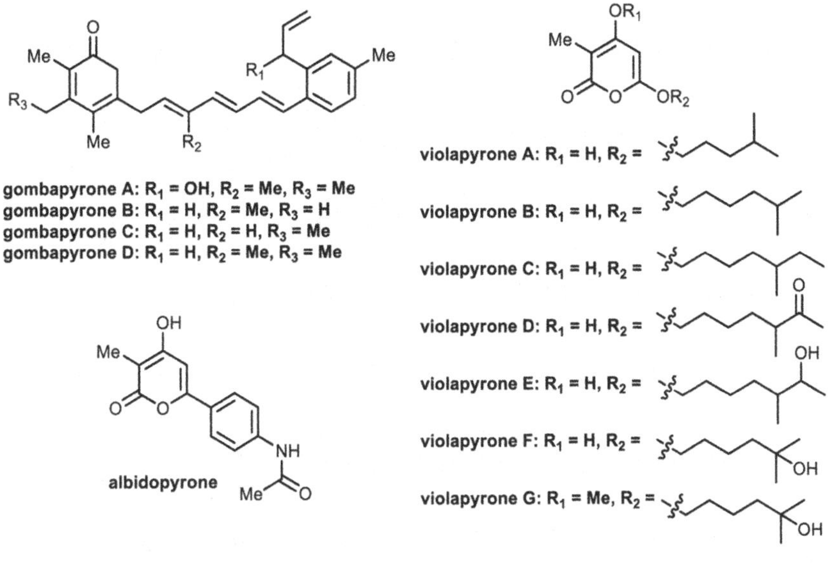

gombapyrone A: R_1 = OH, R_2 = Me, R_3 = Me
gombapyrone B: R_1 = H, R_2 = Me, R_3 = H
gombapyrone C: R_1 = H, R_2 = H, R_3 = Me
gombapyrone D: R_1 = H, R_2 = Me, R_3 = Me

albidopyrone

violapyrone A: R_1 = H, R_2 =
violapyrone B: R_1 = H, R_2 =
violapyrone C: R_1 = H, R_2 =
violapyrone D: R_1 = H, R_2 =
violapyrone E: R_1 = H, R_2 =
violapyrone F: R_1 = H, R_2 =
violapyrone G: R_1 = Me, R_2 =

FIGURE 4.4 α-Pyrone family from actinomycetes.

C-glycoside angucyclines (Figure 4.5) such as urdamycin E, urdamycinone G and E, and dehydroxyaquayamycin were produced by *Streptomyces* sp. BCC 45596 from marine sponge (Supong et al., 2012). These compounds mostly showed cytotoxicity against cancerous (KB, MCF–7, NCI-H187) and non-cancerous (Vero) cells. In addition, urdamycinones E, G, and dehydroxyaquayamycin exhibited strong antimalarial activity with IC_{50} values of 0.053, 0.142, and 2.93 mg/ml and showed activity against *Mycobacterium tuberculosis* with MIC values of 3.13, 12.50, and 6.25 mg/ml, respectively.

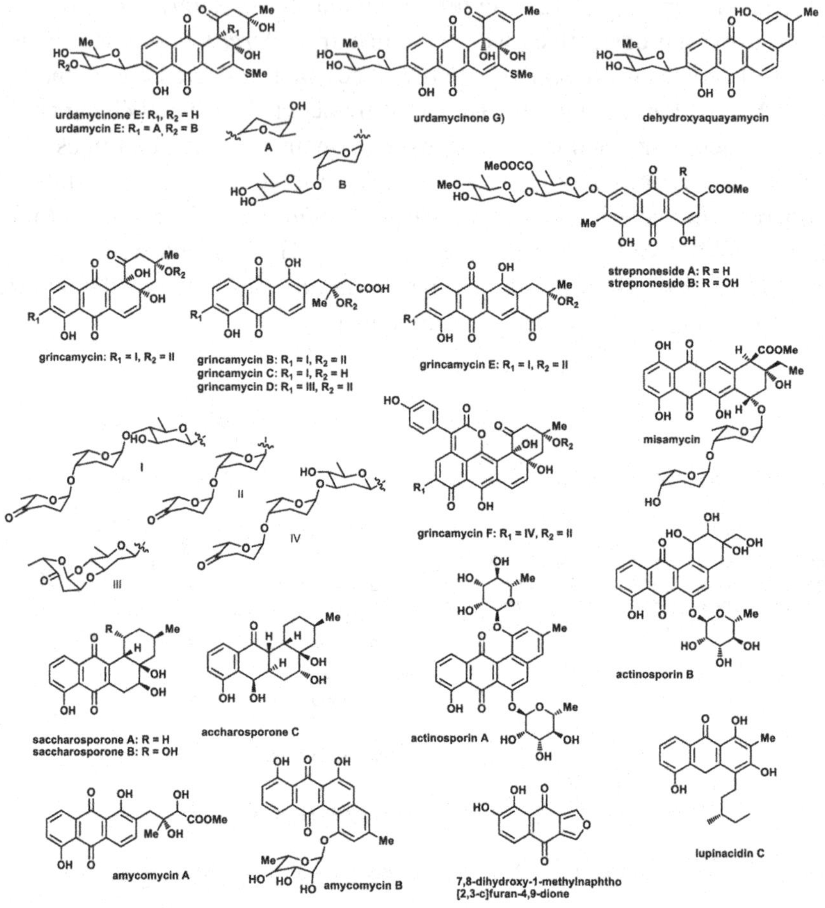

FIGURE 4.5 Quinone glycosides from actinomycetes.

An actinomycete strain WBF–16 isolated from the sediment of soil sample of Wei-Hai Sea in China, produced the anthraquinone glycosides that comprised strepnonesides A and B (Lu et al., 2012). For bioassays, three isolated compounds were tested for cytotoxic activity against HCT116 cell lines. Strepnonesides A and B were exhibited activities with IC_{50} values of 30.2 mM, and 40.2 mM. *Streptomyces lusitanus* SCSIO LR32 isolated from deep-sea produced grincamycin analogs that showed antiproliferative activities with the selected panel of cells; IC_{50} values ranged from 1.1 to 11 μM (Huang et al., 2012). Grincamycin derivatives

(B–E) showed cytotoxicity toward these cell lines with IC_{50} values ranging from 2.1 to 31 μM. It is worth noting that grincamycin F differs from grincamycin primarily in the structure of its enlarged aglycone, consist of a six-membered lactone ring and a hydroxybenzene in addition to the typical angucycline four-ring system.

Anthracycline misamycin isolated from the endophytic *Streptomyces* sp. YIM66403 showed modest antibacterial activity against *S. aureus* with an MIC value 64 μg/ml, but it did not show activity against fungus *Monilia albican* and Gram-negative bacterium, *E. coli*. In addition, misamycin exhibited modest cytotoxicity against the tumor that comprised HL–60, SMMC–7721, MCF–7, A–549, and SW4801, with IC_{50} values of 15.37, 16.34, 25.98, 20.71 and 9.75 μM, respectively (Li et al., 2015).

In addition, the angucycline antibiotics were found in rare actinomycetes including saccharosporones A–C that produced by *Saccharopolyspora* sp. BCC 21906 (Boonlarppradab et al., 2013). Saccharosporones A and B exhibited activity against *Plasmodium falciparum* KI with IC_{50} values 4.1 and 3.9, respectively, but they showed cytotoxic activities against cancer cells. *Actinokineospora* sp. strain EG49 from the marine sponge *Spheciospongia vagabunda*, produced angucycline antibiotics, actinosporins A and B. The angucycline series were found in the insect-derived actinobacterium *Amycolatopsis* sp. HCa1, which produced news amycomycins A and B, and showed cytotoxicity against A375 and KB cells (Gou et al., 2012). The metabolites in the group of anthraquinone, lupinacidin C produced by *Micromonospora lupini* Lupac 08 that isolated from the root nodule of *Lupinus angustifolius* (Igarashi et al., 2011). This compound displayed anti-invasive activity with IC_{50} value 0.019 μg/ml. Isofuranonaphthoquinone, 7,8-dihydroxy-1-methylnaphtho[2,3-c]furan-4,9-dione, was isolated from *Actinoplanes* sp. strain ISO06811 but showed no any bioactivity against bacteria, *B. subtilis* 1A1 and *E. coli* K12 (Zhang et al., 2009).

4.3 TERPENOIDS

Terpenoids are produced by condensation of isopentenyl diphosphate (IPP) and its isomer, dimethylallyl diphosphate (DMAPP). These starting materials are produced via the mevalonate pathway in eukaryotes, archaea, and the cytoplasm of plants, while the methylerythritol phosphate pathway (MEP) is used in prokaryotes and the chloroplasts of plants These metabolites rarely found in the bacteria while some actinomycete strains could produce the terpenoids through the mevalonate pathway (Motohashi et al., 2008).

Oxaloterpins A–E and viguiepinone were isolated from the culture broth of *Streptomyces* sp. KO–3988. The diterpene oxaloterpins A showed anti-bacterial activity against Gram-positive, *B. subtilis* ATCC 43223 and *S. aureus* ATCC 29213 with IC_{50} values of 1.9 and 3.7 nM, respectively (Motohashi et al., 2007). Napyradiomycin SR, 16-dechloro–16-hydroxyna-pyradiomycin C2, 18-hydroxynapyradiomycin A1, 18-oxonapyradiomycin A1, 16-oxonapyradiomycin A2, 7-demethyl SF2415A3, 7-demethyl A80915B, and (R)–3-chloro–6-hydroxy–8-methoxy-R-lapachone were isolated from the *Streptomyces antimycoticus* NT17 which exhibited antibacterial activities (Motohashi et al., 2008). Anti-influenza virus compounds from *Streptomyces* sp. RI18, which was isolated, using the membrane filter method, uncovered a terpene JBIR–68 inhibited influenza virus growth in plaque assays (Takagi et al., 2010). The tricyclic diterpene skeleton cyclooctatin was previously isolated from *Streptomyces melano-sporofaciens*, showed activity for lysophospholypase inhibition (Aoyagi et al., 1992). In addition, this compound was obtained from *Streptomyces* sp. LZ35 (Zhao et al., 2013). Diterpene 17-hydroxycyclooctatin produced by *Streptomyces* sp. MTE4a from the soil sample obtained from Phoenicia, NY which exhibited weak antibacterial activity against *S. aureus* (Kawamura et al., 2011).

Actinomadurol and derivative JBIR–65 produced by *Actinomadura* sp. KC 191 isolated from soil, exhibited significant inhibitory activities comparable to or even stronger than those of ampicillin against *S. aureus* (MIC 0.78 μg/ml), *B. subtilis* (MIC 1.56 μg/ml), *Kocuria rhizophila* (MIC 0.39 μg/ml), *Proteus hauseri* (MIC 0.78 μg/ml), and *Salmonella enterica* (MIC 3.12 μg/ml), but did not inhibit *E. coli* (MIC > 100 μg/ml) (Shin et al., 2016). For biological evaluations of actinomadurol and JBIR–65 using cytotoxicity against various cancer cells (A549, HCT116, SNU638, SK-HEP1, MDA-MB231, and K562), revealed that actinomadurol and JBIR–65 did not inhibit the growth of the tested cancer cell lines, even at high concentrations with $IC_{50} > 100$ μg/ml. In addition, the two compounds did not display anti-fungal activity against *Candida albicans*, *Aspergillus fumigatus*, *Trichophyton rubrum*, and *Trichophyton mentagrophytes* (MIC > 100 μg/ml). The chemical structures of terpene series are shown in Figure 4.6.

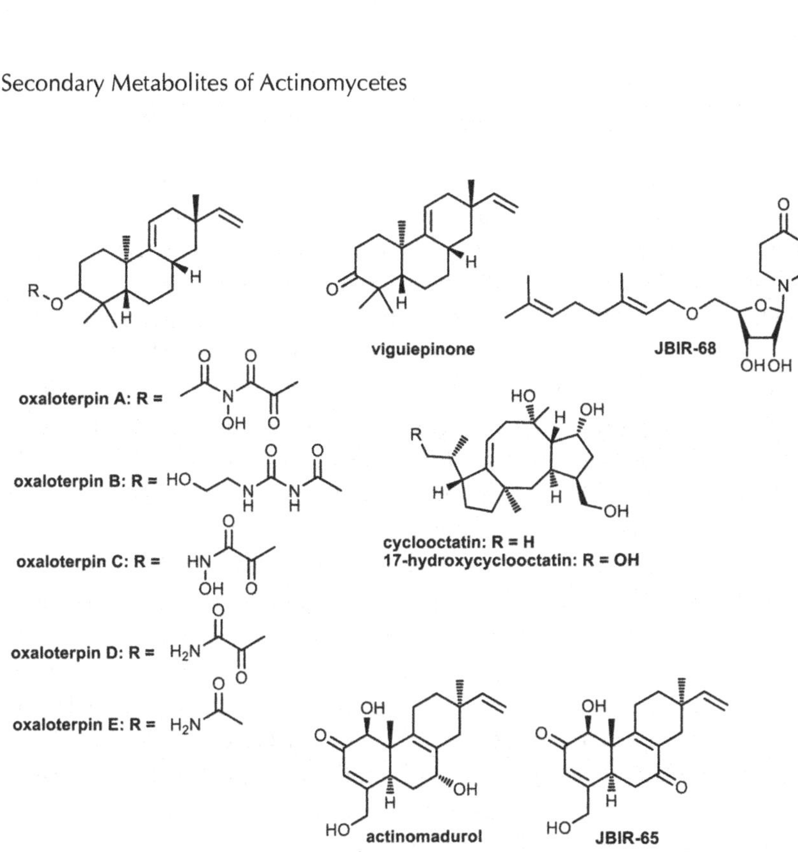

FIGURE 4.6 Terpenoids from actinomycetes.

4.4 PEPTIDES

Cyclohexadepsipeptides, arenamides A-C, were produced by the obligated marine actinomycete, *Salinispora arenicola*. Arenamides A and B blocked the tumor necrosis factor (TNF)-induced activation in a dose- and time-dependent manner with IC_{50} values of 3.7 and 1.7 µM, respectively. In addition, these compounds inhibited prostaglandin E2 (PGE2) and nitric oxide (NO) production with lipopolysaccharide (LPS)-induced RAW 264.7 macrophages. The compounds displayed moderate cytotoxic activity against the human colon carcinoma cell line HCT–116, but did not display activity against cultured RAW cells (Asolkar et al., 2009).

Secondary metabolites in the cultures of the *Streptomyces* sp. Sp080513GE–23 isolated from the sponge was found to possess several unique nonribosomal peptide synthetase (NRPS) genes by screening for biosynthetic genes of secondary metabolites to give two peptides JBIR–34 and JBIR–35. Compounds JBIR–34 and JBIR–35 exhibited weak antioxidant activity with DPPH radical scavenging (IC_{50} values of 1.0 and 2.5 µM, respectively). In addition, JBIR–34 and JBIR–35 did not show cytotoxic activity against several cancer cells and antibacterial activity against Gram-positive, *M. luteus* and Gram-negative, *E. coli*. In addition, phenyl acetylated peptide, JBIR–96, isolated from *Streptomyces* sp. RI051-SDHV6 did not display cytotoxic or antimicrobial activity (Ueda et al., 2011). The nonribosomal peptide producing could be found in the strain BCC 26924, which was identified as *Streptomyces*. The non-ribosomal peptide compound displayed strong activity against *Plasmodium falciparum* (IC_{50} 0.2 µg/ml) and *M. tuberculosis* (MIC 0.12 µg/ml), and this compound was elucidated as heptapeptide cyclomarin C (Intaraudom et al., 2011). A cyclic lipodepsipeptide tumescenamide C has contained a sixteen-membered depsipeptide portion and a methyl-branched acyl group. This compound was isolated from a culture broth of *Streptomyces* sp. KUSC_F05. Tumescenamide C displayed anti-steptomycetes with high selectivity against *Streptomyces* species such as *Streptomyces coelicolor* A3 and *Streptomyces lividans* TK23. This compound did not exhibit antibacterial activity against *B. subtilis*, *S. aureus*, *E. coli* and *P. aeruginosa* (Kishimoto et al., 2012). Sungsanpin is 15-amino-acid peptide, was discovered from a *Streptomyces* sp. isolated from deep-sea sediment collected at Jeju Island, Korea. This sungsanpin inhibited the human lung cancer cell line A549 (Um et al., 2013).

Marine-derived *Streptomyces* sp. strain CMBM0244 was isolated from sediment sample that collected from the South Molle Island, Queensland, produced bioactive peptide mollemycin A. This compound showed antibacterial activity against *S. aureus* ATCC 25293 (IC_{50} 50 nM) and ATCC 9144 (IC_{50} 10 nM), *S. epidermidis* ATCC 12228 (IC_{50} 50 nM), *B. subtilis* ATCC 6051 (IC_{50} 10 nM) and ATCC 6633 (IC_{50} 10 nM), as well *E. coli* ATCC 25922 (IC_{50} 10 nM) and *P. aeruginosa* ATCC 27853 (IC_{50} 50 nM). Mollemycin A displayed moderate activity against *Mycobacterium bovis* (BCG) with IC_{50} 3.2 µM, while it did not exhibit antifungal activity against *C. albicans* ATCC 90028. In addition, this compound exhibited inhibition of malaria parasite against drug sensitive (3D7; IC_{50} 9 nM) and multidrug-resistant (Dd2; IC_{50} 7 nM) *Plasmodium falciparum* (Raju et al., 2014). Anti-infective cycloheptadepsipeptides marformycins A–F were isolated from the deep sea-derived *Streptomyces drozdowiczii* SCSIO 10141 (Zhou et al., 2014). Marformycins A–E were tested for their antibacterial activities against a panel of Gram-positive and Gram-negative bacteria, including *M. luteus*, *Aeromonas hydrophila* subsp. *hydrophila* ATCC 7966,

B. subtilis ATCC 6633, *S. aureus* ATCC 29213 and *E. coli* ATCC 25922. Marformycins A–E only exhibited *in vitro* inhibitory against *M. luteus* with MICs of 0.25, 4.0, 0.25, 0.063 and 4.0 mg/ml, respectively. Marformycins were also evaluated for potential cytotoxic activities against the human breast adenocarcinoma MCF–7, the human glioblastoma SF–268, and the human large-cell lung carcinoma NCI-H460 cell lines. None of the cyclopeptides were found to display cytotoxic activity against any of the three cell lines (IC_{50} >100 mM) (Zhou et al., 2014).

In addition, secondary metabolites in the group of non-ribosomal peptide were found in the rare actinomycetes such as peptidolipins B–F that these compounds formed the lipopeptide. peptidolipins were isolated from marine-derived *Nocardia* sp. Peptidolipins B–F were tested for antibacterial activity against methicillin-sensitive *S. aureus* (MSSA) and methicillin-resistant *S. aureus* (MRSA) (Wyche et al., 2012). Peptidolipins B and E showed activity against MSSA and MRSA at MIC of 64 µg/ml. Peptidolipins C, D, and F have inhibited MSSA and MRSA at MIC value more than 64 µg/ml. In order to identify the activity of peptidolipins B and E were bactericidal or bacteriostatic, a sterile swab was dipped into each well that showed inhibition of bacterial growth and was inoculated on an LB plate. A bactericidal agent would show no growth on the LB plate, while a bacteriostatic agent would show bacterial growth. Thiodepsipeptides were firstly isolated from *Verrucosispora* sp. isolated from a marine sponge (*Chondrilla caribensis*). Thiocoraline analogs, which comprised 22′-deoxythiocoraline and 12′-sulfoxythiocoraline demonstrated significant cytotoxicity against the A549 human cancer cell line with EC_{50} values of 0.13and 1.26 µM, respectively (Wyche et al., 2011). The non-ribosomal peptide structures are in Figure 4.7. Rakicidin A was produced by *Micromonospora* sp. TP-A0860 and showed cytotoxicity against HCT–8 and PANC–1 (Oku et al., 2014). These functionalities should be conserved when exploiting the structure of rakicidin A for drug development and designing bioprobes to study its mode of action.

4.5   ALKALOIDS AND OTHER NITROGEN CONTAINING METABOLITES

Alkaloid is nitrogen containing metabolites that mostly found in plants. In addition, these compounds have been produced by microorganisms such as fungi and actinomycetes. The alkaloid bioactive metabolites mostly found in streptomycetes, but some structures were isolated from rare actinomycetes (Figure 4.8), for example, streptonigrin, and 7-(1-methyl–2-oxopropyl) streptonigrin were produced by *Micromonospora* sp. IM 2670 (Wang et al., 2002). The biological activity of streptonigrin and derivative exhibited very potent and selective cytotoxicity. Streptonigrin displayed toxicity against the SH-SY5Y (IC$_{50}$ 0.05 µM) and the SH-SY5Y–5.6 (IC$_{50}$ 4.6 µM) cell lines, suggesting that apoptosis occurs through a p53-dependent pathway. 7-(1-methyl–2-oxopropyl) streptonigrin showed activity against SH-SY5Y and SH-SY5Y–5.6 cell lines with IC$_{50}$ values 0.9 and 75 µM, respectively.

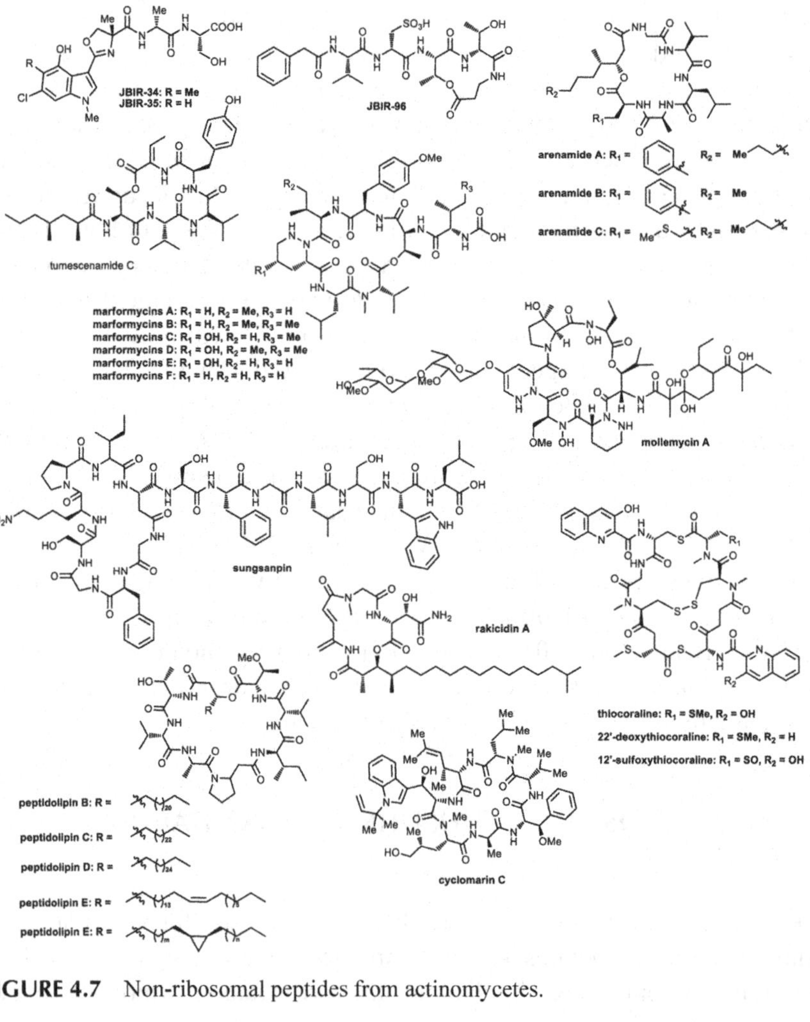

FIGURE 4.7 Non-ribosomal peptides from actinomycetes.

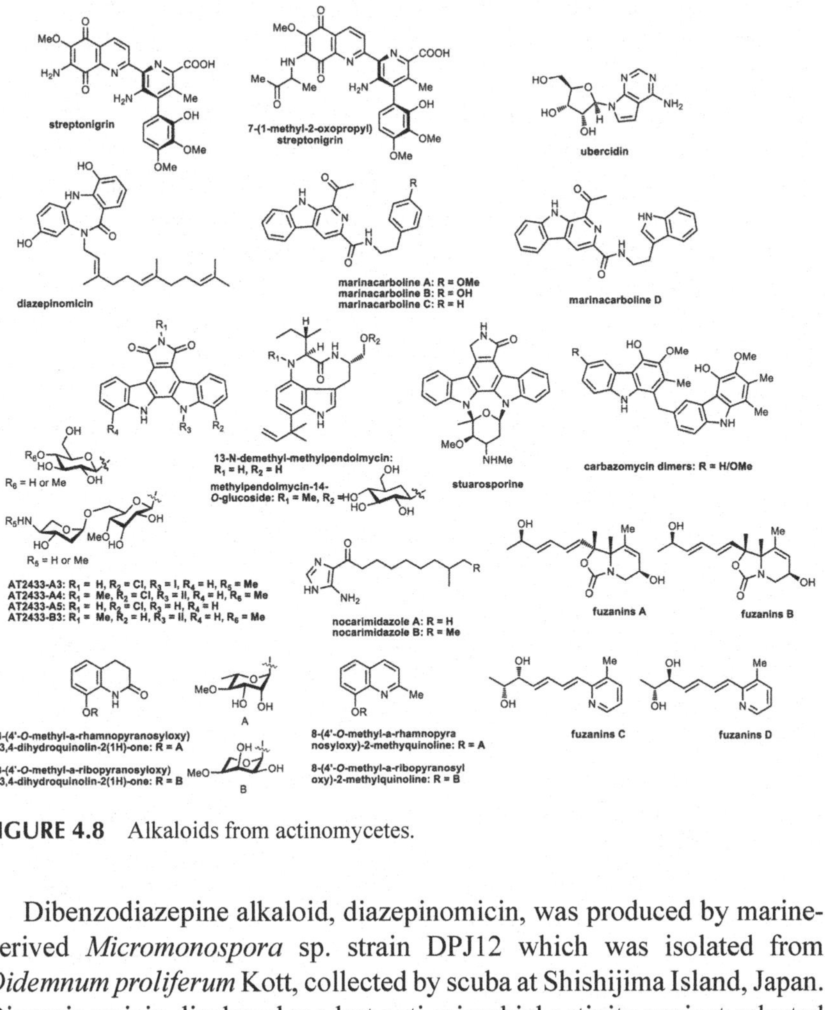

FIGURE 4.8 Alkaloids from actinomycetes.

Dibenzodiazepine alkaloid, diazepinomicin, was produced by marine-derived *Micromonospora* sp. strain DPJ12 which was isolated from *Didemnum proliferum* Kott, collected by scuba at Shishijima Island, Japan. Diazepinomicin displayed modest anti-microbial activity against selected Gram-positive bacteria (Charan et al., 2004). The genus *Kitasatospora* has been an interested source for bioactive metabolites discovery. The nitrogen containing metabolites in the group of carbamate- or pyridine compounds such as fuzanins A, B, C, and D were produced by *Kitasatospora* sp. IFM10917 (Aida et al., 2009). Fuzanin D exhibited activity against human colon carcinoma DLD–1 cell, with an IC_{50} value of 41.2 mM, while uzanins A, B, and C proved to be inactive (IC_{50} >50 mM). Moreover, fuzanin D showed moderate inhibition of Wnt signal transcription (inhibition: 61.7%) along with mild cytotoxicity (27.5%) at 25 mM.

Ubercidin was isolated from the halophilic *Actinopolyspora erythraea* YIM 90600. This compound revealed Pdcd4 stabilizing activity with an IC_{50} of 0.88 μM that was showed its antitumor activity (Zhao et al., 2011).

β-Carboline alkaloids, marinacarbolines A–D, indolactam alkaloids, 13-*N*-demethyl-methylpendolmycin and methylpendolmycin–14-*O*-α-glucoside, and 1-acetyl-β-carboline, methylpendolmycin, and pendolmycin, which were produced by *Marinactinospora thermotolerans* SCSIO 00652, showed anti-malarial activity, especially marinacarboline A and methylpendolmycin–14-*O*-α-glucoside could inhibit *P. falciparum* line Dd2 with IC_{50} values of 1.92 and 5.03 μM, respectively. Marinacarbolines B, C, and D inhibited *P. falciparum* lines 3D7 and Dd2 with IC_{50} values (3.09–16.65 μM) (Huang et al., 2011).

Actinomadura strain could produce the quinolin alkaloids, the compound series which were mostly found in the plant. 8-(4'-*O*-methyl-α-rhamnopyranosyloxy)–3,4-dihydroquinolin-2(1*H*)-one, 8-(4'-*O*-methyl-α-ribopyranosyloxy)–3,4-dihydroquinolin-2(1*H*)-one, 8-(4'-*O*-methyl-α-rhamnopyranosyloxy)–2-methylquinoline and 8-(4'-*O*-methyl-α-ribopyranosyloxy)–2-methylquinoline were isolated from *Actinomadura* sp. BCC 27169 from Thai soil. These compounds were inactive with microbial test and non-cytotoxicity with cell lines (Intaraudom et al., 2014). Indolocarbazole analogs, AT2433-A3, A4, A5, and B3 were isolated from *Actinomadura melliaura* ATCC 39691, a strain isolated from a soil sample collected in Bristol Cove, California. Compounds AT2433-A3 and B3 showed antibacterial activity against *S. aureus* ATCC 6538, *M. luteus* NRRL B–287, and *Mycobacterium smegmatis* ATCC 14468 with MIC values 7.5–120 μM. In addition, the isolated compound from strain ATCC 39691 showed activity against human cancer cell lines (Shaaban et al., 2015). However, the indolocarbazole series, stuarosporine was previously isolated from *Streptomyces* strain.

Nocarimidazoles A and B were produced from the marine-derived *Nocardiopsis* sp. CNQ 115. These compounds formed a 4-aminoimidazole ring combined with a conjugated carbonyl side chain. These compounds were not significantly cytotoxic to A498 and ACHN renal cancer cells (Leutou et al., 2015). Nocarimidazoles A and B showed weak inhibitory activity of the acetylcholinesterase (AChE) enzyme while only nocarimidazole A showed weak inhibitory activities against *Bacillus subtilis* and *Staphylococcus epidermidis* with MIC values of 64 and 64 μg/ml, respectively, while nocarimidazole B displayed only weak activity on *B. subtilis* with an MIC value of 64 μg/ml.

The alkaloids and other nitrogen-containing compounds (Figure 4.9) such as the antimalarial secondary metabolites, carbazomycins A–D were produced by terrestrial *Streptomyces* sp. BCC 26924 from Thai soil. In addition, strain BCC 26924 could be produced the carbazomycin dimmers, which was firstly reported as new alkaloid compounds (Intaraudom et al., 2011). Carbazomycins A–D showed antimalarial (IC_{50} 2.10–10.0 µg/ml) and cytotoxic activities.

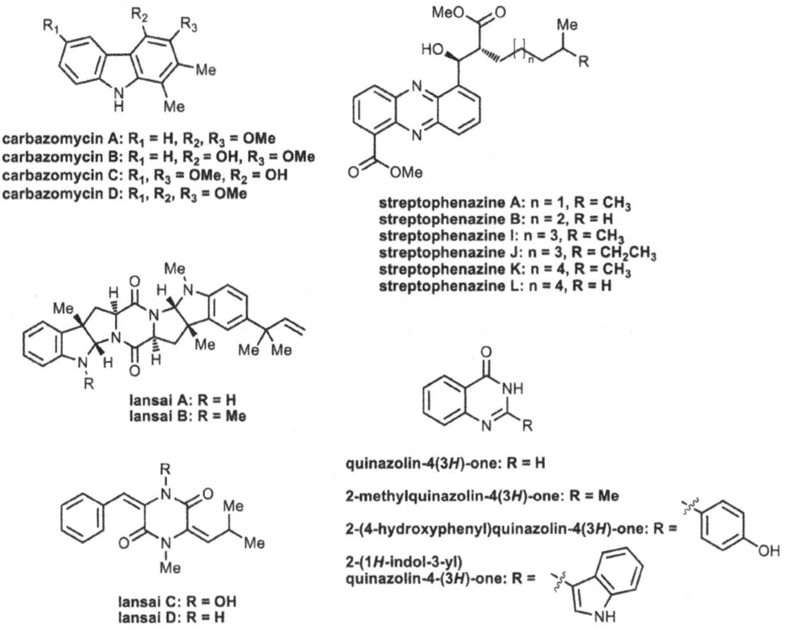

FIGURE 4.9   Nitrogen-containing metabolites from actinomycetes.

But the carbazomycin dimer showed only weak cytotoxicity against cancerous and non-cancerous cell lines. The phenazine derivative is alkaloids that mostly produced by plants. However, streptophenazines were produced by *Streptomyces* sp. BCC21838 that streptophenazines I, J, K, and L firstly reported from this strain. In the biological analysis, streptophenazines exhibited inhibitory activity against *B. cereus* with IC_{50} in a range of 6.25–12.50 mg/ml. In addition, these streptophenazines also exhibited low cytotoxicity against both cancerous (MCF–7, KB, NCI-H187) and non-cancerous (Vero) cells (Bunbamrung et al., 2014). Quinazolinones produced by *Streptomyces* sp. BCC 21795 from sediment, collected at Khao Khitchakut National Park, Chanthaburi Province, Thailand. Quinazolinones exhibited moderate cytotoxic activity (Korn-sakulkarn et al., 2015). *Streptomyces* sp. SUC1 isolated from aerial roots of *Ficus benjamina*, produced bioactive secondary metabolites, lansai A–D. Lansai B showed weakly cytotoxicity against the BC cell line (IC_{50} 15.03 mg/ml), but lansai A, C, and D were inactive (IC_{50} >20 mg/ml). These compounds were also inactive against the fungus *Colletotrichum musae* (MIC >100 mg/ml) (Tuntiwachwuttikul et al., 2008).

4.6 CONCLUSION

Natural products are both a fundamental source of new chemical diversity and an integral component of today's pharmaceutical compendium. Actinomycetes are known for their unprecedented ability to produce novel lead compounds. These bacteria produced diverse chemical structures of bioactive compounds such as polyketides, terpenoids, non-ribosomal peptides, alkaloids, and other nitrogen-containing compounds. The polyketides were the major products from actinomycetes that mostly produced by *Streptomyces* strains, and some of them were produced by the rare actinomycetes, *Saccharopolyspora*, *Actinoplanes*, *Micromonospora*, and related genera. The metabolites from actinomycetes showed diverse biological activities including antibacterial, antifungal, antituberculosis, antimalarial, and anticancer.

KEYWORDS

- actinomycetes
- alkaloids
- antibiotics
- biological activity
- peptides
- polyketides
- rare actinomycetes
- secondary metabolites
- terpenoids

REFERENCES

Abdelmohsen, U. R., Bayer, K., & Hentschel, U., (2014). Diversity, abundance, and natural products of marine sponge-associated actinomycetes. *Nat. Prod. Rep., 31*, 381–399.

Aida, W., Ohtsuki, T., Li, X., & Ishibashi, M., (2009). Isolation of new carbamate- or pyridine-containing natural products, fuzanins A, B, C, and D from *Kitasatospora* sp. IFM10917. *Tetrahedron, 65*, 369–373.

Aoyagi, T., Aoyama, T., Kojima, F., Hattori, S., Honma, Y., Hamada, M., & Takeuchi, T., (1992). Cyclooctatin, a new inhibitor of lysophospholipase, produced by *Streptomyces malanosporofaciens* MI614–43F2. *J. Antibiot., 45*, 1587–1591.

Asolkar, R. N., Freel, K. C., Jensen, P. R., Fenical, W., Kondratyuk, T. P., Park, E. J., & Pezzuto, J. M., (2009). Arenamides A-C, cytotoxic NFKB inhibitors from the marine actinomycete *Salinispora arenicola*. *J. Nat. Prod., 72*, 396–402.

Bérdy, J., (2005). Bioactive microbial metabolites. *J. Antibiot., 58*, 1–26.

Boonlarppradab, C., Suriyachadkun, C., Rachtawee, P., & Choowong, W., (2013). Saccharosporones A, B, and C, cytotoxic antimalarial angucyclinones from *Saccharopolyspora* sp. BCC 21906. *J. Antibiot., 66*, 305–309.

Bull, A. T., Stach, J. E. M., Ward, A. C., & Goodfellow, M., (2005). Marine actinobacteria: Perspectives, challenges, future directions. *Antonie Van Leeuwenhoek, 87*, 65–79.

Bunbamrung, N., Dramae, A., Srichomthong, K., Supothina, S., & Pittayakhajonwut, P., (2014). Streptophenazines I–L from *Streptomyces* sp. BCC21835. *Phytochem. Lett., 10*, 91–94.

Carr, G., Williams, D. E., Díaz-Marrero, A. R., Patrick, B. O., Bottriell, H., Balgi, A. D., et al., (2010). Bafilomycins produced in culture by *Streptomyces* spp. isolated from marine habitats are potent inhibitors of autophagy. *J. Nat. Prod., 73*, 422–427.

Charan, R. D., Schlingmann, G., Janso, J., Bernan, V., Feng, X., & Carter, G. T., (2004). Diazepinomicin, a new antimicrobial alkaloid from a marine *Micromonospora* sp. *J. Nat. Prod., 67*, 1431–1433.

DeBoer, C., Meulman, P., Wnuk, R., & Peterson, D., (1970). Geldanamycin, a new antibiotic. *J. Antibiot., 23*, 442–447.

Donadio, S., Maffioli, S., Monciardini, P., Sosio, M., & Jabes, D., (2010). Antibiotic discovery in the twenty-first century: Current trends and future perspectives. *J. Antibiot., 63*, 423–430.

Dramae, A., Nithithanasilp, S., Choowong, W., Rachtawee, P., Prabpai, S., Kongsaeree, P., & Pittayakhajonwut, P., (2013). Antimalarial 20-membered macrolides from *Streptomyces* sp. BCC33756. *Tetrahedron, 69*, 8205–8208.

Guo, Z. K., Liu, S. B., Jiao, R. H., Wang, T., Tan, R. X., & Ge, H. M., (2012). Angucyclines from an insect-derived actinobacterium *Amycolatopsis* sp. HCa1 and their cytotoxic activity. *Bioorg. Med. Chem. Lett., 22*, 7490–7493.

Hohmann, C., Schneider, K., Bruntner, C., Brown, R., Jones, A. L., Goodfellow, M., et al., (2009). Albidopyrone, a new alpha-pyrone-containing metabolite from marine-derived *Streptomyces* sp. NTK 227. *J. Antibiot., 62*, 75–79.

Huang, H., Yang, T., Ren, X., Liu, J., Song, Y., Sun, A., et al., (2012). Cytotoxic angucycline class glycosides from the deep sea actinomycete *Streptomyces lusitanus* SCSIO LR32. *J. Nat. Prod., 75*, 202–208.

Huang, H., Yao, Y., He, Z., Yang, T., Ma, J., Tian, X., et al., (2011). Antimalarial beta-carboline and indolactam alkaloids from *Marinactinospora thermotolerans*, a deep-sea isolate. *J. Nat. Prod., 74*, 2122–2127.

Igarashi, Y., Yanase, S., Sugimoto, K., Enomoto, M., Miyanaga, S., Trujillo, M. E., et al., (2011). Lupinacidin C, an inhibitor of tumor cell invasion from *Micromonospora lupini*. *J. Nat. Prod., 74*, 862–865.

Intaraudom, C., Dramae, A., Supothina, S., Komwijit, S., & Pittayakhajonwut, P., (2014). 3-Oxyanthranilic acid derivatives from *Actinomadura* sp. BCC27169. *Tetrahedron, 70*, 2711–2716.

Intaraudom, C., Rachtawee, P., Suvannakad, R., & Pittayakhajonwut, P., (2011). Antimalarial and antituberculosis substances from *Streptomyces* sp. BCC26924. *Tetrahedron, 67*, 7593–7597.

Kawamura, A., Iacovidou, M., Hirokawa, E., Soll, C. E., & Trujillo, M., (2011). 17-Hydroxycyclooctatin, a fused 5–8–5 ring diterpene, from *Streptomyces* sp. MTE4a. *J. Nat. Prod., 74*, 492–495.

Keller-Schierlein, W., & Meyet, M., (1983). Isolation and structure elucidation of naphthomycins B and C. *J. Antibiot., 36*, 484–492.

Kishimoto, S., Tsunematsu, Y., Nishimura, S., Hayashi, Y., Hattori, A., & Kakeya, H., (2012). Tumescenamide C, an antimicrobial cyclic lipodepsipeptide from *Streptomyces* sp. *Tetrahedron, 68*, 5572–5578.

Kornsakulkarn, J., Saepua, S., Srijomthong, K., Rachtawee, P., & Thongpanchang, C., (2015). Quinazolinone alkaloids from actinomycete *Streptomyces* sp. BCC 21795. *Phytochem. Lett., 12*, 6–8.

Leutou, A. S., Yang, I., Kang, H., Seo, E. K., Nam, S. J., & Fenical, W., (2015). Nocarimidazoles A and B from a marine-derived actinomycete of the genus *Nocardiopsis*. *J. Nat. Prod., 78*, 2846–2849.

Li, W., Yang, X., Yang, Y., Zhao, L., Xu, L., & Ding, Z., (2015). A new anthracycline from endophytic *Streptomyces* sp. YIM66403. *J. Antibiot., 68*, 216–219.

Lu, C., & Yuemao, S., (2007). A novel ansamycin, naphthomycin K from *Streptomyces* sp. *J. Antibiot., 60*, 648–653.

Lu, C., Li, Y., Deng, J., Li, S., Shen, Y., Wang, H., & Shen, Y., (2013). Hygrocins C-G, cytotoxic naphthoquinone ansamycins from gdmAI-disrupted *Streptomyces* sp. LZ35. *J. Nat. Prod., 76*, 2175–2179.

Lu, Y., Xing, Y., Chen, C., Lu, J., Ma, Y., & Xi, T., (2012). Anthraquinone glycosides from marine *Streptomyces* sp. strain. *Phytochem. Lett., 5*, 459–462.

Motohashi, K., Sue, M., Furihata, K., Ito, S., & Seto, H., (2008). Terpenoids produced by actinomycetes: Napyradiomycins from *Streptomyces antimycoticus* NT17. *J. Nat. Prod., 71*, 595–601.

Motohashi, K., Ueno, R., Sue, M., Furihata, K., Matsumoto, T., Dairi, T., et al., (2007). Studies on terpenoids produced by actinomycetes: Oxaloterpins A, B, C, D, and E, diterpenes from S*treptomyces* sp. KO–3988. *J. Nat. Prod., 70*, 1712–1717.

Ohta, H., Kubota, N. K., Ohta, S., Suzuki, M., Ogawa, T., Yamasaki, A., & Ikegami, S., (2001). Micromonospolides A-C, new macrolides from *Micromonospora* sp. *Tetrahedron, 57*, 8463–8467.

Oku, N., Matoba, S., Yamazaki, Y. M., Shimasaki, R., Miyanaga, S., & Igarashi, Y., (2014). Complete stereochemistry and preliminary structure-activity relationship of rakicidin A, a hypoxia-selective cytotoxin from *Micromonospora* sp. *J. Nat. Prod.*, *77*, 2561–2565.

Pérez, M., Crespo, C., Schleissner, C., Rodríguez, P., Zuniga, P., & Reyes, F., (2009). Tartrolon D, a cytotoxic macrodiolide from the marine-derived actinomycete *Streptomyces* sp. MDG-04-17-069. *J. Nat. Prod.*, *72*, 2192–2194.

Raju, R., Khalil, Z. G., Piggott, A. M., Blumenthal, A., Gardiner, D. L., Skinner-Adams, T. S., & Capon, R. J., (2014). Mollemycin A: An antimalarial and antibacterial glyco-hexadepsipeptide-polyketide from an Australian marine-derived *Streptomyces* sp. (CMB-M0244). *Org. Lett.*, *16*, 1716–1719.

Ritzau, M., Heinze, S., Fleck, W. F., Dahse, H. M., & Grafe, U., (1998). New macrodiolide antibiotics, 11-*O*-monomethyl- and 11,11′-*O*-dimethylelaiophylins from *Streptomyces* sp. HKI–0113 and HKI–0114. *J. Nat. Prod.*, *61*, 1337–1339.

Shaaban, K. A., Elshahawi, S. I., Wang, X., Horn, J., Kharel, M. K., Leggas, M., & Thorson, J. S., (2015). Cytotoxic indolocarbazoles from *Actinomadura melliaura* ATCC 39691. *J. Nat. Prod.*, *78*, 1723–1729.

Sheng, Y., Lam, P. W., Shahab, S., Santosa, D. A., Proteau, P. J., Zabriskie, T. M., & Mahmud, T., (2015). Identification of elaiophylin skeletal variants from the Indonesian *Streptomyces* sp. ICBB 9297. *J. Nat. Prod.*, *78*, 2768–2775.

Shin, B., Kim, B. Y., Cho, E., Oh, K. B., Shin, J., Goodfellow, M., & Oh, D. C., (2016). Actinomadurol, an antibacterial norditerpenoid from a rare actinomycete, *Actinomadura* sp. KC 191. *J. Nat. Prod.*, *79*, 1886–1890.

Stead, P., Latif, S., Blackaby, A. P., Sidebottom, P. J., Deakin, A., Taylor, N. L., et al., (2000). Discovery of novel ansamycins possessing potent inhibitory activity in a cell-based oncostatin M signaling assay. *J. Antibiot.*, *53*, 657–663.

Supong, K., Thawai, C., Choowong, W., Kittiwongwattana, C., Thanaboripat, D., Laosinwattana, C., et al., (2016). Antimicrobial compounds from endophytic *Streptomyces* sp. BCC72023 isolated from rice (*Oryza sativa* L.). *Res. Microbiol.*, *167*, 290–298.

Supong, K., Thawai, C., Suwanborirux, K., Choowong, W., Supothina, S., & Pittayakhajonwut, P., (2012). Antimalarial and antitubercular *C*-glycosylated benz[α] anthraquinones from the marine-derived *Streptomyces* sp. BCC45596. *Phytochem. Lett.*, *5*, 651–656.

Takagi, M., Motohashi, K., Nagai, A., Izumikawa, M., Tanaka, M., Fuse, S., et al., (2010). Anti-Influenza virus compound from *Streptomyces* sp. RI18. *Org. Lett.*, *12*, 4664–4666.

Tuntiwachwuttikul, P., Taechowisan, T., Wanbanjob, A., Thadaniti, S., & Taylor, W. C., (2008). Lansai A–D, secondary metabolites from *Streptomyces* sp. SUC1. *Tetrahedron*, *64*, 7583–7586.

Ueda, J. Y., Izumikawa, M., Kozone, I., Yamamura, H., Hayakawa, M., Takagi, M., & Shin-Ya, K., (2011). A phenylacetylated peptide, JBIR–96, isolated from *Streptomyces* sp. RI051-SDHV6. *J. Nat. Prod.*, *74*, 1344–1347.

Um, S., Kim, Y. J., Kwon, H., Wen, H., Kim, S. H., Kwon, H. C., et al., (2013). Sungsanpin, a lasso peptide from a deep-sea streptomycete. *J. Nat. Prod.*, *76*, 873–879.

Wang, S. L. Y., Jin, X., Xiaoli, X., Hong, H., Francesca, R., Anthony, E. T., et al., (2002). Isolation of streptonigrin and its novel derivative from *Micromonospora* as inducing agents of p53-dependent cell apoptosis. *J. Nat. Prod.*, *65*, 721–724.

Wang, X., Shaaban, K. A., Elshahawi, S. I., Ponomareva, L. V., Sunkara, M., Zhang, Y., et al., (2013). Frenolicins C-G, pyranonaphthoquinones from *Streptomyces* sp. RM–4–15. *J. Nat. Prod., 76*, 1441–1447.

Wang, X., Tabudravu, J., Jaspars, M., & Deng, H., (2013). Tianchimycins A–B, 16-membered macrolides from the rare actinomycete *Saccharothrix xinjiangensis*. *Tetrahedron, 69*, 6060–6064.

Wu, C. Z., Jang, J. H., Ahn, J. S., & Hong, Y. S., (2012). New geldanamycin analogs from *Streptomyces hygroscopicus*. *J. Microbiol. Biotechnol., 22*, 1478–1481.

Wu, C., Tan, Y., Gan, M., Wang, Y., Guan, Y., Hu, X., et al., (2013). Identification of elaiophylin derivatives from the marine-derived actinomycete *Streptomyces* sp. 7–145 using PCR-based screening. *J. Nat. Prod., 76*, 2153–2157.

Wyche, T. P., Hou, Y., Braun, D., Cohen, H. C., Xiong, M. P., & Bugni, T. S., (2011). First natural analogs of the cytotoxic thiodepsipeptide thiocoraline A from a marine *Verrucosispora* sp. *J. Org. Chem., 76*, 6542–6547.

Wyche, T. P., Hou, Y., Vazquez-Rivera, E., Braun, D., & Bugni, T. S., (2012). Peptidolipins B-F, antibacterial lipopeptides from an ascidian-derived *Nocardia* sp. *J. Nat. Prod., 75*, 735–740.

Yang, Y. H., Fu, H. L., Li, L. Q., Zeng, Y., Li, C. Y., He, Y. N., & Zhao, P. J., (2012). Naphthomycins L−N, Ansamycin antibiotics from *Streptomyces* sp. CS. *J. Nat. Prod., 75*, 1409–1413.

Yin, M., Lu, T., Zhao, L. X., Chen, Y., Huang, S. X., Lohman, J. R., et al., (2011). The missing C–17 O-methyltransferase in geldanamycin biosynthesis. *Org. Lett., 13*, 3726–3729.

Zhang, J., Jiang, Y., Cao, Y., Liu, J., Zheng, D., Chen, X., et al., (2013). Violapyrones A-G, alpha-pyrone derivatives from *Streptomyces* violascens isolated from *Hylobates hoolock* feces. *J. Nat. Prod., 76*, 2126–2130.

Zhang, Q., Peoples, A. J., Rothfeder, M. T., Millett, W. P., Pescatore, B. C., Ling, L. L., & Moore, C. M., (2009). Isofuranonaphthoquinone produced by an *Actinoplanes* isolate. *J. Nat. Prod., 72*, 1213–1215.

Zhao, G., Li, S., Wang, Y., Hao, H., Shen, Y., & Lu, C., (2013). 16,17-dihydroxycyclooctatin, a new diterpene from *Streptomyces* sp. LZ35. *Drug Discov. Ther., 7*, 185–188.

Zhao, L. X., Huang, S. X., Tang, S. K., Jiang, C. L., Duan, Y., Beutler, J. A., et al., (2011). Actinopolysporins AC and tubercidin as a Pdcd4 stabilizer from the halophilic actinomycete *Actinopolyspora erythraea* YIM 90600. *J. Nat. Prod., 74*, 1990–1995.

Zhou, X., Huang, H., Li, J., Song, Y., Jiang, R., Liu, J., et al., (2014). New anti-infective cycloheptadepsipeptide congeners and absolute stereochemistry from the deep sea-derived *Streptomyces drozdowiczii* SCSIO 10141. *Tetrahedron, 70*, 7795–7801.

CHAPTER 5

SERRATIOPEPTIDASE: A MULTIFACETED MICROBIAL ENZYME IN HEALTH CARE

SUBATHRA DEVI CHANDRASEKARAN,
JEMIMAH NAINE SELVAKUMAR, and
MOHANASRINIVASAN VAITHILINGAM

School of Biosciences and Technology, VIT University, Vellore, Tamil Nadu, 632014, India

5.1 INTRODUCTION

Naturally occurring substances from bacterial sources are continued to have, a crucial place in drug discovery. The major uses from the source of natural medicines have increased over decades. The highest sources of enzymes are from microbes, plants, and animals. Microbial enzymes have extended much attention and known to exert many advantages. The major mechanism is due to their efficiency in biological environment; enzymes are therapeutically vulnerable. The use of enzymes in therapeutics is as old as the knowledge of their existence. Serratiopeptidase is an enzyme with proteolytic property which is isolated from the bacteria *Serratia marcescens*. Serratiopeptidase is a naturally occurring, physiological agent with no inhibitory effects on prostaglandins and is devoid of gastrointestinal side effects. Serratiopeptidase dissolves blood clots *in vitro* and *in vivo*. It is used to treat many diseases including cardiovascular, arthritis, eye-related problems, and inflammation. The major search for a novel physiological agent with anti-inflammatory potential without enabling any side effects has ended with discovery of an enzyme Serratiopeptidase.

5.2   *SERRATIA MARCESCENS*

An enzyme is a large, biological molecule responsible for the thousands of chemical conversions that sustain life. They are necessary for metabolic reactions such as digestion of food to the synthesis of DNA and everything in between. The microorganism *Serratia marcescens* is part of the kingdom Eubacteria, phylum Proteobacteria, class Gammabacteria, order Enterobacteriales, family Enterobacteriaceae, and genus *Serratia*. It is a gram-negative, flagellated, facultative anaerobe which is considered to be an opportunistic pathogen. The microbe's morphological structure can be seen as short and rod-shaped. *Serratia marcescens* produces a red pigmentation known as prodigiosin. The organism is resistant to many antibiotics making it difficult to treat. Its ability to survive with or without oxygen as well as its mobility makes it all the more difficult. Despite it being a dangerous pathogen, it is continuously produced for numerous commercial benefits.

5.3   SERRATIOPEPTIDASE

Serratiopeptidase is an enzyme with proteolytic potential are produced by Enterobacterium *Serratia* species E–15. Serratiopeptidase is soluble in water and insoluble in alcohol and ether. Serratiopeptidase is white when in powder form. Serratiopeptidase has been discovered to have numerous health benefits and is used for the treatment of a large number of diseases as well as the optimization of multiple pharmaceutical products. *Serratia* peptidase is a powerful anti-inflammatory, used to prevent cancer and tumors, used as alternative medication, and also used for other medical illnesses and side effects. Serratiopeptidase is considered to be a powerful anti-inflammatory enzyme. Inflammation is a general response to the human body when it is injured or attacked by a negative bacteria or virus which means within a few days the pain and inflammation subsides. Traditional medication for inflammation includes many dangers through the conventional treatment. There are two regular medicines for pain and inflammation, and these incorporate steroidal and non-steroidal (NSAIDS) drugs. Steroidal medications have serious symptoms including loss of bone, osteoporosis, expanded glucose, weight pick up caused by a disturbed digestion, joint harm, waterfalls, diminishing of skin, moderate injury recuperating, hypertension, brought down invulnerable framework, passionate clutters, liquid maintenance, and concealment of ordinary adrenal capacities. NSAIDS has been said to cause, gastrointestinal contaminations and confusions, as well as death, interior dying, congestive heart disappointment, harmful impacts, obliteration of joints, and considerably more hazardous symptoms. In this way, serratiopeptidase tablets were utilized as a therapeutic option because of its normal catalysts which dispensed with the dangerous factor. The chemical forestalls swelling and liquid maintenance. Swelling and liquid maintenance is the significant reason for some sicknesses, for example, carpel burrow disorder. Carpel burrow disorder is a provocative issue of the hand and wrist which causes enduring torment and in the long run an inability. In a little scale trial, 65% of the patients were arranged as to demonstrate fruitful clinical changes when treated with serratiopeptidase. Likewise, no huge symptoms were experienced. Research for the clinical utilization of Serratiopeptidase towards sinusitis and bronchitis has been demonstrated fruitful. In both of the cases, the irritation and swelling of the covering of the aviation routes anticipates seepage of bodily fluid. This causes obstacle of the aviation route. Serratiopeptidase diminishes the thickness and consistency of the bodily fluid and enhances its end through bronchopulmonary secretions.

5.4   SERRATIOPEPTIDASE AS THERAPEUTICS

Pain being one of the most troubling aspects of inflammation is produced by cellular chemical reactions as part of the body's natural inflammatory healing response. Chronic pain, on the other hand, can be detrimental to healing. Swelling caused by irritation can make tissue press against touchy nerves and cause pain. Serratiopeptidase proteins can deplete the liquid in aggravated territories and additionally decrease the pain straightforwardly by hindering the arrival of agony initiating amines from tissues. To the extent the proteins connect to malignancy is concerned, a promising cure is continually being investigated. This ensures the current safe framework, as well as makes it more grounded to plan for more serious phases of the disease until the point that it is totally wiped out. Serratiopeptidase medications are naturally occurring enzymes, and they do not irritate the digestive system. Prescription of more than 25 years in Europe, Japan, and Asia, Serratiopeptidase is likewise used to upgrade the execution of anti-infection agents. Bacterial diseases are ending up increasingly impervious to treatment because of a self-created biofilm which encourages them ward off anti-infection agents. Specialists express that Serratiopeptidase can improve the adequacy of anti-infection agents, for example, ampicillin, ciclacillin, cephalexin, minocycline, and cefotiam. It helps in separating bigger particles into littler ones which brings down the thickness. At the point when the sputum is of lower thickness, it is all the more effectively expelled from the respiratory tract. This permits agreeable and less relentless relaxing. Notwithstanding the compounds capacity to separate sputum, it is likewise viewed as a protein-processing chemical which separates "non-living" matter in the human body. Serratiopeptidase conflicts with this development in light of the fact that the chemical processes these non-living tissues and particles while allowing living tissue to sit unbothered. Moreover, serratiopeptidase additionally diminishes the blood, expel blood clumps and battle phlebitis/thrombophlebitis. The catalyst has been looked into to likewise be fit for enhancing course by decreasing levels of dead tissue. The protein has corrective applications, battles develop of fibrin, help endless sinusitis, and effectively treats fibrocystic bosom ailment. Utilization of serratiopeptidase keeps on developing and advance into elective fields. Russian analysts in quantum radio-material science have built up a plasma-construct PC framework that capacities with respect to low-power rotating attractive field to produce biovitality. In result, there is an improvement of the frameworks bioactivity and execution.

5.5   RESEARCH DEVELOPMENTS IN SERRATIOPEPTIDASE

Previous studies on screening and molecular characterization of *Serratia marcescens* VITSD2 with GenBank accession number: KC961637, have shown producing maximum serratiopeptidase. The main aim of the study was to enhance the production of serratiopeptidase from *Serratia marcescens* VITSD2. The study also emphasized the bioactivity of enzyme. Certainly, strain improvement for the selected strain was carried out by random mutagenesis. When compared to wild-type strain mutant showed prominent findings throughout the study. Culture condition for the growth and production were optimized. Among the different carbon and nitrogen sources tested starch and beef extract was found to be the best with maximum yield with enzyme activity of 2921units/mL and 2924 U/mL respectively. The enzyme showed an optimum activity of 2377 U/mL at pH–6.5 and 2906 U/mL at 25°C. Maximum serratiopeptidase production during fermentation was obtained after 24 h incubation with 1% inoculum in the medium at 25°C and yielded 3255 U/mL. Tryptophan stimulated the production with 2857 U/mL. Maximum serratiopeptidase production was detected after 24 h incubation with 2564 U/mL. The results of the present study clearly indicate that the yield of serratiopeptidase was found maximum by varying the cultural conditions by the mutant strain. Serratiopeptidase was purified using gel filtration chromatography. The molecular weight of the protease was determined by using SDS-PAGE, and it was found to be 66 KD The antibacterial activity of the produced serratiopeptidase showed moderate activity against *Staphylococcus aurous* (17 mm) and *Escherichia coli* (15 mm). HPLC chromatogram for purified enzyme produced from *Serratia marcescens* VITSD2 indicates the presence of serratiopeptidase and the retention time peak was found to be 3.02 min. The growth kinetics and optimization of fermentation conditions for serratiopeptidase were investigated, where the production of serratiopeptidase was carried out in trypticase soya broth, and the enzyme was partially purified using ammonium sulfate precipitation and dialysis. The specific activity was determined by casein hydrolysis assay and was found to be 12.00, 21.33, and 25.40 units/mg for crude, precipitated, and dialyzed samples (Subathra et al., 2013).

Rocha and Brennan (1973) isolated *Serratia marcescens* on a new medium-commercial deoxyribonuclease agar. Koki et al. (1984) in their experiment showed that serratiopeptidase has a remarkable record of safety from decades of use of several peoples. Noboru et al. (1984) reported the characterization of the precursor of *Serratia marcescens* serine protease. Peter and Michael (1988) described a method to identify suitable crystallization condition *de novo* for the purification of a recombinant protein variant. It has been proved that the enzyme from *Serratia marcescens* consists of two polypeptide chains that form a complex. Mazzone et al. (1990) proved that serratiopeptidase is a proteolytic enzyme. Serratiopeptidase has been gaining wide acceptance in Europe and Asia as a potent analysis and anti-inflammatory drug. Mammone et al. (1990) showed that serratiopeptidase and drain fluid from the inflamed area can reduce swelling and pain. Serratiopeptidase also reduces pain through its ability to release of pain-inducing amines from inflamed tissues. The recommended dosage of serratiopeptidase is 10 mg to 30 mg a day for arthritis, sinusitis, fibrocystic breast swelling, bronchitis, carpel tunnel syndrome, and cardiovascular problems, 20 µg a day.

Tang et al. (1991) explained the mechanism behind the action of Serratiopeptidase enzyme. The same mechanism makes it possible for peptidases to inactivate virus, by pruning the viral proteins necessary for infectivity. Serratiopeptidase is commercially obtained from the purification and characterization of four proteases from *Serratia marcescens* by ammonium sulfate precipitation, chromatography, SDS–PAGE. Sugai et al. (1996) suggested the purification of endopeptidase is effective by column chromatography.

Glen et al. (2001) explained the crystal structure of Anthranilate Synthase (AS) from *Serratia marcescens*. A mesophilic bacterium has been solved in the presence of its substrates, chorismate, and glutamine and one product, glutamate, at 1.95A° and with its bound feedback inhibitor, tryptophan, at 2.4A° in comparison with the AS structure from the hyperthermophile S*ulfolobus solfataricus*.

A study on the partial characterization of an alkaline protease produced by a non-pigmenting *Serratia* spp. was done. The characteristics of extracellular alkaline protease produced by non-pigmenting *Serratia* spp. were compared with proteases from pigmenting *Serratia* spp. The non-pigmented strain was isolated directly from the soil. It was discovered that enzyme has maximum activity at a pH of 9.0. However, it was also stable at pH 6–10. When these results were compared to the protease produced by the pigmenting strains, the alkaline protease was noted to have reduced enzymatic stability.

An exoprotease produced by *Serratia marcescens* ATCC 25419 was studied. In this experiment, three exoproteases were discovered, and they were labeled one, two, and three respectively. Labeling the cells with radioactive amino acids revealed no intracellular proteins that could be precipitated with antibodies raised against purified exoproteases. The interactions of some substances with exoprotease were noted. These substances included tosyl-L-lysine, chloromethyl ketone, phenethyl alcohol, and procaine. The microheterogeneity of the isolated exoforms was revealed by anion exchange chromatography, and sodium dodecyl sulfate-polyacrylamide gel electrophoresis was also observed in the samples pulse labeled with radioactive amino acids. Discovery of four different types of proteases from culture filtrates of *Serratia marcescens* from a patient with severe corneal ulcer was studied. The sample was purified using ammonium sulfate precipitation, ion exchange chromatography, and gel filtration chromatography. The proteases were differentiated from each other using polyacrylamide gel electrophoresis. The molecular weights of the purified proteases were estimated to be 56×10^3, 60×10^3 and

73×10^3 (designated as 56K, 60K, and 73K respectively). Upon isoelectric focusing, 73K was separated into 73Ka and 73Kb. 5.3, 4.4, 5.8 and 7.3 were found to be the isoelectric points of 56K, 60K, 73Ka and 73Kb, respectively. 56K and 60K were classified as metalloenzymes as they were inactivated by EDTA at pH 5.0 and were reactivated by zinc ions. No carbohydrates, cysteine, and cystine were detected in them. 73K was classified as a thiol protease. Amino acid compositions, partial amino acid sequence logical and immunological, enzyme logical and immunological properties revealed that these four enzymes are distinct from each other (Matsumoto et al., 1984).

A study was done concerning an extracellular protease produced by *Serratia marcescens* when grown on skim milk agar. This extracellular protease was isolated by ethanol precipitation. Further purification was done such as salt fractionation, DEAE-cellulose ion exchange chromatography, and gel filtration chromatography on agarose P–100. The optimum pH for proteolytic activity on casein was found to be at pH 6.0 to 9.0, and optimum temperature was found to be 45°C. Diisopropylfluorophosphate, ρ-chloromercuribenzoate, and dithiothreitol did not affect the proteolytic activity of the extracellular protease. It was given metalloprotease classification due to its inactivation by metal-ion chelators and reactivation by ferrous ions. A study based on a metalloprotease which is induced by chitin in a new chitinolytic bacterium *Serratia* sp. was done. Strain KCK, which was purified and characterized during the study exhibited a broad pH activity range (pH 5.0–8.0) as well as thermostability. Its deduced amino acid sequence showed high similarity to those of bacterial zinc binding metalloproteases and a well-conserved serralysin family motif. Pre-treatment of chitin with the metalloproteinase promoted chitin degradation by chitinase A, which suggests that metalloprotein participates and facilitates chitin degradation by this microorganism.

5.6 FERMENTATIVE PRODUCTION OF SERRATIOPEPTIDASE

The effects of aeration and agitation rates on serratiopeptidase production in a fermentor were studied. The organism used was *Serratia marcescens* NRRL B–23112. Consistent assessment of serratiopeptidase yield, the broke up oxygen fixation, maltose use, biomass, SRP yield, and pH were done all through the aging procedure. An unsettling of 400 rpm gave the most extreme serratiopeptidase generation of 11580 EU/mL with a particular SRP profitability of 78.8 EU/g/h. This was around 58% higher than the shake-jar level. The volumetric mass exchange coefficient (KLa) for the ideal aging framework was observed to be 11.3 h^{-1}. Motor examination done on the aging procedure demonstrated that SRP generation was blended development related (Pansuriya et al., 2011). An experiment concerning the fermentation of *Serratia marcescens* ATCC 13880 was done. It was done to study the biosynthesis of the organism; various physiological parameters were optimized. The advanced media contained (g/L) 10.0 maltose as carbon source, 10.0 ammonium sulfate as inorganic nitrogen source, 10.0 peptone as natural nitrogen source, 10.0 glycerine, 10.0 dihydrogen phosphate, 10.0 sodium acetic acid derivation as inorganic salt source, 10.0 sodium bicarbonate and 10.0 ascorbic corrosive as stabilizer in 1 L refined water. The ideal pH was observed to be 7.0. 37°C was observed to be the ideal temperature and ideal time was observed to be 24 hours. This streamlined medium created 27.36 IU/mL of serratiopeptidase contrasted with 17.97 IU/mL in basal medium. 52kD was the sub-atomic weight of the refined serratiopeptidase (Badhe et al., 2009).

5.7   SERRATIOPEPTIDASE AND INFECTIOUS DISEASES

Serratiopeptidase is a single enzyme which has been responsible for preventing inflammations. This study involved sixteen albino rats who were randomly divided into four groups. They were fed and sheltered in normal conditions. Acute inflammation was artificially induced using an injection. Diclofenac was orally regulated to one rodent in each gathering while serratiopeptidase was controlled to the next three. Diclofenac or refined water was administered twice per day. Notwithstanding the little example estimate, the oral group of serratiopeptidase was demonstrated successful. In the part of treatment for irritation, serratiopeptidase tablets are usually utilized because of their limited symptoms in contrast with NSAIDS and steroidal medications.

The serratiopeptidase tablets are typically created and ingested orally as an enteric covered tablet. This examination explored the likelihood of creating topical arrangements of serratiopeptidase as balms and gels. PEG-GMO gel definition of Serratia peptidase was produced and tried on a prompted ear edema technique in mice. Serratiopeptidase is ordinarily utilized as a part of the treatment of incendiary and horrendous swelling, in Germany and other European nations. It was likewise uncovered that the vast majority of the exploration around there was of European birthplace. A twofold visually impaired investigation was directed by German scientists, in which, a gathering of patients were regulated serratiopeptidase on post-agent day. The gathering directed serratiopeptidase indicated 50 % diminishment of swelling and furthermore turned out to be quickly torment free contrasted with the controls. By the tenth day, no agony was seen in the patients regulated serratiopeptidase. The impact of serratiopeptidase in the treatment of fibrocystic bosom infection was contemplated. In a twofold visually impaired, two gatherings were haphazardly shaped from an aggregate of 70 patients grumbling of bosom engorgement. One gathering was controlled serratiopeptidase, while the other was managed a fake treatment. Around 86% of the serratiopeptidase controlled patients indicated direct to checked change. There was change in span (immovability), bosom torment and bosom swelling. No unfriendly reactions were noted, and in this way, it was inferred that serratiopeptidase was sheltered and viable. The impact of serrapeptidase in patients with carpal passage disorder (CTS) was considered. A clinical trial with twenty patients with CTS was led. After gauge electrophysiological thinks about, these patients were given 10 mg of serrapeptidase twice day by day with introductory short course of nimesulide. Following a month and a half, clinical, and electrophysiological reassessment was finished. 65% of the cases controlled with serratiopeptidase demonstrated critical clinical change; this was additionally fortified by change in electrophysiological parameters. Four cases indicated repeat of CTS. No huge reactions were noted. Therefore it was reasoned that serratiopeptidase could be utilized to treat CTS yet additionally considers are required.

The impact of catalyst Danzen (Serratiopeptidase, Takeda Chemical Industries) on patients with bosom engorgement was assessed. An aggregate of 70 patients with bosom engorgement were incorporated into the trail. The patients were isolated into two gatherings. One gathering was directed Danzen, and the other gathering was controlled a fake treatment. Patients directed Danzen indicated change of bosom torment; bosom swelling and in span contrasted with the individuals who regulated the fake treatment. While 85.7% of the patients getting serratiopeptidase had "Direct to Marked" change, just 60.0% of the patients accepting fake treatment had a comparative level of change. "Stamped" change was found in 22.9% of the treatment gathering and 2.9% of the fake treatment gathering. These distinctions were measurably noteworthy. No antagonistic symptoms were watched. Serratiopeptidase has numerous clinical uses as a calming operator (especially for post horrendous swelling), for fibrocystic bosom illness, for bronchitis. Other than lessening aggravation, one of Serratiopeptidase most significant advantages is decrease of torment, because of its capacity to hinder the arrival of torment actuating amines from aroused tissues. Doctors all through Europe and Asia have perceived the mitigating and agony blocking advantages of this normally happening substance and are utilizing it in treatment as a contrasting option to NSAIDs. The learning of generation, cleansing, and portrayal of serratiopeptidase is essential for ad-libbing the action and the business estimation of the protein. The controlled aging of *Serratia* sp. secretes serratiopeptidase compound in the exceptionally specific medium. The recuperation procedure includes different sorts of filtration, fixation, and ventures to make chemical helpful for pharmaceutical applications lastly dried to a fine free streaming powder shape. The restorative utilization of serratiopeptidase is notable and all around archived. Late Japanese licenses even recommend that oral serratiopeptidase may treat or forestall such popular sicknesses as AIDS, Hepatitis B and C. Serratiopeptidase lessens swelling, enhances microcirculation and expectoration of sputum. Serratiopeptidase, when expended in unprotected shape, is annihilated by corrosive in the stomach. Be that as it may, enteric layer of tablet empower the chemical to go through the stomach unaltered and assimilate in the digestive system. Serratiopeptidase decreases post-agent edema at infusion destinations. Serratiopeptidase lessens inner tissue edema and aggravation caused at post-agent taking care of. Diminishment in edema decreases odds of crack at tissue and additionally hazard in the event of plastic surgery and join dismissal. Serratiopeptidase has mucolytic action in sinuses, ear depressions, and calming movement in upper respiratory tract organs and aides in speedier determination, better anti-toxin bioavailability, and quicker cure rates. Serratiopeptidase is utilized as a part of intense, excruciating aggravated dermatitis. Serratiopeptidase helps in better control over dental diseases and irritation.

5.8   SERRATIOPEPTIDASE AND RESPIRATORY AILMENTS

The effect of administering serratiopeptidase to patients suffering from chronic sinusitis was evaluated. Due to this thickening of the mucus, it is expelled less frequently. Japanese researchers administered serratiopeptidase 30 mg/day orally for four weeks. They evaluated the viscosity and elasticity of the mucus in their adult patients. The viscosity had decreased, thus improving expulsion of bronchopulmonary secretions. This indicates that serratiopeptidase can be used in the treatment of chronic sinusitis (Majima et al., 1988). The effects of using serratiopeptidase to treat patients with chronic sinusitis were studied. A clinical study was done in which 140 patients suffering from acute or chronic nose, throat, and ear pathologies were administered either a placebo or active serratiopeptidase.

The group that received active serratiopeptidase showed significant decrease in severity of pain, difficulty in swallowing, nasal obstruction. Patients with conditions like laryngitis, catarrhal rhinopharyngitis and sinusitis were also administered serratiopeptidase and showed a rapid betterment in their symptoms after 3–4 days. The efficacy of the treatment was assessed to be good for 97.3 % of the patients administered serratiopeptidase compared to 21.9% administered the placebo (Mazzone et al., 1990). An experiment to study the effect of serratiopeptidase on sputum and symptoms in patients with chronic airway diseases was conducted. The trail was an open-labeled type with non-treatment control group. The patients were randomly split into two groups. One group was administered serratiopeptidase 30 mg/day for 4 weeks while the second group was not administered serratiopeptidase. The patients' sputum samples were collected in the morning; part of each sputum sample was weighed and then completely dried and reweighed. Thus it was concluded that serratiopeptidase could be used to treat chronic airway diseases.

5.9   SERRATIOPEPTIDASE AGAINST BIOFILM FORMATION

The impact of serratiopeptidase in treating prosthetic disease was anticipated. There is different strategy for bacterial protection from against microbial resistance however the most critical one is biofilm development. The examination demonstrated that proteolytic catalysts could altogether improve the action of anti-infection agents against biofilms. Under different conditions, serratiopeptidase incredibly improved the action of ofloxacin. The effects of serratiopeptidase in the treatment of infection caused by slime forming bacteria around an implanted orthopedic device were evaluated. It was demonstrated that serratiopeptidase was successful against biofilm framing microorganisms and improved the adequacy of the anti-infection agents utilized.

5.10   SERRATIOPEPTIDASE AND ENHANCED ANTIBIOTIC ACTIVITY

The effects of administering cephalexin, an antibiotic in conjunction with serratiopeptidase, a proteolytic enzyme or a placebo was studied and evaluated. The test sample was 93 patients suffering from either chronic rhinitis with sinusitis, chronic relapsing bronchitis or perennial rhinitis. The group administered serratiopeptidase showed significant improvement in nasal stuffiness, coryza, rhinorrhea, and the para-nasal sinus shadows (Perna et al., 1988).

5.11   CLINICAL PHARMACOLOGY

After oral organization, serratiopeptidase is assimilated through the digestive tract and transported straightforwardly into the circulatory system. Being a peptide there would be a high inclination of this bioactive molecule to experience enzymatic debasement in the gastrointestinal tract and low layer porousness because of the hydrophilic idea of peptides and proteins. Consequently, these components could prompt low bioaccessibility of this protein when utilized restoratively. The intestinal ingestion of serratiopeptidase was evaluated by measuring its focus in plasma, lymph, and concentrate of incendiary tissue of rats by sandwich compound immunoassay (EIA) strategy. Serratiopeptidase was additionally identified in carrageenan-incited incendiary tissue in creatures at fixations higher than that in plasma. It was deduced in the investigation that serratiopeptidase is consumed from the digestive tract, conveyed to the provocative site by blood or lymph. In this manner, showing that orally controlled serratiopeptidase is consumed from intestinal tract and achieves dissemination in an enzymatically dynamic frame (Moriya et al., 1994; Moriya et al., 2003).

5.12   EVIDENCE-BASED CLINICAL INFORMATION

Serratiopeptidase has been used in traumatic inflammation, traumatic swelling sinusitis, rhinitis, laryngitis, bronchitis, periodontitis, pericoronitis of wisdom tooth.

5.13 CONCLUSION

Serratiopeptidase have been utilized as a part of fruitful treatment of numerous ailments. Serratiopeptidase, when directed for provocative ailments or to anticipate plaque development on the courses, are very much endured. Because of its absence of symptoms and mitigating abilities, serratiopeptidase is a consistent choice to supplant destructive NSAIDs. Subsequently, analysts have stepped toward discovering help for incendiary infection sufferers. In such a manner, the general public can be enormously profited by the restorative estimation of serratiopeptidase which is an extraordinary request. The chemical eyes a future treatment for a large number of fatal and developing clinical conditions to which almost no or no options are accessible. With this advanced approach, serratiopeptidase portrayal and serious examination on it can pioneer future control of the catalyst for improvement of its execution.

KEYWORDS

- anti-inflammatory drug
- bacteria
- biofilm
- cancer
- enzyme
- immune system
- infectious diseases
- inflammation
- microbes
- pain
- *Serratia marcescens*
- serratiopeptidase
- therapeutics

REFERENCES

Badhe, R. V., Nanda, R. K., Kulkarni, M. B., Bhujbal, M. N., Patil, P. S., & Badhe, S. R., (2009). Media optimization studies for serratiopeptidase production from *Serratia marcescens* ATCC 13880. *Hindustan Antibiot. Bull., 51*(1–4), 17–23.

Majima, Y., Inagaki, M., Hirata, K., Takeuchi, K., Morishita, A., & Sakakura, Y., (1988). The effect of an orally administered proteolytic enzyme on the elasticity and viscosity of nasal mucus. *Arch. Otorhinolaryngol., 244*(6), 355–359.

Mammone, G., Labkovsky, E. E. M., Billings, P. C., Rahil, J., Pratt, R. F., & Knox, J. R., (1990). Crystallographic structure of a phosphonate derivative of the *Enterobacter cloacae* P99 cephalosporinase mechanistic interpretation of a β-lactamase transition state analog. *Biochem., 33*, 6762–6772.

Matsumoto, K., Maeda, H., Takata, K., Kamata, R., & Okamura, R., (1984). Enzymes determining the order of the nucleotide bases DNA strand sequencing. *J. Bacteriol., 157*(1), 225–232.

Mazzone, A., Catalani, M., Costanzo, M., Drusian, A., Mandoli, A., Russo, S., et al., (1990). Evaluation of *Serratia peptidase* in acute or chronic inflammation of otorhinolaryngology pathology: A multicentre, double-blind, randomized trial versus placebo. *J. Int. Med. Res., 18*(5), 379–388.

Mazzone, K., Banumann, U., Bauer, M., Letoffe, S., Delepelaire, P., & Wandersman, C., (1990). Crystal structure of a complex between *Serratia marcescesns* metallo-protease and an inhibitor from *Erwinia chrysanthemi*. *J. Mol. Biol., 248*, 653–661.

Moriya, N., Nakata, M., Nakamura, M., Takaoka, M., Iwasa, S., Kato, K., & Kakinuma, A., (1994). Intestinal absorption of serratiopeptidase (TSP) in rats. *Biotechnol. Appl. Biochem., 20*, 101–108.

Moriya, N., Shoichi, A., Yoko, H., Fumio, H., & Yoshiaki, K., (2003). Intestinal absorption of serratiopeptidase and its distribution to the inflammation sites. *Japan Pharmacol. Therap., 31*, 659–666.

Noboru, M., Kyuji, K., & Ryoichi, O., (1984). Purification and characterization of proteases from a clinical isolate of *Serratia marcescens*. *J. Bacteriol., 157*(1), 255–232.

Pansuriya, R. C., & Singhal, R. S., (2010). Evolutionary operation (EVOP) to optimize whey independent serratiopeptidase production from *Serratia marcescens* NRRL B–23112. *J. Microbiol. Biotechnol., 20*, 950–957.

Perna, L., Majima, Y., Inagaki, M., Hirata, K., & Takeuchi, K., (1988). The effect of an orally administered proteolytic enzyme on the elasticity and viscosity of nasal mucus. *Arch. Otorhinolaryngol., 244*(6), 355–359.

Rocha, V., & Brennan, E. F., (1973). Purification and partial characterization of *Serratia marcescens* tryptophan synthetase. *J. Bacteriol., 134*(3), 950–957.

Spraggon, G., & Yanofsky, C., (2001). The structures of anthranilate synthase of *Serratia marcescens* crystallized in the presence of its substrates. *J. Biochem., 98*(11), 6021–6026.

Subathra, D. C., Renuka, E., Joseph, H. S., Jemimah, N. S., & Mohana, S. V., (2013). Screening and molecular characterization of *Serratia marcescens* VITSD2: A strain producing optimum serratiopeptidase. *Frontiers Biol., 2*(6), 632–639.

Sugai, M., Fujiwara, T., Akiyama, T., Ohara, M., Komatsuzawa, H., Inoue, S., & Suginaka, H., (1996). Purification and molecular characterization of Coly cylglycine endopeptidase produced by *Staphylococcus capitis* Epkl. *J. Bacteriol., 179*, 1193–1202.

Tang, U., Heijl, L., Dahlen, V., & Sundin, Y., (1991). A 4-quadrant comparative study of periodontal treatment using tetracycline-containing drug delivery fibers and scaling. *J. Clin. Periodontol., 18*, 111–116.

CHAPTER 6

PRODUCTION OF BACTERIORHODOPSIN, AN INDUSTRIALLY IMPORTANT MEMBRANE PROTEIN OF HALOPHILIC ARCHAEA: A BIOTECHNOLOGICAL CHALLENGE

PRADNYA PRALHAD KANEKAR[1], SNEHAL OMKAR KULKARNI[1], and SAGAR PRALHAD KANEKAR[2]

[1]Department of Biotechnology, Modern College of Arts, Science and Commerce, Shivajinagar, Pune–411005, Maharashtra, India

[2]Department of Biotechnology, Modern College of Arts, Science and Commerce, Ganeshkhind, Pune–411007, Maharashtra, India

6.1 INTRODUCTION

Among the natural microbial resources, extremophiles that live under stress conditions are known to be a rich source of novel biomolecules, biomaterials, and secondary metabolites. Halophiles are salt-loving organisms inhabiting extreme saline environments. They are capable of adapting to changes in osmotic pressure to protect from denaturing of cell proteins. Halophilic archaea form a major component of the prokaryotic world in marine and terrestrial ecosystems, indicating that organisms from this domain might be playing an important role in global energy cycles. Their unusual properties make them a potentially valuable resource in the development of novel biotechnological processes. In general, halophilic archaea have been studied so far from the biodiversity point of view, evolution aspects and mainly biotechnological potential in the production of extremozymes, exopolysaccharide, PHA, carotenoids (pigments) and bacteriorhodopsin (BR).

6.2 HALOPHILIC ARCHAEA

Halophilic archaea are the microorganisms that live only in the presence of high concentration of salt and thus are well adapted to saturating sodium chloride concentrations. They are found mainly in extreme saline environments like hypersaline lakes, natural hypersaline brines, solar salt pans, saline lakes, salted food products, ancient salt deposits, rock salt mines and rarely from the low saline environment like sea water. They require at least 1.5 M NaCl (approximately 9% NaCl) for their growth and optimal growth usually occurs at 3.5–4.5 M NaCl (20–25 % NaCl) (Das Sarma and Arora, 2001). They can grow up to saturation (approximately 37 % NaCl concentration) (Grant et al., 2001). They are Gram-negative, pleomorphic, and contain polar lipids in their membrane. Their high densities in saline water often lead to pink or red coloration of water, called a red' bloom. This pink/red coloration is due to the presence of C_{50} carotenoid pigments in haloarchaea. According to Oren (2009), salt-saturated lakes and salterns of crystallizer ponds turn red due to microbial bloom. The carotenoids of bacterioruberin group produced by haloarchaeal strains appear to be the main factor causing the characteristic red color of hypersaline brines worldwide. Kanekar et al. (2012) have described different aspects of haloarchaea as taxonomy, physiology, and applications.

The haloarchaea are classified under the Domain Archaea, Phylum Euryarchaeota, class Halobacteria, orders Halobacteriales, Haloferacales, and Natrialbales, Families, Halobacteriaceae, Haloarculaceae, Halococcaceae, Haloferaceae, Halorubraceae, and Natrialbaceae (Gupta et al., 2015, 2016). Many novel species of halophilic archaea have been isolated from Great Salt Lake Utah, USA, Owens Lake in California and from the Dead Sea. The Dead Sea is a terminal lake between Israel and Jordon. The Dead Sea and the Great Salt Lake have been extensively studied during last 20 years for isolation of many novel species of halophilic archaea, *Halobacterium gomorrense, Haloarcula marismortui,* and *Haloferax volcanii.* One of the more unusual shaped haloarchaea is the 'square haloarchaeon of Walsby.' Walsby described in 1980 the presence of square-shaped haloarchaea in high concentrations in a hypersaline pond near the Red Sea. It was cultivated in 2004 using a very low nutritional solution to allow growth along with high salt concentration and classified under genus *Haloquadratum* (Salt Square). The cells of 'square haloarchaea of Walsby' are extremely thin around 0.15 μm and square-shaped (Burns et al., 2004a, 2007).

Culturable halophilic archaea have been isolated from rock salt mines of up to millions of years of age (195–250 million years) representing haloarchaeal genera of *Haloarcula, Halobacterium, Halococcus*, and *Natronobacterium*. Viable halophilic archaea were reported for the first time from ancient rock salt by Reiser and Tasch (1960) and Dombrowski (1963). Haloarchaeal strains of *Haloarcula* and *Halorubrum* were reported from rock salt in Winsford salt mine by Norton et al. (1993). Presence of *Halococcus salifodinae* in Permo-Triassic salt deposit was detected by Tan-Lotter et al. (1999). McGenity et al. (2000) proposed the origin of halophilic microorganisms and haloarchaea in ancient rock salt deposits. Permian alpine salt deposit and Middle-Late Eocene rock salt were studied as a source of haloarchaea, *Halococcus dombrowskii* sp. nov. and *Halobacterium* sp. respectively. Thus haloarchaea are known as the most primitive organisms on the earth.

A few researchers over the globe have studied isolation and cultivation of haloarchaea from solar salterns. Archaeal biodiversity in the crystallizer pond of solar saltern was reported by Benlloch et al. (2001) using culture dependant as well as culture-independent methods. Ochsenreiter et al. (2002) reported a diversity of archaea in hypersaline environments using molecular-phylogenetic and cultivation studies. Diversity studies of halo-archaea in the Australian crystallizer ponds were carried out by Burns et al. (2004b). Asha et al. (2005) isolated and cultivated halophilic archaea from solar salterns located in Peninsular Coast of India. Dave and Desai (2006) worked on microbial diversity of marine salterns near Bhavnagar, Gujarat, India. Ahmad et al. (2008) studied phylogenetic analyses of halophilic archaea from salt pan sediments of Mumbai, India. Halophilic archaebac-terium has been isolated from salt pan situated in Kandla, Gujarat, India (Akolkar et al., 2008). Diversity of halophilic microorganisms in solar salterns of Tamil Nadu, India was examined by Manikandan et al. (2009). Sabet et al. (2009) reported characterization of halophiles isolated from solar salterns in Baja, California, Mexico. Thomas et al. (2012) reported a novel haloarchaeal lineage distributed in hypersaline marshy environment of Little and Great Rann of Kutch, India. Karthikeyan et al. (2013) reported halocin producing extreme haloarchaeon *Natrinema* sp. BTSH10 isolated from salt pans of Kanyakumari District, Tamil Nadu, India. Isolation of halophilic archaea from sea water and soil samples from salt production sites of Mumbai, Maharashtra, India have been documented by Digaskar et al. (2015).

Various novel haloarchaeal species have been isolated from different solar salterns all over the world, *Halogeometricum borinquense* gen. nov. sp. nov., from solar saltern of Cabo Rojo, Puerto Rico (Montalvo-Rodriguez et al., 1998); *Haloarcula quadrata* sp. nov., from brine pool in Sinai, Egypt (Oren et al., 1999); and *Haloferax alexandrinus* sp. nov., from solar saltern in Alexandria, Egypt (Asker and Ohta, 2002). Solar salterns harbor a diverse community of halophilic microorganisms that represent the three domains of life: Archaea, bacteria, and Eukarya (Oren, 2002). The archaeal members of the order *Halobacteriales*, family *Halobacteriaceae*, are significant components of salt pans (Benlloch et al., 2002); *Haloquadratum walsbyi* gen. nov. sp. nov., from Australia and Spain (Burns et al., 2007); *Haloferax larsenii* sp. nov., from solar salterns in China (Xu et al., 2007); *Halopelagius inordinatus* gen. nov. sp. nov., *Halogranum rubrum* gen. nov. sp. nov., from Rudong and Haimen marine solar saltern, China respectively (Cui et al., 2010a,b); *Halorussus rarus*gen. nov., sp. nov., from Taibei marine solar salterns, China (Cui et al., 2010c); *Haloplanus salinus* sp. nov. from Chinese solar saltern (Qui et al., 2013); *Halorubrum rutilum* sp. nov. from Yangjiang marine solar saltern, China (Yin et al., 2015); *Haloarchaeobius amylolyticus* sp. nov., from Chinese marine saltern (Yuan et al., 2015) have been described.

Few studies have shown that haloarchaea are not restricted to hypersaline ecosystems but also inhabit in low saline environments. Rodriguez–Valera et al. (1979) first time reported presence of haloarchaea from seawater. Elshahed et al. (2004) reported halophilic archaea from low salinity, sulfide, and sulfur-rich spring. Purdy et al. (2004) isolated novel haloarchaea from non-extreme ecosystem like salt marsh sediments with salinity close to seawater. Recently Kanekar et al. (2015, 2017) also reported isolation of haloarchaeon *Halostagnicola larsenii* from rock pit seawater and *Haloferax larsenii* from rock showing red coloration, Rock Garden, Malvan, District Sindhudurga, West Coast of India.

6.3   BACTERIORHODOPSIN AND ITS APPLICATIONS

Certain species of haloarchaea can carry out a light-driven synthesis of adenosine triphosphate (ATP). This occurs without chlorophyll pigment.

In these species of haloarchaea, retinal-containing protein BR is present which acts as a light-driven proton pump capable of producing proton gradient across the membrane and is used for production of ATP. BR is named so because of its analogy with the vertebrate visual pigment 'rhodopsin.' A unique feature of haloarchaea is a presence of purple membrane (PM) as a specialized region of the cell membrane. It contains a two-dimensional crystalline lattice of a chromoprotein, BR. Production of BR was first described from *Halobacterium halobium* isolated from salterns of San Francisco Bay, USA (Oesterhelt and Stoeckenius, 1971) and further studied by Khorana (1988) and Shand and Betlatch (1991, 1994). BR contains a protein moiety (Bactrio-opsin) and a covalently bound chromophore (retinal) which acts as a light dependant transmembrane proton pump (Krebs and Khorana, 1993).

The retinal protein, BR is the major photosynthetic protein of a haloarchaeon *Halobacterium salinarum*. It converts the energy of green light (500–600 nm, max 568 nm) into an electrochemical proton gradient, which in turn is used for ATP production by ATP synthase. *Halobacterium salinarum* can use both, oxygen, and light as energy sources because of presence of BR in PM. The most efficient source for *Halobacterium salinarum* is oxygen; however, under conditions of low aeration, these cells synthesize BR in their PM and produce energy in the form of ATP.

BR is a 25 kDa transmembrane protein mainly found in the PM of *Halobacterium salinarum*. The protein:lipid ratio of PM is 75:25. The only protein present in PM is BR which forms a hexagonal 2-D crystal consisting of BR trimer. BR is composed of seven alpha helices that arrange in a homotrimer within a hexagonal 2-D crystalline array in a PM. A molecule of retinal chromophore is bound to Lys 216 by a Schiff base. The retinal molecule is carotenoid like molecule that can absorb light energy and catalyze the formation of proton motive force. This retinal gives BR purple color. When the BR is excited by light of suitable wavelength, a complex photocycle is initiated that lasts up to 10 milliseconds. During this process, the retinal group undergoes isomerization from all-*trans* to *13-cis*-isomer. This transformation results in the translocation of protons to the outside surface of the membrane. The retinal molecule then returns to its more stable form, all-*trans* isomer in the dark along with the uptake of protons from the cytoplasm, this completes the cycle. As the protons accumulate on the outer surface of membrane, a proton motive force increases until the membrane is sufficiently 'charged' to drive ATP synthesis through the activity of proton-translocating ATPase.

Oesterhelt and Stoeckenius (1971, 1974) reported that when grown under low oxygen tension in light, *Halobacterium halobium* produces distinct patches in its plasma membrane. These patches were named PM because of their characteristic color. This color is due to the retinal component of BR, which is the sole protein found in PM. They reported isolation of PM by differential and sucrose gradient centrifugation after lysis of cells by dialysis against distilled water. Oesterhelt and Stoeckenius (1973) and Oesterhelt and Hess (1974) studied the proton pumping of BR present in the PM of *H. halobium*. They reported that during proton pumping action of PM, BR undergoes a light-induced cyclic reaction involving a proton release on one side of the membrane and proton uptake on another side. A light-induced transient shift of the absorption maxima to 412 nm and concomitant release and uptake of protons have been observed in PM. Racker and Stoeckenius (1974) observed that isolated PM, when incorporated into lipid vesicles, has been shown to act as a light-driven proton pump. Lozier et al. (1975) reviewed BR as a light-driven proton pump in *Halobacterium halobium*. Knowledge of the primary structure of BR was necessary to study the mechanism of proton pump. The primary structure of BR containing 248 amino acids residues was elucidated by Khorana et al. (1979). Ovchinnikov et al. (1979) studied the complete structural elucidation of BR and described the structural basis of the functioning of BR. Singh et al. (1980) studied the photoelectric conversion by BR in charged synthetic membranes. Photoelectroactivity of oriented PM layers attached to an ion exchange film has been studied by the researchers. The photocurrents as high as 20 μA cm^{-2} were obtained from this study. Kouyama et al. (1987) studied the activity of BR in *Halobacterium halobium* cell envelopes, and it was found that the concentration of magnesium ions greatly enhanced the light-induced pH change. The results suggested that some divalent cations act as a buffer against a large increase in the internal pH and that internal pH was important factor in determining activity of BR. It was also noted that high levels of proton pump activity was maintained in a wide range of external pH, at least between 4.5–9.4.

Kouyama et al. (1988) studied the structure and function of BR by preparing BR containing membrane vesicles and demonstrated that BR exhibits a high proton pump activity within a wide pH range (pH 4–9.5). The three-dimensional structure of BR was studied by fluorescence energy transfer techniques. It was shown that the retinal chromophore is located 10A below a surface of PM. The dynamic structure of BR was investigated by fluorescence depolarization techniques. It was shown that the retinal chromophore is tightly buried in the pocket within a protein.

Molecular basis of BR production has been investigated by a few researchers. Dunn et al. (1981) identified the BR gene in a 5.3 kilo-base restriction endonuclease fragment isolated from *Halobacterium halobium* DNA, using a cloned cDNA fragment as a probe. Betlach et al. (1984), Leong et al. (1988) and Gropp et al. (1993, 1994) reported that the bacterio-opsin (*bop*) gene which encodes the structural protein bacterio-opsin is located in the *bop* gene cluster with at least three other genes, *brp*: bacterio-opsin related protein; *bat*: bacterio-opsin activator gene and the *blp*: bacterio-opsin linked product gene. Betlach et al. (1989) studied the regulation of the bacterio-opsin gene of halophilic archaea. Shand and Betlach (1991) studied expression of the *bop* gene cluster of *Halobacterium halobium* by varying oxygen tension and light/dark condition. Phototrophic chamber which delivers 30,000 lux light to the cultures was used to grow *Halobacterium halobium* cultures. The culture was grown alternatively in high oxygen tension and low oxygen tension mode, in presence of light and in dark condition. They observed that *bat* gene expression is induced under conditions of low oxygen tension and *bat* gene product activates *bop* gene expression and light induces the *brp* transcription which stimulates *bat* transcription. It was thus reported that expression of *bop* gene cluster is induced by low oxygen tension and by light. Shand and Betlach (1994) reported the bop gene cluster expression in BR overproducing mutants of *H. halobium*. Xu et al. (2005) reported detection of *bop* gene in a novel strain of *Natrinema* isolated from Ayakekum Lake, China. Lynch et al. (2012) first time identified BR gene (GI: 445742603) in the genome of *Haloferax mucosum*. Recently, Hsu et al. (2013) reported a few haloarchaeal strains, *Halobacterium salinarum*, *Haloarcula marismortui*, *Haloquadratum walsbyi*, and haloalkaliphilic *Natronomonas pharaoni* to contain a *bop* gene.

6.3.1   PROPERTIES AND APPLICATIONS OF BR

Unlike the other proteins of halophilic archaea, BR is stable in absence of salts and retains its photochemical activity over longer periods. It is active at temperatures from 0–45°C and in pH range of 1–11 and tolerates the temperature over 80°C in water and 140°C in dry condition. BR is stable in sunlight for years and resistant to hydrolysis by many proteases. It can be stored by immobilizing on glass plates or embedding in polymers. They have high quantum efficiency of converting light into a state change, ability to form thin films in polymers and gels. These properties of BR have enabled multiple usages in different applications.

6.3.2 APPLICATIONS IN CONVERSION OF SOLAR ENERGY AND GENERATION OF ELECTRICITY

BR has been used as a light-induced electricity generator. The electric current was generated by illuminating oriented PM sheets containing light-driven proton pump, BR. BR is used in photoelectric, photochromic, or energy converting element.

Thavasi et al. (2008) reported the feasibility of BR as bio-photosensitizer in excitonic solar cells. They studied the wild-type BR and triple mutant BR in combination with wide gap semiconductor TiO_2 as efficient light harvester in solar cell and successfully prepared the bio-sensitized solar cell (BSSC) based on the wild-type BR and triple mutant BR adsorbed on nanocrystalline TiO_2 film. Yen et al. (2011) have maximized the solar energy conversion of BR by tailoring plasmonic and electrostatic field effects. They have explored the plasmonic filed enhancement of current production from BR by maximizing the blue light effect, leading to current enhancement. Trivedi et al. (2011) have reviewed different applications of BR. According to them, use of BR has been patented for collection of solar energy into direct current. Shiu et al. (2013) described the procedure for measurement of photocurrent activity of isolated PM solution. The PM containing BR was drop coated on ITO glass and dried at room temperature. The photocurrent activity measurements were taken using photocell consisted of two ITO glass electrodes between which PM film was sandwiched. Xenon lamp (140 W) with 560 nm pass filter was used as a light source to illuminate photocell and photocurrents were measured.

6.3.3 OPTICAL APPLICATIONS

Researchers have used BR with photoelectric properties as photochromic material for optical applications. BR is used in optical applications as for optical lenses, optical grating, pattern generator, beam reshaper, data storage carrier, fraud-proof optical data carrier, security ink, optical switches, ultra-sensitive optical biosensors. Hampp (2000) reviewed applications of BR as optoelectronic material.

6.3.4 BIOTECHNOLOGICAL AND MEDICAL APPLICATIONS

BR is used as a diagnostic material for detection of respiratory-related viral nucleoprotein (influenza virus) with high sensitivity and specificity. It can also be used in therapeutics for treatment of degenerated retinal blindness, eye disorders, retinal disorders, malignant tumors, and vaccine therapy. BR is used in the devices for sensing and communicating nerve cell ending in humans and neural signals involving building neural bridge, neural bridge switch. Thus it has applications in neuro-stimulation devices and sensing devices. Its role in drug delivery, transport, and release of drug, cell signaling and induction of cell apoptosis in neoplastic cells is also patented. Its pharmaceutical applications like transport and delivery of immunogenic species is patented. BR is used in optical circuits, optical fibers, which may be used in biomedical applications and inserted in the body as transmissive media.

6.3.5 OTHER APPLICATIONS

BR films are used as a short-term memories, in particular in holographic application. It is also used in detection of quality of air, water, and soil. Fimia et al. (2003) reported preparation of PM–polyacrylamide films as holographic recording material. The holographic parameters of those films were measured. Jin et al. (2008) reviewed applications of BR as electronic conduction medium for biomolecular electronics. Trivedi et al. (2011) have reviewed application of BR for legal document storage in writable or rewritable formats. BR can also be used for filtration of water, purification of water by reverse osmosis and has biosensor applications (Trivedi et al., 2011).

6.4 BR PRODUCING HALOPHILIC ARCHAEA

Only a few strains of halophilic archaea possess BR in their cell membrane. Although a major studies on BR have been carried out using the candidate organism *Halobacterium salinarum* (formerly called as *Halobacterium halobium*), research appears to be directed to explore new strains of haloarchaea for production of BR. *Haloarcula japonica* was the second organism reported for BR production (Yatsunami et al., 2000) followed by *Natrinema* (Xu et al., 2005), *Halorubrum sodomense* (Mohan Raj and Vatsala, 2009), *Haloferax mucosum* (Lynch et al., 2012), *Halostagnicola larsenii* (Kanekar et al., 2015) and *Haloferax larsenii* (Kanekar et al., 2017) (Table 6.1).

TABLE 6.1   BR Producing Haloarchaea

Name of haloarchaeon	Source of the organism/ presence of bop gene	References
Halobacterium halobium	Bay of San Francisco, California, USA	Oesterhelt and Stoeckenius, 1971
Halobacterium sp.	Dead Sea	Oren and Shilo, 1981
Haloarcula japonica	Japanese saltern soil	Yatsunami et al., 2000
Natrinema	Ayakekum Lake, China	Xu et al., 2005
Halorubrum sodomense	Dead Sea	Mohan Raj and Vatsala, 2009
Haloferax mucosum	BR gene (GI: 445742603)	Lynch et al., 2012
Haloarcula marismortui	*bop* gene	Hsu et al., 2013
Haloquadratum walsbyi	*bop* gene	Hsu et al., 2013
Natronomonas pharaoni	*bop* gene	Hsu et al., 2013
Halostagnicola larsenii	Rock pit sea water, Rock Garden, Malvan, West Coast of India	Kanekar et al., 2015
Haloferax larsenii	Red rocky substrate, Rock Garden, Malvan, West Coast of India	Kanekar et al., 2017

6.5   PRODUCTION OF BR

Only some haloarchaea have been reported to contain BR in their cell membrane. In natural saline environments, the red-pigmented haloarchaea impart red color to the ecosystem they populate with high density. BR is a chromophore associated integral membrane protein originally discovered in extremely halophilic archaeon *Halobacterium halobium* (now called as *Halobacterium salinarum*) by Oesterhelt and Stoeckenius (1971). The production of PM allows the haloarchaeal strain to switch from oxidative phosphorylation to photosynthesis as a means of energy production

(Oesterhelt and Stoeckenius, 1973). According to Rogers et al. (1978), the haloarchaea synthesize PM when dissolved oxygen concentration becomes too low to sustain aerobic ATP production in the presence of light. Oren and Shilo (1981) observed a red colored dense bloom of halobacteria in the Dead Sea. The bloom was found to contain 0.4 nmol per mg protein BR. The haloarchaeon, *Halobacterium* was isolated from this red bloom. The isolate of *Halobacterium* also contained BR. Javor (1983) found 2.2 nmol/l BR in the haloarchaeal community in the crystallizer ponds of the salterns of Baja, California, Mexico. The cells of *H. salinarum* were mainly found in amino acids rich, frequently oxygen-poor, concentrated brines (Woese et al., 1990). *H. salinarum* is most frequently employed for BR production. Under suitable conditions, *H. salinarum* produces large quantity of BR. The color of the culture broth then changes from red-pink to purple.

Excitation of BR by light results in release of protons from the cell to outer medium (Oesterhelt and Stoeckenius, 1973). This leads to generation of proton gradient which may drive production of adenosine triphosphate (ATP). The generation of chemical energy in form of ATP was demonstrated by Danon and Stoeckenius (1974). The BR containing cells can survive under starvation conditions, in light (Brock and Peterson, 1976). Thus production of BR is likely to be enhanced under stress conditions. With this additional energy generating mechanism, the cells may utilize it when conventional energy sources are limited, e.g., when organic substrates are lacking, or there is a limitation of fermentable substrates (Hartmann et al., 1980).

After a gap of about 10 years, studies on production of BR were diverted to expression of genes coding BR (*bop*) in *Escherichia coli* or other organisms. Hildebrandt et al. (1989, 1991) exploited *Schizosaccharomyces pombe* as a expression host for natural BR gene (*bop*) and production of functional BR in heterologous host. Pompejus et al. (1993) studied production of BR by expressing synthetic gene in *Escherichia coli*. A gene (*bos*) coding for bacterioopsin (BO) which is an apoprotein of BR was assembled from chemically synthesized oligonucleotides. A new method of repeated rounds of insertion mutagenesis was developed for expression of gene in *E. coli*. The yield of BO was 30–50 mg pure protein/L culture medium. BO could be used for reconstitution of fully functional BR.

Deshpande and Sonar (1999) studied synthetic pathway of production of BR by investigating the role of mRNA and membrane proteins in cellular metabolism by *in vivo* introductions of through liposome fusion. The experimental evidence indicated that the formation of BO acts as a trigger for lycopene conversion to β-carotene in retinal biosynthesis.

Shimano et al. (2009) used cell-free expression method which is a promising method for production of membrane proteins. A dialysis based *Escherichia coli* cell-free system was employed for production of BR. The system in the dialysis mode was supplemented with a combination of a steroid detergent (digitonin, cholate or 3-[cholamidopropyl) dimethylammonio]-2-hydroxyl-1-propanesulfonic acid, CHAPSO) and egg phosphatidyl choline, which resulted in high yield (0.3–0.7 mg/mL reaction mixture) of the fully functional BR.

Abarghooi et al. (2012) attempted overproduction of BR in *E. coli*. They designed a synthetic gene and cloned in PET30a for overexpression of BR gene through fusion to the mystic. The synthetic BR was separated from mystic by trypsin cleavage at the factor of Xa site between mystic and BR. The induction temperature and medium were found to influence quality and quantity of recombinant BR production.

Kanekar (2014) explored halophilic archaea isolated from Andaman Islands, India for production of BR. Kanekar et al. (2015) have reported flask culture production of BR using 11 strains of *Halostagnicola larsenii* with BR yield ranging from 0.035 to 0.258 g/L in Sehgal and Gibbon's (SG) medium containing 4.28 M NaCl. Kanekar et al. (2017) also have reported production of BR (0.137 g/L) by another haloarchaeon *Haloferax larsenii* RG3D.1 isolated from red rocky substrate, Rock Garden, Malvan, West Coast of India.

6.6   OPTIMIZATION OF BR PRODUCTION

The yield of BR is associated with cell mass since it is a transmembrane protein. Optimization of culture conditions and nutritional parameters, therefore, becomes necessary to increase the yield of BR. Statistical optimization of production of BR using *Halobacterium salinarum* was extensively studied by Ghasemi et al. (2008). Various carbon and nitrogen sources, temperature, and pH were studied by one factor at a time approach. The corn steep powder, peptone, and yeast extract were found to be the most suitable sources and 38°C and 7.5 as the optimal temperature and pH respectively. The authors further optimized concentration of carbon and nitrogen sources by Taguchi method for production of BR. The optimization studies resulted in enhanced yield of 191.7 mg/L of BR. Manikandan et al. (2009) optimized growth media to obtain high cell density of halophilic archaeon *Halobacterium salinarum*. The media components namely potassium chloride (6.35 g/L), magnesium sulphate (9.70 g/L), gelatin (13.38 g/L), soluble starch (12.0 g/L), and artificial seawater with 20% (w/v) NaCl were found to be optimum by response surface methodology yielding 2.4 fold increases in biomass.

Optimization of physical parameters as illumination, agitation speed, temperature, and nitrogen source was studied using *Halobacterium salinarum* for production of BR by Seyedkarimi et al. (2015). The authors used fractional factorial design and found that peptone from meat was the most suitable nitrogen source, temperature as 39°C, agitation speed of 150 rpm and light intensity of 6300 lux for production of BR with the yield of 196 mg/L. Shiu et al. (2015) investigated effect of complex nutrients on growth and production of BR by *Halobacterium salinarum*. Tryptone was found to be the best nitrogen source for growth and BR production.

6.7   DEVELOPING BIOREACTORS FOR BR PRODUCTION

Very few researchers attempted large-scale production of BR using bioreactor. Lee et al. (1998) reported production of BR using cell recycle culture of *H. halobium* R1 (ATCC 29341). The batch culture was carried out in 2.5 L jar fermenter containing 1.3 L medium. Sterilized air was supplied at 1.3 vvm. A halogen lamp was used as a light source. Cell recycle was carried out in the same fermenter equipped with an external membrane filter module. The cell concentration and BR concentration obtained using cell recycle culture was 30.3 dry cell weight/l and 282 mg/L in 10 days with the productivity of 1.15 mg/L/h. Ghasemi et al. (2008) studied bioreactor fermentations using *H. salinarum* PTCC 1685 in 15 L stirred tank fermenter. The fermenter was run at temperature, 38°C; agitation speed, 150 rpm; aeration rate, 1.5vvm; working volume, 4 L of the optimized medium and initial pH 7.5. Under optimal conditions, maximum 234.6 mg/L BR production was observed. Shiu et al. (2015) reported repeated batch cultivation of *H. salinarum* for obtaining high yield of BR. Bubble column photobioreactor of 1 L capacity with working volume 800 ml was used for repeated batch cultivation. The height of the column was 15 cm. The air was supplied from the bottom of the column to generate bubbles and also provide mixing. The reactor was supplied by 4 fluorescent 40

W lamps and 1 LED lamp for illumination. Half tryptone medium was employed for BR production. The photobioreactor was placed in 37°C water bath and stirred with magnetic stir bar at 50 rpm. The aeration rate was 1 vvm till 48 h then decreased to 0.1 vvm till end of cultivation. For repeated batch culture, the cell pellet was collected by centrifugation and re-suspended with the same volume of fresh half tryptone medium. A cell density of 4.8 g/L with 201.8 mg/L BR concentration was achieved after 210 h after 5 batches of repeated cultivation.

6.8   COMMERCIAL PRODUCTION OF BR

Commercial production of BR was initially started in Spain by the company COBELL. At present, Munich Innovative Materials (MIB) GmbH, Marburg, Germany founded in 1997, is the only manufacturer of BR in technical quantities worldwide. The products include high-performance recording media for optical information processing, and data analysis, specialty, and custom-made BR variants with single to multiple amino-acid replacements from laboratory to technical quantities.

For quite many years after its discovery, BR was being produced on laboratory scale. A low-cost technology was developed at Fraunhofer Institute for Interfacial Engineering and Biotechnology, for production of BR (https://www.fraunhofer.de/en/press/research-news/2019/january/producing-vaccines-without-the-use-of-chemicals..html). A continuously operating bioreactor was developed, and the cost of production was reduced by replacing peptone with yeast extract in the growth medium. In the down-stream processing, centrifugation, and dialysis steps were replaced by a multistage cross-flow filtration technique leading to a simplified operation scheme and minimization of yield losses. The spent medium was reused for production of BR, thus saving the cost of production.

6.9   ISOLATION OF PURPLE MEMBRANE AND PURIFICATION OF BR

Separation and purification of BR has been a tedious jobs that being a transmembrane protein. Traditionally, the PM purification procedure was mainly based on steps of centrifugation, homogenization, dialysis, and the lengthy sucrose gradient ultracentrifugation. Oesterhelt and Stoeckenius

(1974) first time reported use of differential and sucrose density gradient centrifugation for the isolation of PM from *H. halobium* and its fractionation into red and PM. Becher and Cassim (1975) reported an improved method for isolation of PM of *H. halobium*. Absorption spectra of PM was used as a criteria to determine purity of PM solutions. The ratio of protein aromatic amino acid absorbance at 280 nm to chromophore absorbance at 567 nm was also considered to show increase in the purity of PM preparations. Miercke et al. (1989) studied purification of BR by isolating PM sucrose density gradient centrifugation and then solubilizing PM in Triton–X–100. BR was purified by size exclusion chromatography using 3-[cholamidopropyl) dimethylammonio]–2-hydroxyl–1-propanesulfonic acid (CHAPSO) detergent at pH 5.0. Tan and Birge (1996) reported use of series of alkyl ammonium surfactants in solubilization of BR.

Huh et al. (2010) described a separation of BR from PM of *H. salinarum* using a laminar flow extraction system in a microfluidic device. The PM was isolated using sucrose density gradient ultracentrifugation and PM residing in 40% (w/v) sucrose layer was used for microfluidic system. The microfluidic device was prepared using soft lithography, and replica molding methods and the material for replica molding was Polydimethylsiloxane (PDMS). The microfluidic separation system was carried out using two-phase aqueous system (ATPS) and the ionic liquid two-phase system (ILTPS). Shiu et al. (2013) described isolation of PM from *H. salinarum* using aqueous two-phase system (ATPS) based on the polyethylene glycol (PEG) and potassium phosphate solution. The results showed that PM was completely recovered from the interface of PEG-phosphate ATPS with BR purity of 94.1%. Thus lengthy and tedious sucrose density gradient ultracentrifugation which was commonly employed for purification of BR can now be replaced by rapid ATPS purification process.

6.10   PRODUCTION OF BR: A BIOTECHNOLOGICAL CHALLENGE

A large amount of BR is in demand for research and development, but only small amount is available due to difficulty in culturing *Halobacterium* cells. So far some investigators have attempted to study BR production by *H. salinarum* strains in batch mode or continuous culture method, but less attention is being paid to optimize the nutritional and environmental conditions for obtaining high yield of BR.

Although BR is a novel biomolecule and has many applications encompassing energy, electronics, biomedical, and pharmaceutical sectors, its production is not easy. The growth period of haloarchaea is long (7–8 days), the spent medium has high concentration of salts leading to the generation of high chloride-containing waste and equipment like fermenters, bioreactors, centrifuge are likely to face pitting/corrosion problems due to high contents of salt in the medium. BR being a membrane protein, its yield is associated with the cell mass. The lysis of cells is comparatively easy because the haloarchaeal cells get lysed in water. However, purification of BR from membrane becomes tedious due to requirement of ultracentrifugation or sucrose gradient ultracentrifugation. In the light of these difficulties, research efforts were therefore directed towards expressing *bop* genes in *E. coli* or *Schizosaccharomyces pombe*. Heterologous expression of synthetic genes has also been tried. Cell-free synthesis of BR has been a novel approach.

So far, only the candidate organism *Halobacterium salinarum* has been employed for production of BR. In recent past, researchers over the globe have explored other haloarchaeal genera, e.g., *Haloarcula*, *Halorubrum*, *Haloferax*, and *Halostagnicola*. However, all these are flask culture or laboratory scale studies. Developing bioreactor and downstream processing are still awaiting success.

Open ponds system has been a suitable technique for obtaining mass culture of haloarchaea. Developing suitable bioreactors and centrifuges of corrosion resistant material are still the technological problems to be addressed. Perhaps this is the reason why exploitation of haloarchaea for production of BR has been lagging behind. Production of BR thus becomes a biotechnological challenge.

6.11 CONCLUSION AND FUTURE PERSPECTIVES

Production of BR has been studied using the candidate organism *Halobacterium salinarum*. Optimization of nutritional and environmental factors, isolation of PM and BR also has been studied using *H. salinarum*. Only some other species of halophilic archaea have been recently reported to produce BR or contain *bop* genes. BR is a very versatile biomolecule having applications in energy, bioelectronics, pharmaceuticals, and biomedical fields. Hence there is a demand for this industrially important biomolecule. Production of BR is, therefore, being studied globally. Since the growth of haloarchaea is slow (7 to 8 days), the production process becomes costly.

Due to high concentration of NaCl in the growth medium, bioreactors or fermenters and centrifuges of salt resistant material are to be developed to obtain mass culture of the haloarchaeal strain. Since BR is associated with the cell membrane, purification procedure is tedious involving ultracentrifugation. These are the biotechnological challenges in production of BR.

Haloarchaea are unique living entities on the earth, and more efforts are needed to isolate new strains yielding high biomass and BR. A number of saline environments are yet to be explored for obtaining haloarchaea. Although the haloarchaea require at least 1.5 M (9%) NaCl for growth, certain strains have been shown to survive and grow in low saline environments like seawater. To avoid use of high content of salt and subsequently generation of high chloride containing wastewaters, such strains could be explored. Nutritional requirement of haloarchaea is low and hence low carbon, and nitrogen-containing renewable substrates could be explored and optimized.

Development of bioreactors of salt-resistant material can be dealt with by collaboration with material engineers and instrumentation engineers. For application of BR in various fields like bioenergy, electronics, biomedical, and pharmaceutical sectors, a team of physicists, chemists, engineers, and microbiologists is required. International collaborations to finally obtain a good yield of BR and its applications are envisaged.

KEYWORDS

- aqueous two-phase system
- bacteriorhodopsin
- cell membrane proteins
- generation of electricity
- *Halobacterium salinarum*
- halophilic archaea
- *Halostagnicola larsenii*
- hypersaline environments
- light-driven proton pump
- optoelectronic material
- photocycle
- photochromic material
- proton gradient
- purple membrane
- retinal pigment
- solar salterns
- sucrose density gradient ultracentrifugation

REFERENCES

Abarghooi, F., Babaeipour, V., Rajabi, M. H., & Mofid, M. R., (2012). Overproduction of bacteriorhodopsin in *E. coli* as pharmacological targets. *Res. Pharm. Sci., 7*, S473.

Ahmad, N., Sharma, S., Khan, F. G., Kumar, R., Johri, S., Abdin, M. Z., & Qazi, G. N., (2008). Phylogenetic analyses of Archaeal ribosomal DNA sequences from salt pan sediments of Mumbai, India. *Curr. Microbiol., 57*, 145–152.

Akolkar, A. V., Deshpande, G. M., Raval, K. N., Durai, D., Nerurkar, A., & Desai, A. J., (2008). Organic solvent tolerance of *Halobacterium* sp. SP1 (1) and its extracellular protease. *J. Basic Microbiol., 48*, 421–425.

Asha, K. R. T., Vinitha, D. A. J., Sehgal, G. K., Manjusha, W. A., Sukumaran, N., & Selvin, J., (2005). Isolation and cultivation of halophilic Archaea from solar salterns located in peninsular coast of India. *Int. J. Microbiol., 1*, 1–6.

Asker, D., & Ohta, Y., (2002). *Haloferax alexandrinus* sp. nov., an extremely halophilic canthaxanthin-producing archaeon from a solar saltern in Alexandria (Egypt). *Int. J. Syst. Evol. Microbiol., 52*, 729–738.

Becher, B. M., & Cassim, J. Y., (1975). Improved isolation procedures for the purple membrane of *Halobacterium halobium*. *Prep. Biochem., 5*, 161–178.

Benlloch, S., Acinas, S. G., Anton, J., Lopez-Lopez, A., Luz, S. P., & Rodriguez-Valera, F., (2001). Archaeal biodiversity in crystallizer ponds from a solar saltern: Culture versus PCR. *Microb. Ecol., 41*, 12–19.

Benlloch, S., Lopez-Lopez, A., Casamayor, E. O., Ovreas, L., Goddard, V., Daae, F. L., et al., (2002). Prokaryotic genetic diversity throughout the salinity gradient of a coastal solar saltern. *Environ. Microb., 4*, 349–360.

Betlach, M. C., Shand, R. F., & Leong, D. M., (1989). Regulation of the bacterio-opsin gene of a halophilic archaebacterium. *Can.. J. Microbiol., 35*, 134–140.

Betlach, M., Friedman, J., Boyer, H. W., & Pfeifer, F., (1984). Characterization of a halo-bacterial gene affecting bacterio-opsin gene expression. *Nucl. Acids Res., 12*, 7949–7959.

Brock, T. D., & Peterson, S., (1976). Some effects of light on the viability of rhodopsin-containing halobacteria. *Arch. Microbiol., 109*, 199–200.

Burns, D. G., Camakaris, H. M., Janssen, P. H., & Dyall-Smith, M. L., (2004a). Cultivation of Walsby's square haloarchaeon. *FEMS Microbiol. Lett., 238*, 469–473.

Burns, D. G., Camakaris, H. M., Janssen, P. H., & Dyall-Smith, M., (2004b). Combined use of cultivation dependant and cultivation-independent methods indicates that members of most haloarchaeal groups in an Australian crystallizer ponds are cultivable. *Appl. Environ. Microbiol., 79*, 5258–5265.

Burns, D. G., Janssen, P. H., Itoh, T., Kamekura, M., Li, Z., Jensen, G., et al., (2007). *Haloquadratum walsbyi* gen. nov. sp. nov., the square haloarchaeon of Walsby, isolated from saltern crystallizers in Australia and Spain. *Int. J. Syst. Evol. Microbiol., 57*, 387–392.

Cui, H. L., Gao, X., Sun, F. F., Dong, Y., Xu, X. W., Zhou, Y. G., et al., (2010b). *Halogranum rubrum* gen. nov., sp. nov., a halophilic archaeon isolated from a marine solar saltern. *Int. J. Syst. Evol. Microbiol., 60*, 1366–1371.

Cui, H. L., Gao, X., Yang, X., & Xu, X. W., (2010c). *Halorussus rarus* gen nov., sp. nov., a new member of the family *Halobacteriaceae* isolated from a marine solar saltern. *Extremophiles, 14*, 493–499.

Cui, H. L., Li, X, Y., Gao, X., Xu, X. W., Zhou, Y. G., Liu, H. C., et al., (2010a). *Halopelagius inordinatus* gen. nov., sp. nov., a new member of the family *Halobacteriaceae* isolated from a marine solar saltern. *Int. J. Syst. Evol. Microbiol., 60,* 2089–2093.

Danon, A., & Stoeckenius, W., (1974). Photophosphorylation in *Halobacterium halobium. Proc. Natl. Acad. Sci., USA, 71,* 1234–1238.

Das Sarma, S., & Arora, P., (2001). Halophiles. In: *Encyclopedia of Life Sciences* (pp. 1–9). Macmillan Press, Nature Publishing Group, London.

Dave, S. R., & Desai, H. B., (2006). Microbial diversity at marine salterns near Bhavnagar, Gujarat, India. *Curr. Sci., 90,* 497–500.

Deshpande, A., & Sonar, S., (1999). Bacteriorhodopsin-triggered retinal biosynthesis is inhibited by bacteriorhodopsin formation in *Halobacterium salinarum. J. Biol. Chem., 274,* 23535–23540.

Digaskar, V. U., Thombre, R. S., & Oke, R. S., (2015). Screening of extremely halophilic archaea for its biotechnological potential. *Int. Jour. Pharma. Biosci., 6*(B), 811–819.

Dombrowski, H., (1963). Bacteria from Paleozoic salt deposits. *Ann. NY Acad. Sci., 108,* 453–460.

Dunn, R., McCoy, J., Simsek, M., Majumdar, A., Chang, S. H., Rajbhandary, U. L., & Khorana, H. G., (1981). The bacteriorhodopsin gene. *Proc. Natl. Acad. Sci., USA, 78,* 6744–6748.

Elshahed, M. S., Najar, F. Z., Roe, B. R., Oren, A., Dewers, T. A., & Krumholz, L. R., (2004). Survey of Archaeal diversity reveals an abundance of Halophilic Archaea in a low salt, sulfide- and sulfur-rich spring. *Appl. Environ. Microbiol., 70,* 2230–2239.

Fimia, A., Acebal, P., Murciano, A., Blaya, S., Carretero, L., Ulibarrena, M., & Aleman, R., (2003). Purple membrane-polyacryamide films as holographic recording materials. *Optics Express, 11,* 3438–3444.

Ghasemi, M. F., Shodjai-Arani, A., & Moazami, N., (2008). Optimization of bacteriorhodopsin production by *Halobacterium salinarum* PTCC 1685. *Process Biochem., 43,* 1077–1082.

Grant, W. D., Kamemkura, M., McGenity, T. J., & Ventosa, A., (2001). Class III Halobacteria class, nov. In: Boone, D. R., Castenholz, R. W., & Garrity, G. M., (eds.), *Bergey's Manual of Systematic Bacteriology* (Vol. 1, pp. 294–334). Springer, New York.

Gropp, F., & Betlach, M. C., (1994). The *bat* gene of *Halobacterium halobium* encodes a trans-acting oxygen inducibility factor. *Proc. Nat. Acad. Sci., USA, 91,* 5475–5479.

Gropp, F., Gropp, R., & Betlach, M. C., (1993). A fourth gene in the *bop* gene cluster of *Halobacterium halobium* is co-regulated with the *bop* gene. *Syst. Appl. Microbiol., 16,* 716–724.

Gupta, R. S., Naushad, S., & Baker, S., (2015). Phylogenomic analyses and molecular signatures for the class Halobacteria and its two major clades: A proposal for division of the class Halobacteria into an emended order *Halobacteriales* and two new orders, *Haloferacales* ord. nov. and *Natrialbales* ord. nov., containing the novel families *Haloferacaceae* fam. nov. and *Natrialbaceae* fam. nov. *Int. J. Syst. Evol. Microbiol., 65,* 1050–1069.

Gupta, R. S., Naushad, S., Fabros, R., & Adeolu, M. A., (2016). Phylogenomic reappraisal of family-level divisions within the class Halobacteria: Proposal to divide the order *Halobacteriales* into the families *Halobacteriaceae, Haloarculaceae* fam. nov., and *Halococcaceae* fam. nov., and the order *Haloferacales* into the families *Haloferacaceae* and *Halorubraceae* fam. nov. *Antonie Van Leeuwenhoek, 109,* 565–587.

Hampp, N. A., (2000). Bacteriorhodopsin: Mutating a biomaterial into an optoelectronic material. *Appl. Microbiol. Biotechnol., 53*, 633–639.

Hartmann, R., Sickinger, H. D., & Oesterhelt, D., (1980). Anaerobic growth of halobacteria. *Proc. Natl. Acad. Sci., USA, 77*, 3821–3825.

Hildebrandt, V., Polakowski, F., & Buldt, G., (1991). Purple fission yeast: Overexpression and processing of the pigment bacteriorhodopsin in *Schizosaccharomyces pombe. Photochem, Photobiol., 54*, 1009–1016.

Hildebrandt, V., Ramezani-Rad, M., Swida, U., Wrede, P., Grzesick, S., Prinke, M., & Buldt, G., (1989). Genetic transfer of the pigment bacteriorhodopsin in the eukaryote *Schizosaccharimyces pombe. FEBS Lett., 243*, 137–140.

Hsu, M. F., Yu, T. F., Chou, C. C., Fu, H. Y., Yang, C. S., & Wang, A. H. J., (2013). Using *Haloarcula marismortui* bacteriorhodopsin a fusion tag for enhancing and visible expression of integral membrane proteins in *Escherichia coli. PLoS One, 8*, 1–11.

Huh, Y. S., Jeong, C. M., Chang, H. N., Lee, S. Y., Hong, W. H., & Park, T. J., (2010). Rapid separation of bacteriorhodopsin using a laminar-flow extraction system in a microfluidic device. *Biomicrofluidics, 4*, 014103 1–10.

Javor, B. J., (1983). Planktonic standing crop and nutrients in a saltern ecosystem. *Limnology and Ocenography, 28*, 153–159.

Jin, Y., Honig, T., Ron, I., Friedman, N., Sheves, M., & Cahen, D., (2008). Bacteriorhodopsin as an electronic conduction medium for biomolecular electronics. *Chem. Society Rev., 37*, 2422–2432.

Kanekar, P. P., Kanekar, S. P., Kelkar, A. S., & Dhakephalkar, P. K., (2012). Halophiles - taxonomy, diversity, physiology, and applications. In: Satyanarayana, T., Johri, B., & Prakash, A., (eds.), *Microorganisms in Environmental Management* (pp. 1–34). Springer, Netherlands.

Kanekar, P. P., Kulkarni, S. O., Dhakephalkar, P. K., Kulkarni, K. G., & Saxena, N., (2017). Isolation of an halophilic, bacteriorhodopsin-producing archaeon, *Haloferax larsenii* RG3D.1 from the rocky beach of Malvan, West Coast of India. *Geomicrobiol. J., 34*, 242–248.

Kanekar, P. P., Kulkarni, S. O., Shouche, Y., Jani, K., & Sharma, A., (2015). Exploration of a haloarchaeon, *Halostagnicola larsenii*, isolated from rock pit sea water, West Coast of Maharashtra, India, for production of bacteriorhodopsin. *J. Appl. Microbiol., 118*, 1345–1356.

Kanekar, S. P., (2014). *Biodiversity and Biotechnological Exploration of Halophiles Isolated From Andaman Islands and Lonar Lake, India* (pp. 1–219). PhD Thesis, Savitribai Phule Pune University, Pune, India.

Karthikeyan, P., Bhat, S. G., & Chandrasekaran, M., (2013). Halocin SH10 production by an extreme haloarchaeon *Natrinema* sp. BTSH10 isolated from salt pans of South India. *Saudi Journal of Biological Sciences, 20*, 205–212.

Khorana, H. G., (1988). Bacteriorhodopsin, a membrane protein that uses light to translocate protons. *J. Biol. Chem., 263*, 7439–7442.

Khorana, H. G., Gerber, G. E., Herlihy, W. C., Gray, C. P., Anderegg, R. J., Nihei, K., & Biemann, K., (1979). Amino acid sequence of bacteriorhodopsin. *Proc. Natl. Acd. Sci., USA, 76*, 5046–5050.

Kouyama, T., Kinosita, K., & Ikegami, A., (1988). Structure and function of bacteriorhodopsin. *Adv. Biophys., 24*, 123–175.

Kouyama, T., Kouyama, A. N., & Ikegami, A., (1987). Bacteriorhodopsin is a powerful light-driven proton pump. *Biophys. J., 51*, 839–841.

Krebs, M. P., & Khorana, H. G., (1993). Mechanism of light-dependent proton translocation by bacteriorhodopsin. *J. Bacteriol., 175*, 1555–1560.

Lee, S. Y., Chang, H. N., Um, Y. S., & Hong, S. H., (1998). Bacteriorhodopsin production by cell recycle culture of *Halobacterium halobium*. *Biotech. Lett., 20*, 763–765.

Leong, D., Pfeifer, F., Boyer, H., Betlach, M., (1988). Characterization of a second gene involved in bacterio-opsin gene expression in a halophilic archaebacterium. *J. Bacteriol., 170*, 4903–4909.

Lozier, R. H., Bogomolni, R. A., & Stoeckenius, W., (1975). Bacteriorhodopsin: A light-driven proton pump in *Halobacterium halobium*. *Biophys. J., 15*, 955–962.

Lynch, E. A., Langille, M. G. I., Darling, A., Wilbanks, E. G., Haltiner, C., Shao, K. S. Y., et al., (2012). Sequencing of seven haloarchaeal genomes reveals patterns of genomic flux. *PLoS One, 7*, e41389.

Manikandan, M., Kannan, V., & Pasic, L., (2009). Diversity of microorganisms in solar salterns of Tamil Nadu, India. *World J. Micro Biotech., 25*, 1001–1017.

McGenity, T. J., Gemmell, R. T., Grant, W. D., & Stan-Lotter, H., (2000). Origins of halophilic microorganisms in ancient salt deposits. *Env. Microbiol., 2*, 243–250.

Miercke, L. J. W., Ross, P. E., Stroud, R. M., & Dratz, E. A., (1989). Purification of bacteriorhodopsin and characterization of mature and partially processed forms. *J. Biol. Chem., 264*, 7531–7535.

Mohan, R. S., & Vatsala, T. M., (2009). Tryptophan inhibits bacteriorhodopsin formation in *Halorubrum sodomense* A01. *J. Basic Microbiol., 49*, 304–309.

Montalvo-Rodriguez, R., Vreeland, R. H., Oren, A., Kessel, M., Betancourt, C., & Garriga, J. L., (1998). *Halogeometricum borinquense* gen. nov., sp. nov., a novel halophilic archaeon from Puerto Rico. *Int. J. Syst. Evol. Microbiol., 48*, 1305–1312.

Norton, C. F., McGenity, T. J., & Grant, W. D., (1993). *Archaeal halophiles* (halobacteria) from two British salt mines. *J. Gen. Microbiol., 139*, 1077–1081.

Ochsenreiter, T., Pfeifer, F., & Schleper, C., (2002). Diversity of archaea in hypersaline environments characterized by molecular phylogenetic and cultivation studies. *Extremophiles, 6*, 247–274.

Oesterhelt, D., & Hess, B., (1974). Reversible photolysis of the purple complex in the purple membrane of *Halobacterium halobium*. *Eur. J. Biochem., 37*, 316–326.

Oesterhelt, D., & Stoeckenius, W., (1971). Rhodopsin-like protein from the purple membrane of *Halobacterium halobium*. *Nature New Biology, 233*, 149–154.

Oesterhelt, D., & Stoeckenius, W., (1973). Functions of a new photoreceptor membrane. *Proc. Natl. Acad. Sci., USA, 70*, 2853–2857.

Oesterhelt, D., & Stoeckenius, W., (1974). Isolation of the cell membrane of *Halobacterium halobium* and its fractionation into red and purple membrane. *Method Enzymol., 31*, 667–678.

Oren, A., & Shilo, M., (1981). Bacteriorhodopsin in a bloom of halobacteria in the Dead Sea. *Arch. Microbiol., 130*, 185–187.

Oren, A., (2002). Molecular ecology of extremely halophilic archaea and bacteria. *FEMS Microbiol. Ecol., 39*, 1–7.

Oren, A., (2009). Microbial diversity and microbial abundance in salt-saturated brines: why are the waters of hypersaline lakes red? *Natural Resources and Environmental Issues, 15*, 247–255.

Oren. A., Ventosa, A., Gutierrez, M. C., & Kamekura, M., (1999). *Haloarcula quadrata* sp. nov., a square, motile archaeon isolated from a brine pool in Sinai (Egypt). *Int. J. Syst. Evol. Microbiol., 49*, 1149–1155.

Ovchinnikov, Y. A., Abdulaev, N. G., Feigina, M. Y., Kiselev, A. V., & Lobanov, N. A., (1979). The structural basis of the functioning of bacteriorhodopsin: An overview. *FEBS Letters, 100*, 219–224.

Pompejus, M., Friedrich, K., Teufel, M., & Fritz, H. J., (1993). High yield production of bacteriorhodopsin via expression of a synthetic gene in *E. coli. Eur. J. Biochem., 211*, 27–35.

Purdy, K. J., Cresswell-Maynard, T. D., Nedwell, D. B., McGenity, T. J., Grant, W. D., Timmis, K. N., & Embley, T. M., (2004). Isolation of haloarchaea that grows at low salinities. *Environ. Microbiol., 6*, 591–595.

Qui, X. X., Zhao, M. L., Han, D., Zhang, W. J., & Ciu, H. L., (2013). *Haloplanus salinus* sp. nov., an extremely halophilic archaeon from a Chinese marine solar saltern. *Arch. Microbiol., 195*, 799–803.

Racker, E., & Stoeckenius, W., (1974). Reconstitution of purple membrane vesicles catalyzing light-driven proton uptake and adenosine triphosphate formation. *J. Biol. Chem., 249*, 662–663.

Reiser, R., & Tasch, P., (1960). Investigation of the viability of osmophile bacteria of great geological age. *Trans Kans. Acad. Sci., 63*, 31–34.

Rodriguez-Valera, F., Ruiz-Berraquero, F., & Ramos-Cormenzana, A., (1979). Isolation of extreme halophiles from seawater. *Appl. Environ. Microbiol., 38*, 164–165.

Rogers, P. J., & Morris, C. A., (1978). Regulation of bacteriorhodopsin synthesis by growth rate in continuous cultures of *Halobacterium halobium. Arch. Microbiol., 119*, 323–325.

Sabet, S., Diallo, L., Hays, L., Jung, W., & Dillon, J. G., (2009). Characterization of halophiles isolated from solar salterns in Baja California, Mexico. *Extremophiles, 13*, 643–656.

Seyedkarimi, M. S., Aramvash, A., & Ramezani, R., (2015). High production of bacteri-orhodopsin from wild-type *Halobacterium salinarum. Extremophiles, 19*, 1021–1028.

Shand, R. F., & Betlatch, M. C., (1991). Expression of the bop gene cluster of *Halobacterium halobium* is induced by low oxygen tension and by light. *J. Bacteriol., 173*, 4692–4699.

Shand, R. F., & Betlatch, M. C., (1994). *Bop* gene cluster expression in bacteriorhodopsin overproducing mutants of *Halobacterium halobium. J. Bacteriol., 176*, 1655–1660.

Shimano, K., Goto, M., Kikukawa, T., Miyauchi, S., Shirouzu, M., Kamo, N., & Yokoyama, S., (2009). Production of functional bacteriorhodopsin by an *Escherichia coli* cell-free protein synthesis system supplemented with steroid detergent and lipid. *Protein Sci., 18*, 2160–2171.

Shiu, P. J. R., Ju, Y. H., Chen, H. M., & Lee, C. K., (2015). Effect of complex nutrients and repeated batch cultivation of *Halobacterium salinarum* on enhancing bacteriorhodopsin production. *J. Microb. Biochem. Technol., 7*, 289–293.

Shiu, P. J., Ju, Y. H., Chen, H. M., & Lee, C. K., (2013). Facile isolation of purple membrane from *Halobacterium salinarum* via aqueous-two-phase-system. *Protein Expression and Purification, 89*, 219–224.

Singh, K., Korenstein, R., Lebedeva, H., & Caplan, S. R., (1980). Photoelectric conversion by bacteriorhodopsin in charged synthetic membranes. *Biophys. J., 31*, 393–402.

Stan-Lotter, H., McGenity, T. J., Legat, A., Denner, E. B. M., Glaser, K., Stetter, K. O., & Wanner, G., (1999). Very similar strains of *Halococcus salifodinae* are found in geographically separated Permo-Triassic salt deposits. *Microbiol., 145*, 3565–3574.

Tan, E. H. L., & Birge, R. R., (1996). Correlation between surfactant/micelle structure and the stability of bacteriorhodopsin in solution. *Biophys. J., 70*, 2385–2395.

Thavasi, V., Lazarova, T., Filipek, S., Kolinski, M., Querol, E., Kumar, A., et al., (2008). Study on the feasibility of bacteriorhodopsin as bio-photosensitizer in excitonic solar cell: a first report. *J. Nanosci. Nanotechnol., 8*, 1–9.

Thomas, M., Pal, K. K., Dey, R., Saxena, A. K., & Dave, S. R., (2012). A novel haloarchaeal lineage widely distributed in the hypersaline marshy environment of Little and Great Rann of Kutch in India. *Curr. Sci., 103*, 1078–1084.

Trivedi, S., Choudhary, O. P., & Gharu, J., (2011). Different proposed applications of bacteriorhodopsin. *Recent Patents on DNA and Gene Sequences, 5*, 35–40.

Woese, C. R., Kandler, O., & Wheelis, M. L., (1990). Towards a natural system of organisms: Proposal for the domains Archaea, Bacteria, and Eucarya. *Proc. Natl. Acad. Sci., USA, 87*, 4576–4579.

Xu, X. W., Wu, M., & Huang, W. D., (2005). Isolation and characterization of a novel strain of *Natrinema* containing a *bop* gene. *J. Zhejiang Univ. Sci.,* B6, 142–146.

Xu, X. W., Wu, Y. H., Wang, C. S., Oren, A., Zhou, P. J., & Wu, M., (2007). *Haloferax larsenii* sp. nov., an extremely halophilic archaeon from a solar saltern. *Int. J. Syst. Evol. Microbiol., 57*, 717–720.

Yatsunami, R., Kawakami, T., Ohtani, H., & Nakamura, S. A., (2000). Novel bacteriorhodopsin like protein from *Haloarcula japonica* strain TR–1: Gene cloning, sequencing, and transcript analysis. *Extremophiles, 4*, 109–114.

Yen, C. W., Heyden, S. C., Dreaden, E. C., Szymanski, P., & El-Sayed, M. A., (2011). Tailoring plasmonic and electrostatic field effects to maximize solar energy conversion by bacteriorhodopsin, the other natural photosynthetic system. *Nano Lett., 11*, 3821–3826.

Yin, S., Wang, Z., Xu, J. Q., Xu, W. M., Yuan, P. P., & Cui, H. L., (2015). *Halorubrum rutilum* sp. nov., isolated from a marine solar saltern. *Arch. Microbiol., 197*, 1159–1164.

Yuan, P. P., Sun, X. J., Liang, X., Chen, X. J., Han, D., Zhang, W. J., & Cui, H. L., (2015). *Haloarchaeobius amylolyticus* sp. nov., isolated from a marine solar saltern. *Arch. Microbiol., 197*, 949–953.

CHAPTER 7

PROPERTIES AND BIOTECHNOLOGICAL APPLICATIONS OF β-GLUCURONIDASES

NASSER EL-DIN IBRAHIM[1], WEILAN SHAO[2], and KESEN MA[1]

[1]*Department of Biology, University of Waterloo, ON, N2L 3G1, Canada*

[2]*School of Environment, Jiangsu University, Zhenjiang, China*

7.1 INTRODUCTION

Glucuronic acid is an acid form of glucose, which is obtained by the oxidation of the sixth carbon of the glucose (C6) to a carboxylic group. However, the oxidation reaction occurs after glucose is transformed into uridine diphosphate (UDP) glucose or UDP-α-D-glucose (UDPG), and is catalyzed by an UDP-glucose dehydrogenase (UGDH). Depending on the stereochemical configuration at the first carbon of the molecule (C1), glucuronic acid is classified as α- and β-glucuronic acid. In β-glucuronic acid, the hydroxyl group attached to C1 is on the same side of the pyranose ring as the carboxyl group on C6. The β-form is the predominant one (~64%) in the free glucuronic acid, which can be formed either after the dephosphorylation of glucuronic acid-1-phosphate derived from UDP-glucuronic acid (UDPGA), an oxidized form of UDPG, or by the hydrolysis of UDPGA. In contrast, the non-free acid has a predominant α-form that exists as UDPGA. Glucuronic acid has two stereoisomers or enantiomers based on the configuration at C5, D-glucuronic acid and L-glucuronic acid, where the most physiologically predominant configuration is D-glucuronic acid. Glucuronic acid is first identified in urine, and its name is from the Greek word that means "sweet urine".

Glucuronic acid is rarely found as a free sugar acid in organisms; however, it is a crucial building block of many vital macromolecules in human, animal, plant, and microbial cells. Glucuronic acid in the form of UDPGA is polymerized with other sugar moieties to form polysaccharides structures or can be conjugated with other aglycon compounds (lipids and proteins) (Dutton, 1980). The linkage between glucuronic acid and other compounds is called glycosidic bond, which takes place between the hydroxyl group (-OH) on the anomeric carbon of hemiacetal form of glucuronic acid and hydroxyl (-OH), amine ($-NH_2$) or thiol (-SH) groups on the other compounds in a process called glucuronidation or glucurono-sylation (Figure 7.1).

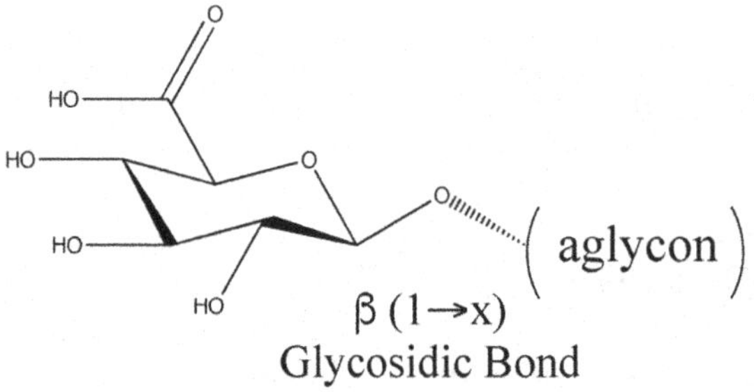

FIGURE 7.1   Glycosidic bond in the β-D-glucuronides.

The vital macromolecules that contain glucuronic acid are widely spread in all organisms, for example in animals it is found in the glyco-calyx, a glycoprotein coat that covers some cells, and in the proteoglycans (mucopolysaccharides) which include chondroitin sulfate (CS), heparin, mucoitin sulfate, hyaluronic acid, and intercellular matrix. Heparan sulfate (HS) and heparin are repeating disaccharide unit of α-D-(1,4)-glucosamine and β-D-(1,4)-glucuronic or α-L-(1,4)-iduronic acid with different sulfa-tions. Several biological processes are mediated by HSs, e.g., effects of chemokines, selectins, and coagulation. CS is a linear sulfated polysac-charide of glycosaminoglycan (GAG) family. CS structure is composed of repeating disaccharide unit consisting of glucuronic acid (GlcA) and N-acetylgalactosamine (GalNAc). CS is widely distributed in various cell membrane and extracellular matrix.

In plants, glucuronic acid presents in gums including Arabic (18% glucuronic acid) and xanthan gums. It is also shown that the down-regulation of UGDH catalyzing the key step in biosynthesis of UDPGA results in severe development problem in *Arabidopsis thaliana*, as the UDPGA is considered as a common precursor for the synthesis of many sugars including arabinose, xylose, galacturonic acid, and apiose (Reitsma et al., 2007; Reboul et al., 2011; Kolarova et al., 2014). In bacteria, glucuronic acid is found in their glycocalyx and polysaccharides. It is clear that glucuronic acid plays a key role in the metabolism of all three domains of life including human, animals, plants, and microorganisms.

Glycosyl hydrolases (EC 3.2.1.x) are key enzymes participated in the carbohydrate metabolism of all the domains of life (Archaea, Bacteria, and Eukarya). The hydrolysis of glucuronic acid derivatives can be catalyzed by β-glucuronidase (EC 3.2.1.31) that plays key physiological roles in human, animals (invertebrates such as insects, nematodes, gastropods, and almost all tissues of vertebrates), plants, and some bacteria (Kyle et al., 1992; Fior et al., 2009) (Figures 7.2 and 7.3). This enzyme specifically catalyzes the hydrolysis of β-glucuronosyl-*O*-bonds in glucuronide molecules in different animal tissues such as liver, kidney, spleen, intestinal epithelium, endocrine, and reproductive organs to liberate the β-glucuronic acid moieties (Dutton, 1980). It also plays a role in the regulation of glucuronidation of xenobiotics and endogenous compounds (Himeno et al., 1974).

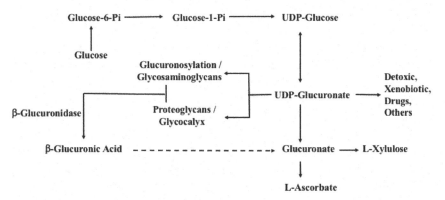

FIGURE 7.2 Glucuronic acid pathway.

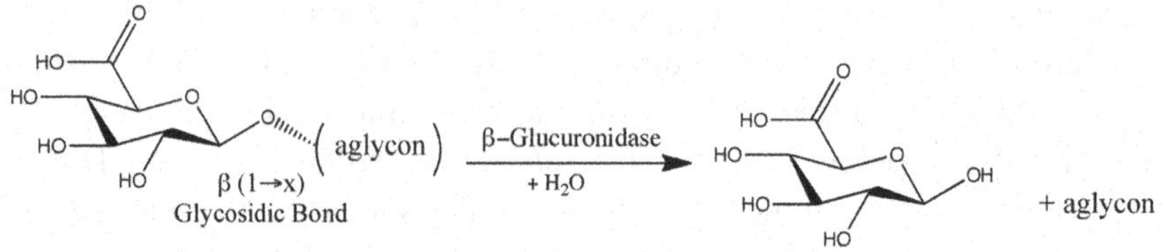

FIGURE 7.3 Hydrolysis of β-D-glucuronides by β-glucuronidase. β-glucuronidase hydrolyzes the glycosidic bond between the glucuronic acid and aglycon moiety in the β-D-glucuronides.

7.2 MOLECULAR PROPERTIES OF β-GLUCURONIDASES

GUSB (glucuronidase, beta) cDNAs encoding for β-glucuronidase have been identified in human, mouse, and rat (Nishimura et al., 1986; Oshima et al., 1987; Gallagher et al., 1988; Powell et al., 1988; Miller et al., 1990). Human GUSB gene is 21-kb in length, and it contains 12 exons ranging from 85 to 376 bp and it has been mapped at chromosome bands 7q11.21 and 7q11.22 with two pseudogenes at chromosome bands 5p13 and 5q13 (Tsui et al., 1988; Miller et al., 1990; Speleman et al., 1996).

Human GUSB is a homotetrameric enzyme that plays an important role in the degradation of GAGs that contain glucuronic acid (Paigen, 1979). Inherited deficiency of GUSB activity results in the accumulation of GAGs in lysosomes, a lysosomal storage disease known as mucopolysaccharidosis type VII (MPS VII) or Sly syndrome that has been reported in humans, dogs, mice, and cats (Sly et al., 1973; Birkenmeier et al., 1989; Haskins et al., 1991; Gitzelmann et al., 1994).

In plants, it was mistakenly reported for a long time that plant has no GUSB despite the GUSB activity has been identified since 1932, initially as baicalinase from *Scutellaria baicalensis*, and later as GUSB (Miwa, 1932; Levvy, 1954; Levvy and Marsh, 1959). Activities of GUSB have been reported in rye (*Secale cereale*) primary leaves (Schulz and Weissenböck, 1987), where it is proposed to have a role in lignin biosynthesis (Anhalt and Weissenböck, 1992). This enzymatic activity is also reported in maize developing kernels (Muhitch, 1998). Moreover, endogenous GUSB activity is present in many other plants such as fruit wall, seed coat, endosperm, embryos (Hu et al., 1990) and during male gametophyte development of potato, tobacco, and tomato (Plegt and Bino, 1989).

The regulation of the *gus* (glucuronidase) gene expression has been studied in many model plants such as *Arabidopsis thaliana, Oryza sativa, Nicotiana tabacum* and *Zea mays* (Sudan et al., 2006). As a result, an improved protocol is developed for eliminating the interfering of the endogenous GUSB activity in the transformed plants (Alwen et al., 1992) especially with the plant of known high endogenous GUSB activity such as rapeseed (Abdollahi et al., 2011).

Bacterial GUSB activity has been identified in some species, including those from gastrointestinal tract, for example *Escherichia coli*, Entero-bacteriaceae, *Lactobacillus gasseri, Staphylococcus* sp., *Clostridium perfringens, Staphylococcus aureus, Thermotoga maritina*, Bacteroides, Bifidobacterium, Eubacterium, and Ruminococcus (Hawksworth et al., 1971; Hill et al., 1971; Jefferson et al., 1986; Jefferson, 1989; McBain and Macfarlane, 1998; Akao, 1999; Nelson et al., 1999; Akao, 2000; Russell and Klaenhammer, 2001; Shimizu et al., 2002).

On the other hand, GUSB activity is unable to be identified in many commonly known bacteria such as *Agrobacterium, Azospirillum, Bradyrhizobium, Pseudomonas*, and *Rhizobium* (Wilson et al., 1992) nor some fungi such as *Aspergillus, Neurospora, Saccharomyces, Schizosaccharomyces*, and *Ustilago* (Wilson et al., 1995). With few exception of thio-β-glucuronides, all β-glycosidic bond between glucuronic acid and conjugated aglycone in the glucuronides can be hydrolyzed by GUSB (Stoeber, 1957; Wilson et al., 1992). GUSB is encoded by the *uidA* gene that is later known as *gusA* gene of the *gus* operon that is located at 36.5 min on the *E. coli* genome (Jefferson et al., 1986; Blattner et al., 1997).

One of the intensively characterized GUSB genes is the *gusA* of *E. coli* that encodes for GUSB, a widely used reporter system for transformants (Jefferson et al., 1986; Jefferson et al., 1987). Initially, it is constructed for the expression in the roundworm nematode, *Caenorhabditis elegans* (Jefferson et al., 1987), and due to its unique properties it is then used as a main plant expression report system, and especially GUSB can be translocated across membranes (Jefferson et al., 1987).

In *E. coli*, the *gusA* gene that encodes for the GUSB is a part of the *gusRABC* operon, in which *gusR* encodes for the transcriptional repressor while *gusB* encodes for inner membrane glucuronide transporter with the assistance by an outer membrane protein encoded by *gusC* gene with an unknown mechanism (Wilson et al., 1992; Liang et al., 2005). Similarly, an open reading frame (ORF) is identified in *Ruminococcus gnavus* E1, which is encoded for a protein with GUSB activity and it has 61% similarity to that of *E. coli* GUSB. This ORF is identified as a part of an operon without any identifiable glucuronides transporter genes (Beaud et al., 2005). However, in *Lactobacillus gasseri* ADH, the *gusA* gene is transcribed as a single or monocistronic mRNA, and its encoded protein have 39% similarity to *E. coli* gusA protein but without any identifiable operon structure similar to that in *E. coli* (Russell and Klaenhammer, 2001).

7.3   CATALYTIC PROPERTIES OF β-GLUCURONIDASES

Human GUSB is a lysosomal enzyme that is translated initially as 80-kDa glycoprotein monomer precursor containing 653 amino acids, and then is post-translationally modified into 629 amino acid residues peptide (subunit), which has a mass of 75-kDa as monomer and can form as a 330-kDa tetramer. Many post-translational modifications take place on GUSB precursor, including cleavage of the N-terminus signal sequence, N-linked glycosylation, trimming of carbohydrate chains and removing 18 residues from the C-terminus of the protein. It was proposed that such a precursor is to serve as a retention signal at Endoplasmic Reticulum (ER), where GUSB targeted to both ER and lysosome (Tabas and Kornfeld, 1980; Rosenfeld et al., 1982; Kornfeld, 1986; Islam et al., 1993; Shipley et al., 1993). The active form of bacterial GUSB is homotetrameric enzyme that hydrolyzes β-D-glucuronides. The enzyme has a subunit of 601 amino acid residues with a mass of 68 kDa as a monomer and 272 kDa as a tetramer. Both human and *E. coli* GUSBs exhibit their catalytic activities in their tetramers forms (Jain et al., 1996), and their homodimer forms are also reported to be active. *E. coli* GUSB is even reported to be active as aggregates of octamer (dimer of tetramer) or hexadecylmer (16-mer) (Gehrmann et al., 1994; Burchett et al., 2015). A kinetics study on the tetramerization of *E. coli* GUSB has revealed that there are two rate-limiting steps during its folding and assembly, step 1 at monomer-dimer assembly and step 2 at dimer-tetramer assembly (Matsuura et al., 2011). Both human and *E. coli* GUSBs contain three structural domains, the sugar binding domain, immunoglobulin-like beta-sandwich domain and $(\alpha/\beta)_8$ TIM barrel domain (Jain et al., 1996). TIM barrel domains are a characteristic feature of many glycosyl hydrolases where the active site residues are found to be near the

C-termini of domain III and are localized within a large cleft that occurs at the interface of two monomers of the dimer. This is consistent with the earlier report that GUSB is functionally active as a dimer or tetramer but not as a monomer (Flores and Ellington, 2002).

Superimpose of both human and *E. coli* structures reveal that the active site residues of both GUSBs are completely superimposable. The only remarkable difference of the human enzyme is the absence of the loop (Ser–360:Glu–378) near the catalytic site of *E. coli* GUSB. On the other hand, human GUSB has lysosomal targeting motif in each monomer located near the active site vicinity (Hassan et al., 2013, 2015).

Carbohydrate-Active enZyme database (CAZy) has 5 classes of the enzymes that regulate breakdown and adding carbohydrate moieties, including glycosidases and transglycosidases, which are known as glycoside hydrolases (GH). GHs are divided into 135 families based on experimental characterization and protein sequences (http://www.cazy. org/) (Cantarel et al., 2009). Most of the known GUSBs belong to GH family 2, while few occur in GH families 1, 30 and 79. *Homo sapiens, E. coli, Lactobacillus gasseri* ADH and *Lactobacillus brevis* RO1 GUSBs belong to GH family 2 (Russell and Klaenhammer, 2001; Kim et al., 2009). Enzymes belonging to GH families 1, 30 and 79 have GUSB activity and other enzymatic activities or highly specific to certain substrates. For example, Klotho GUSB of GH family 1 is capable of hydrolyzing steroid β-glucuronides and removal of sialic acids or has sialidase activities (Tohyama et al., 2004; Cha et al., 2008). GUSB of GH family 30 that are characterized from *Lactobacillus brevis* subsp. *coagulans* shows catalytic activity on baicalin (flavone glycoside) (Sakurama et al., 2014). Also, baicalin-GUSB of GH family 79, which has very high specificity to baicalin from *S. baicalensis* (Sasaki et al., 2000; Zhang et al., 2005). Another example of GH family 79 is the mammalian heparanase that has an endo-β-D-glucuronidase and can hydrolyze HS proteoglycans (Gandhi et al., 2012; Peterson and Liu, 2013).

Two mechanisms are proposed for the catalysis of GHs, including GUSBs. Based on the products stereochemistry at the anomeric center, it is either "inverted" or "retained" relative to that of the substrate glycoside (Koshland, 1953). In inversion mechanism, the anomeric center uses concomitant general acid catalysis and general base catalysis mechanisms, the former mechanism to facilitate breaking the glycosidic bond and separating the aglycone leaving group, and the later mechanism to facilitate H_2O molecule attack at the anomeric center. While in retention mechanism, the anomeric center participates in a two-steps catalytic reaction. Initially, a general acid/base catalyst (residue) facilitates the separation of the leaving group while a second residue would stabilize the formed enzyme-intermediate complex. The second step of the reaction would be a general acid/base catalyst facilitating nucleophilic H_2O molecule attack on the anomeric center to break down the enzyme-intermediate to products with retained stereochemistry at the anomeric center. The two mechanisms require two acidic residues participating directly in the catalysis. One amino acid acts as a catalytic nucleophile and the other functions as an acid-base catalyst or the proton donor. It is proposed that the GUSB active site has a carboxylic acid and a carboxylate anion (usually two glutamic acid residues) as the catalytic functional groups (Wang and Touster, 1972; Islam et al., 1999; Speciale et al., 2014). GUSBs of *Homo sapiens*, *E. coli*, and *Staphylococcus* sp. RLH1 show the following conserved active site residues, glutamic acid–1 (E1) which acts as a nucleophilic residue at positions 451, 394 and 396 respectively, and glutamic acid–2 (E2) which acts as an acid-base catalyst at positions 540, 504 and 508 respectively, and tyrosine (Y) that has important catalytic role at 504, 468 and 471 respectively (Arul et al., 2008). The amino acid sequences identity between the human and *E. coli* GUSBs is 42.6% while that of human and *Staphylococcus* sp. RLH1 is 41.8%. The catalytic reactions take place through an oxocarbenium ion-like transition state and conformation distortion of the pyranoside ring favorite a significant bond breakage to the departing (leaving) group and limited bond formation to the attacking nucleophile (Figure 7.4) (Zechel and Withers, 2000; Speciale et al., 2014). It is proposed that the reaction catalyzed by human GUSB takes place in a two-step mechanism (Wong et al., 1998). The nucleophilic attack on the sugar anomeric carbon by one of the carboxylated residues occurs in the active site in the first step. Secondly, the second carboxylated residue of the active site acts as a general base in the presence of water molecule and attacks at the anomeric carbon, resulting in the cleavage of glycosidic bond in the oxocarbenium transition state with retention of the anomeric configuration. A hydrophobic interaction between pyranose rings and aromatic side chain of the aromatic residues in the vicinity of the active site may play a role in the stabilization of the transition states (Meyer and Schulz, 1997; Amaya et al., 2004).

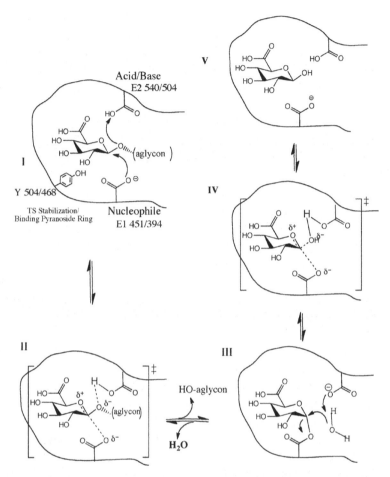

FIGURE 7.4 Catalytic mechanism of β-glucuronidase. I: the active site resides, Glu1 (E1), Glu2 (E2) and Tyr (Y) positioned according to *Homo sapiens/E. coli* structures. II: Transition State (TS) after binding of substrate to the active site. III: Binding of a H_2O molecule to the TS, IV: Bonds rearrangement in the TS, V: Final stage after the cleavage of the glycosidic bond.

7.4 APPLICATIONS OF β-GLUCURONIDASES

7.4.1 HUMAN AND ANIMAL β-GLUCURONIDASES

Human GUSB gene is located on the long arm (q) of chromosome 7 at position 21.11 (7q21.11), precisely from base pair number 65,960,683 to

65,982,313 on chromosome 7. The human GUSB gene has a length of 21.6 kb and contains 11 introns, and 12 exons and the transcripted mRNA has 1953 bp in length that encodes a 651 amino acid residues precursor polypeptide. Several post-translational modifications occur during the GUSB gene translation and after its biosynthesis by membrane-bound ribosomes. GUSB has dual localizations, where it is translocated to ER and then to lysosome. The early post-translational processing step includes cleavage of the N-terminus signal peptide (22 residues) (Rosenfeld et al., 1982), and N-linked glycosylation (Tabas and Kornfeld, 1980; Kornfeld, 1986) where the molecular weight of the peptide formed is 78 kDa (Brot et al., 1978). The latter processes steps in the lysosome, including trimming of carbohydrate chains and removes 18 amino acid residues from the C-terminus of the premature protein to form the mature monomer subunit with a molecular weight of 75 kDa (Islam et al., 1993). The protein structure of human GUSB contains three domains, domain 1 is a jelly roll-like domain containing the lysosomal targeting residues, domain 2 is an immunoglobulin domain, and domain 3 is a TIM barrel domain that contains the active site (Jain et al., 1996; Hassan et al., 2013).

The human GUSB is found in many organs like the spleen, kidney, lung, bile, serum, urine, and others (Wakabayashi, 1970; Paigen, 1989). In addition, GUSB has been reported in many body fluids such as blood, ascitic, pleural, spinal, ventricular, and vaginal, synovial fluid and sometimes it is present in these fluids at higher concentrations in some patients (Boyland et al., 1957; Weissmann et al., 1971). It degrades glucuronic acid-containing GAGs, like HS, CS, and dermatan sulfate (Sly et al., 1973). Elevated levels of GUSB activity have been reported in different patients with diseases such as renal and kidney diseases, epilepsy, breast, larynx, testes, urinary tract infections, children leprosy and liver injury (Bank and Bailine 1965; Roberts et al., 1967; Paigen, 1989; Nandan et al., 2007; George, 2008). Also, elevated activity of GUSB is reported in different types of tumors (Boyland et al., 1957). These elevated GUSB activity levels present in the tumor environment are proposed to be due to the over expression of GUSB (Sperker et al., 2000), which is released from necrotic tumor tissues (Bosslet et al., 1995) or tumor-infiltrating immune cells (Juan et al., 2009). Furthermore, GUSB activity is shown to be a sensitive biomarker for organophosphorus insecticide exposure (Ueyama et al., 2010). The richest source of GUSB has been reported to be the preputial gland of the female rat, where it represents 7% of the total soluble protein (Keller and Touster, 1975).

Mucopolysaccharidosis VII (MPS VII) or Sly syndrome is an auto-somal recessive and the most common lysosomal storage disorders (LSD) that is characterized by the deficiency of GUSB activity, resulting in mental retardation due to the partially degradation of CS, dermatan sulfate, and HS and subsequently accumulated in the lysosomes of many tissues, including brain, eventually leading to cellular and organ dysfunction (Sly et al., 1973; Vogler et al., 2005).

MPS VII has various consequences based on its severity. It can cause mental retardation, hepatosplenomegaly, and skeletal dysplasia. In the most severe phenotype, it causes *Hydrops fetalis* that is characterized by an accumulation of fluid, or edema, in at least two fetal compartments and often fetus cannot survive beyond few months, while in a milder phenotype it causes later onset and normal intelligence. Similarly, MPS VII has been reported in different animals such as canine, feline, and murine species (Birkenmeier et al., 1989; Haskins et al., 1991; Sands and Birkenmeier, 1993; Gitzelmann et al., 1994).

Nearly 55 different mutations in the GUSB gene have been identified in MPSVII patients (Khan et al., 2016). These mutations include point mutations, deletions, missense mutations, splice-site mutations, nonsense mutations, and rearrangements. Among these mutations, 49 unique Sly syndrome mutations that subdivided into 36 missense mutations, 6 nonsense mutations, 5 deletions and 2 splice site alterations (Tomatsu et al., 2009). With the highest mutation percentage, the missense mutations such as Leu176Phe, Pro415Leu, Pro408Ser, Ala619Val, Arg216Trp, Arg382Cys, and Arg477Trp are highly frequent (Tomatsu et al., 1990; Tomatsu et al., 1991) while Leu176Phe is the most common mutation among the MPS VII patients (Vervoort et al., 1996). In a very informa-tive work using Molecular Dynamic Simulation, Khan et al. (2016) show how these mutations resulted in malfunction or alteration of the GUSB activity. For instance, the Leu176Phe mutation implies changing leucine residue at the 176th position, a highly conserved residue, to phenylalanine with a bulky hydrophobic phenyl group. This leads to increasing in the hydrophobicity, decreasing in the residual contact distance from 2.7 Å to 1.8 Å and consequently, increasing in hydrogen bond strength and stability (Khan et al., 2016).

Drugs inside the body are considered as foreign substances or xenobi-otics, and they are metabolized inside the body through two phases, I, and II. In phase II, the drugs become more water soluble and hence more excretable by conjugation with different compounds including glucuronic acid in a process called glucuronidation that takes place by large number of Uridine 5'-diphospho-glucuronosyltransferase (UDP-glucuronosyltransferase, UGTs) (Brand et al., 2010; Jancova et al., 2010). Glucuronidation has been assigned as the major reason for many cases of the multidrug resistance, especially for the drug used in the management of epilepsy, schizophrenia, depression, HIV infection, hypertension, organ transplantation, hypercholesterolemia, osteoporosis, and some cancer types (Mazerska et al., 2016). The reverse process to the glucuronidation is the catalytic removal of the glucuronic acid by the activity of GUSB. The balance between these two processes has crucial role in the drug action efficiency on many diseases and applications. In some instances, glucuronidation acts as a detoxification mechanism of some drugs such as antitumor drugs Irinotecan/Camptothecin–11 (Topoisomerase I inhibitor for colorectal cancer) (Cummings et al., 2003), Belinostat (Histone deacetylase inhibitor for hematological malignancies) (Dai et al., 2011), Etoposide (Topoisomerase II inhibitor for many types of cancer) (O'Dwyer et al., 1985) and Epirubicin (Topoisomerase II inhibitor for breast and bladder cancer) (Harris et al., 2003) and therefore increasing β-glucuronidase activity is beneficial. On the other hand, glucuronidation may act as an activation mechanism of some drugs such as antitumor drugs Tamoxifen (Antiestrogen for estrogen-dependent breast cancer, (White, 2003) and Acridinone (Antitumor) (Pawlowska et al., 2013). In such circumstances, the use of GUSB inhibitors is beneficial.

Tumor and bacteria have mysterious connections with unknown mechanism(s). It is reported that many bacterial species with selective manner localize and proliferate in tumors vicinity after the systemic administration. These reports including *E. coli* (Yu et al., 2004; Cheng et al., 2008), *Clostridium* (Moese and Moese, 1964), *Bifidobacterium* (Kimura et al., 1980; Yazawa et al., 2000) and *Salmonella* (Clairmont et al., 2000; Forbes, 2010). This extraordinary phenomenon attracts many scientific researchers to establish "delivery vehicle" for enzyme prodrug tumor therapy using bacteria and this approach called "bacteria directed enzyme prodrug therapy" or BDEPT (Forbes, 2010). This BDEPT system has unique potential advantage features including that bacteria can grow in tumors for many weeks (Stritzker et al., 2007; Zhang et al., 2010), which means we will have sustained expression of therapeutic enzymes. Second, the bacteria colonization in tumor seems to be "universal" and not specific to certain tumor types (Cheng et al., 2008; Jiang et al., 2010). Third feature is that the bacterial colonization in tumors can reach high density or high tumor/normal tissue ratios, make this bacterial delivery vehicle system more powerful than that alternative targeted therapies methods such as antibodies (Reddy et al., 2011; Hess and Neri, 2014) and nanoparticles (Zhang et al., 2014) and that will be reflected on the toxicity of the treatment therapy. Intravenous injection of *E. coli* DH5α in tumor-bearing mice resulting in colonization of the *E. coli* selectively in the tumors with tumor/normal tissue colonization ratios of around 10,000-fold after 4 days of injection and prolongs up to 18 days (Zhang et al., 2010).

However, the only limitation of BDEPT approach is how bacterial enzyme contents can be released into the necrotic regions of the tumors (Westphal et al., 2008). Engineering the bacterial delivery vehicles to secret their therapeutic enzymes or induction of bacterial lysis showed significant improvement in tumor therapy using BDEPT (Loeffler et al., 2007; Liu et al., 2008; Loeffler et al., 2009; Jiang et al., 2010; Jiang et al., 2013; Jeong et al., 2014).

An example for the BDEPT application in tumor therapy is the use of Camptothecin–11 (CPT–11) in colorectal adenocarcinoma. CPT–11 is a plant alkaloid that can be hydrolyzed in the human body into SN–38, a topoisomerase I inhibitor and induces tumor cell apoptosis (Staker et al., 2002). However, SN–38 rapidly conjugated to glucuronic acid to form SN–38G (Figure 7.5), which is not active as topoisomerase I poison. To reactivate SN–38G to SN–38, Hsieh et al. (2015) have engineered *E. coli* with constitutive expression of highly active GUSB and enhanced treatment is shown with these engineered bacteria, especially if the bacterial lysate is injected into the intratumoral region (Hsieh et al., 2015). Similar work using 9-aminocamptothecin (9AC) has been carried out on CL1–5 human lung tumor in mice (Cheng et al., 2008).

In a similar way, a new class of Prodrugs has been synthesized as an efficient way to target specifically the tumor tissues and decrease the cytotoxicity of the drug itself. Prodrug consists of three parts fused together, glucuronic acid, linker molecule, and the active drug. The prodrug itself is inactive, which means less toxic, once administrated it is hydrolyzed at the tumor by elevated GUSB activity to release the active drug *in situ*. The linker plays an important role in keeping sufficient space between the glucuronide moiety and the antitumor drugs that helps to allow an easy recognition of the enzymatic substrate by GUSB. Spontaneously, after the removal of glucuronic acid, the linker is decomposed and unleashes the active drug. This type of drugs called GUSB-responsive prodrugs (Tranoy-Opalinski et al., 2014).

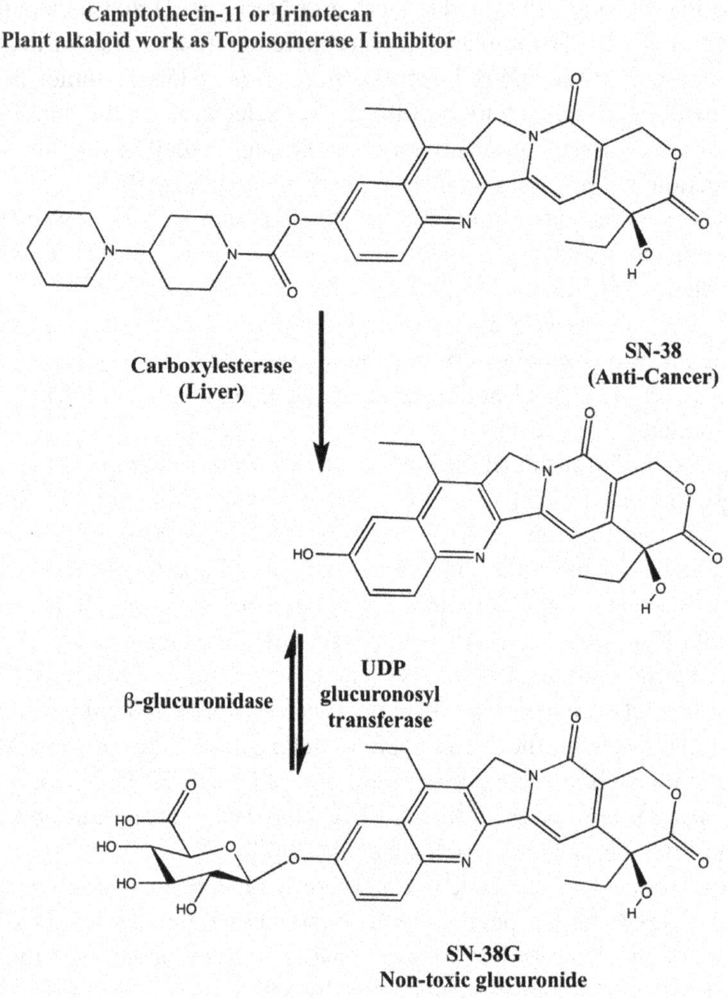

FIGURE 7.5 *E. coli* directed β-glucuronidase CPT–11 prodrug therapy. Irinotecan/Camptothecin–11/CPT–11 is hydrolyzed by liver carboxyesterase to anticancer active form SN–38, however, it is rapidly converted back to the glucuronidated inactive form SN–38G. Presence of modified bacteria with *E. coli* β-glucuronidase can reactivate SN–38G to SN–38 *in situ* of the tumor (modified from Hsieh et al., 2015).

There are elevated levels of GUSB activity in many types of cancers. It is always very important to conduct an assessment of this activity that is used as a tumor biomarker and as a crucial factor for the drug efficiency (Sperker et al., 2000). This activity assessment is carried out by using glucuronidated fluorescence dyes or "Probes" that are hydrolyzed "*in situ*" by the elevated GUSB activity at the tumor site in a similar way to the activation of the prodrugs that mentioned above. Using the positron emission tomography (PET), glucuronide probes are developed to pursue more optimized protocols for personalized cancer therapy and imaging depending on GUSB activity. These probes such as fluoresce in di-β-D-glucuronide probe (FDGlcU) (Tzou et al., 2009), FITC-TrapG, and NIR-TrapG probes (Chopra, 2004; Cheng et al., 2012; Su et al., 2014), [124]I-phenolphthalein-glucuronide probe ([124]I-PTH-G) (Tzou et al., 2009), 1-O-(4-(2-fluoroethyl-carbamoyloxymethyl)–2-nitrophenyl)-O-β-D-glucopyronuronate methyl ester (^{18}F-FEAnGA-Me) (Antunes et al., 2012) and [124]I-TrapG (Su et al., 2014). These reports on the glucuronidated prodrugs and probes and tumor GUSB are very promising and may lead to the development of more personalized medicine, imaging, and clinical diagnosis approaches.

On contrary to the above examples, where the higher GUSB activity is beneficial, it is not desirable in some medical conditions to have high levels of the GUSB activity, so it is importance to have GUSB inhibitors. Due to the increase in GUSB activity in renal and kidney diseases, epilepsy, breast, larynx, testes, urinary tract infections, children leprosy and liver injury, it is important to have inhibitors that can decrease or eliminate the undesirable activity of human or gut flora GUSBs associated with some diseases. For instance, CPT–11 or irinoteacan as prodrug for colon and pancreatic cancers causes severe diarrhea and gastrointestinal tract damage that may lead to dose reduction or a complete termination of the therapy (Stringer et al., 2008). However, using a low-potency GUSB inhibitor, there is a reduction in the gastrointestinal tract toxicity associated with irinotecan in rats and mice (Fittkau et al., 2004; Wallace et al., 2010). Some natural compounds are known to inhibit the GUSB activity such as D-saccharic acid–1,4-lactone and some natural glucuronides (Narita et al., 1993). Potent GUSB inhibitors have been also identified or synthesized for pharmaceutical applications such as bisindole (Khan et al., 2014), thiadiazole, benzimidazole, 6-chloro-2-aryl–1H-imidazo[4,5-b]pyridine derivatives and 2,5-disubstituted–1,3,4-oxadiazoles as a benzimidazole derivative (Taha et al., 2015; Zawawi et al., 2015; Salar et al., 2016; Taha et al., 2016).

7.4.2   PLANT β-GLUCURONIDASE

It is probably the first reported GUSB activity in plant. Baicalinase from *Scutellaria baicalensis* is known to have GUSB activity, and it is confirmed after further characterization (Miwa, 1932; Levvy, 1954; Levvy and Marsh, 1959). Following research on plant GUSB or "endogenous" GUSB that plays key roles in development of the primary leaves of rye plant (*Secale cereale*), the same is found to be in the pedicel of maize kernels during its developing and lignin biosynthesis (Schulz and Weissenböck, 1987; Anhalt and Weissenböck, 1992; Muhitch, 1998). Endogenous GUSB activity in plant is reported in fruit wall, seed coat, endosperm, embryos, and during male gametophyte development of many plants (Plegt and Bino, 1989; Hu et al., 1990). Inhibition of endogenous GUSB activity leads to seedling developmental problems such as arresting the growth and inhibiting root-hair development in tobacco (*Nicotiana tabacum*) and parasitic angiosperm, *Cuscuta pentagona* (Sudan et al., 2006; Schoenbeck et al., 2007). Calli and somatic embryos from *Coffea arabica* and *Coffeaca nephora,* zygotic embryo from *Capsicum chinense* and microspores and microspore-derived embryos from rapeseed *Brassica napus* all are reported to have transiently expressed endogenous GUSB activity (Sreenath and Naveen, 2004; Solis-Ramos et al., 2010). In addition to developmental role of endogenous GUSB, it is reported to have a protection role against the oxidative stress in *Scutellaria baicalensis* cells (Matsuda et al., 1995; Morimoto et al., 1998; Sasaki et al., 2000). GUSB from *Arabidopsis thaliana* (AtGUS) is reported to hydrolyze glucuronic acids from carbohydrate chains of arabinogalactan-protein (Eudes et al., 2008; Konishi et al., 2008; Anbu and Arul, 2013). Arabinogalactan-proteins are a family of proteoglycans that involved in many important physiological processes (Fincher et al., 1983). This spatiotemporal expression of endogenous GUSB probably is the main reason for a long time debate about plant GUSB existence. The regulation details of the GUSB gene has been studied in *Arabidopsis thaliana, Oryza sativa, Nicotiana tabacum* and *Zea mays* (Sudan et al., 2006; Eudes et al., 2008).

For applications in plant biotechnology, the *E. coli* GUSB gene is used extensively as a report gene during the plant transformation. It is used with different types of promotors including constitutive or tissue-specific promoters, for instance, Cauliflower Mosaic Virus (CaMV) 35S which is considered to be a very strong constitutive and the mostly used promoter in plant biotechnology. *E. coli* GUSB can be glycosylated in plant in the ER when the GUSB gene is fused to Patatin signal peptide that leads to considerable reduction of enzyme activity (Iturriaga et al., 1989). To overcome this glycosylation, *E. coli* GUS gene has been modified by site-directed mutagenesis which can be especially used in protein targeting (Farrell and Beachy, 1990). On the other hand, the level of GUSB gene expressed in the transgenic plant is enhanced dramatically after introduction of plant intron inside the *E. coli* GUSB gene, thanks to the mysterious intron-mediated enhancement (IME). IME controls both the amount and the expression site, so it is called Intron Dependent Spatial Expression (IDSE) (Vancanneyt et al., 1990; Gianì et al., 2009; Gallegos and Rose, 2015; Laxa, 2016).

Since there are two GUSB activities in the transgenic tissues, one bacterial from *E. coli* and the other is endogenous; more attention should be paid during measuring the GUSB activities specially with plants rich in endogenous GUSB activity such as rapeseed (Abdollahi et al., 2011). There are two major differences between *E. coli* and endogenous GUSB activities. Optimal pH for *E. coli* GUSB is 7–8 while that for endogenous GUSB is 4–5, and the other difference is the sensitivity to the saccharic acid 1,4-lactone inhibitor present in the enzyme assay mixture, which significantly decreases the endogenous GUSB activity (Kosugi et al., 1990; Abdollahi et al., 2011). An improved protocol for eliminating the endogenous GUSB activity interfering with the transformed plants has been developed (Alwen et al., 1992; Hänsch et al., 1995). An alternative solution to overcome this activity interference or false positive results is proposed by using directed-evolution modified *E. coli* GUSB gene. Measuring the GUSB activity after heat treatment at 70°C for 30 min or at 80°C for 10–30 min allows the differentiation between the GUSB activity of the modified *E. coli* one and that from the endogenous one and eliminates the false positive results (Xiong et al., 2007; Xiong et al., 2011). It is worth to mention that the GUSB activity has been reported in some thermophilic bacteria, such as *Thermotoga maritima* (Salleh et al., 2006). The *gus* gene of *T. maritima* is cloned and overexpressed, and the recombinant GUSB shows a broad pH range from 4.5 to 7.5, and has a maximum activity at pH 6.5 and thermal stability with a half-life of 3 h at 85°C, in contrast to the *E. coli* enzyme, which has a half-life of 2.6 h at 50°C. *T. maritime* GUSB gene could be used as a reporter gene to replace the *E. coli* one and hence the GUSB activity could be measured at higher temperature to overcome false positive results. Another disadvantage of using *E. coli* GUSB gene as a reporter gene is that some plants are known to have endogenous GUSB inhibitor(s), as in the case of leaves and roots of some genotypes of wheat (Ramadan et al., 2011).

7.4.3   MICROBIAL β-GLUCURONIDASE

Microbial GUSB hydrolyzes β-glucuronides as well as β-galacturonides, and it is the first enzyme of the Hexuronide-Hexuronate pathway in *E. coli*, which has a gene regulation mechanism independent from that of the other enzymes of the pathway (Didier-Fichet and Stoeber, 1968; Baudouy-Robert et al., 1970). GUSB can be induced by β-glucuronide/glucuronate and fructuronate (Novel et al., 1974).

GUSB assays are extensively performed in various fields such as plant biotechnology, diagnostic purposes in clinical, environmental samples, biotransformation of pharmaceutical compounds and others (Ender et al., 2017; Kaleem et al., 2017). Different discontinuous assays are available for histochemical, spectrophotometric or fluorometric analysis as well as continuous spectrophotometric and fluorometric assays (Aich et al., 2001; Briciu-Burghina et al., 2015). Also, a rapid on-site detection protocol for the detection of *E. coli* in surface water by measuring its GUSB activity has been developed (Heery et al., 2016). A modified GUSB gene has been created to overcome the limitation of GUSB during tissue fixation in the histochemical analysis due to the loss of activity by glutaraldehyde or formaldehyde (Naleway, 1992). Also, the GUSB gene from *Lactobacillus gasseri* ADH has been modified to be used as a reporter gene in acidophilic, lactic acid and non-acidophilic bacterial hosts (Callanan et al., 2007).

β-glucuronidase from filamentous fungus *Penicillium purpurogenum* Li–3 is reported to have a highly selective hydrolyzing property on glycyrrhizin, and it is active as a homodimer and catalyzes the direct conversion to glycyrrhetic acid mono-glucuronid (Kaleem et al., 2017; Xu et al., 2017). Glycyrrhizin is the principal active ingredient of liquorice root, *Glycyrrhiza glabra* that has some medicinal applications. This β-glucuronidasehas enhanced structure stability and more resistance to deactivation by denaturing when expressed in *Pichia pastoris* GS115 than the wild type due to glycosylation inside the *Pichia* cells (Zou et al., 2013).

GUSB from *Staphylococcus* sp. RLH1 has 10-fold higher sensitivity compared to *E. coli* one for an unknown mechanism. The secondary structural comparison reveals increased random coil percentage, the presence of catalytic residues in loops and higher solvent accessibility of active site residues, which may be the reason for this activity difference (Arul et al., 2008).

7.5   CONCLUSION AND FUTURE PERSPECTIVES

GUSB plays a crucial role in all organisms that possess this activity. In human, the balance between glucuronidation and GUSB plays a pivotal role in the medical applications of GUSB. GUSB-responsive prodrugs and GUSB-responsive imaging are very promising areas of applications, which may open the door for more personalized medicine applications.

Both probes for GUSB-responsive prodrugs or GUSB-responsive imaging reported so far have two very important advantages in tumor treatment, particularly for solid tumors. The first advantage is the low toxicity of "inactive" prodrugs and "non-imaging" probes in the normal tissues, however, in tumor tissues or what we called in this chapter "*in situ*," both prodrugs and probes are activated after their hydrolysis with GUSB. The second is the activation selectivity at tumor site due to high content of GUSB. These two advantages should encourage more research in cancer therapy using this approach to increase its efficiency and to increase its repertoires for targeting different types of tumors and maybe even metastasis one.

In plant biotechnology, GUSB gene shows a superior property as a reporter gene; however, the endogenous GUSB interference may be minimized by using GUSB gene from extremely thermophilic microorganisms, which should have much higher thermostability. Consequently, it increases the demand of screening the extremophilic organisms for GUSB with improved properties for many different applications. Discovering novel GUSB enzymes from thermophilic or extremely thermophilic bacteria that are thermostable at temperatures up to 100°C and/or active at higher pH may ensure the complete elimination of the interference of endogenous GUSB enzyme activity present in the host organisms/tissues. Furthermore, the new thermophilic GUSBs could be resistance to the endogenous inhibitors that could be present in some organisms/tissues. For biofuel biotechnology, the new thermophilic GUSBs enzymes may also be used for degradation of microalgae cell wall for biofuel production.

KEYWORDS

- β-glucuronidase
- β-glucuronidase-responsive prodrugs
- bacteria directed enzyme prodrug therapy
- FEAnGA-Me
- FITC-TrapG
- glucuronic acid
- glucuronidation
- glycoside hydrolase
- GUSB gene
- intron-mediated enhancement
- mucopolysaccharidosis type VII
- PTH-G
- reporter gene
- sly syndrome
- UDP-glucuronosyltransferase

REFERENCES

Abdollahi, M. R., Memari, H. R., & Van Wijnen, A. J., (2011). Factor affecting the endogenous β-glucuronidase activity in rapeseed haploid cells: how to avoid interference with the gus transgene in transformation studies. *Gene, 487*(1), 96–102.

Aich, S., Delbaere, L. T., & Chen, R., (2001). Continuous spectrophotometric assay for beta-glucuronidase. *Biotechniques, 30*(4), 846–850.

Akao, T., (1999). Purification and characterization of glycyrrhetic acid mono-glucuronide beta-D-glucuronidase in *Eubacterium* sp. GLH. *Biol. Pharm. Bull., 22*(1), 80–82.

Akao, T., (2000). Competition in the metabolism of glycyrrhizin with glycyrrhetic acid mono-glucuronide by mixed *Eubacterium* sp. GLH and *Ruminococcus* sp. PO1-3. *Biol. Pharm. Bull., 23*(2), 149–154.

Alwen, A., Benito, M. R. M., Vicente, O., & Heberle-Bors, E., (1992). Plant endogenous beta-glucuronidase activity: How to avoid interference with the use of the *E. coli* beta-glucuronidase as a reporter gene in transgenic plants. *Transgenic Res., 1*(2), 63–70.

Amaya, M. A. F., Watts, A. G., Damager, I., Wehenkel, A., Nguyen, T., Buschiazzo, A., et al., (2004). Structural insights into the catalytic mechanism of trypanosoma cruzi trans-Sialidase. *Structure, 12*(5), 775–784.

Anbu, P., & Arul, L., (2013). Beta glucuronidase activity in early stages of rice seedlings and callus: A comparison with *Escherichia coli* beta glucuronidase expressed in the transgenic rice. *Int. J. Biotech. Mol. Biol. Res., 4*(4), 52–59.

Anhalt, S., & Weissenböck, G., (1992). Subcellular localization of luteolin glucuronides and related enzymes in rye mesophyll. *Planta, 187*(1), 83–88.

Antunes, I. F., Haisma, H. J., Elsinga, P. H., Van Waarde, A., Willemsen, A. T., Dierckx, R. A., & De Vries, E. F., (2012). *In vivo* evaluation of 1-O-(4-(2-fluoroethyl-carbamoyloxymethyl)–2-nitrophenyl)-O-beta-D-glucopyronuronate: a positron emission tomographic tracer for imaging beta-glucuronidase activity in a tumor/inflammation rodent model. *Mol. Imaging, 11*(1), 77–87.

Arul, L., Benita, G., & Balasubramanian, P., (2008). Functional insight for β-glucuronidase in *Escherichia coli* and *Staphylococcus* sp. RLH1. *Bioinformation, 2*(8), 339–343.

Bank, N., & Bailine, S. H., (1965). Urinary beta-glucuronidase activity in patients with urinary tract infection. *New England Journal of Medicine, 272*(2), 70–75.

Baudouy-Robert, J., Didier-Fichet, M. L., Jimeno-Abendano, J., Novel, G., Portalier, R., & Stoeber, F., (1970). Modalities of induction of 6 1st enzymes degrading hexuronides and hexuronates in *Escherichia coli* K12. *C. R. Acad. Sci. Hebd. Seances Acad. Sci. D., 271*(2), 255–258.

Beaud, D., Tailliez, P., & Anba-Mondoloni, J., (2005). Genetic characterization of the beta-glucuronidase enzyme from a human intestinal bacterium, *Ruminococcus gnavus*. *Microbiology, 151*(7), 2323–2330.

Birkenmeier, E. H., Davisson, M. T., Beamer, W. G., Ganschow, R. E., Vogler, C. A., Gwynn, B., et al., (1989). Murine mucopolysaccharidosis type VII. Characterization of a mouse with beta-glucuronidase deficiency. *J. Clin. Invest., 83*(4), 1258–1266.

Blattner, F. R., Plunkett, G., 3rd, Bloch, C. A., Perna, N. T., Burland, V., Riley, M., et al., (1997). The complete genome sequence of *Escherichia coli* K–12. *Science, 277*(5331), 1453–1462.

Bosslet, K., Czech, J., & Hoffmann, D., (1995). A novel one-step tumor-selective prodrug activation system. *Tumor Targeting, 1*, 45–50.

Boyland, E., Gasson, J. E., & Williams, D. C., (1957). Enzyme activity in relation to cancer. *Br. J. Cancer, 11*(1), 120–129.

Brand, W., Boersma, M. G., Bik, H., Hoek-van den Hil, E. F., Vervoort, J., Barron, D., et al., (2010). Phase II metabolism of hesperetin by individual UDP-glucuronosyltransferases and sulfotransferases and rat and human tissue samples. *Drug Metab. Dispos., 38*(4), 617–625.

Briciu-Burghina, C., Heery, B., & Regan, F., (2015). Continuous fluorometric method for measuring beta-glucuronidase activity: Comparative analysis of three fluorogenic substrates. *Analyst, 140*(17), 5953–5964.

Brot, F. E., Bell, C. E., & Sly, W. S., (1978). Purification and properties of β-glucuronidase from human placenta. *Biochemistry, 17*(3), 385–391.

Burchett, G. G., Folsom, C. G., & Lane, K. T., (2015). Native electrophoresis-coupled activity assays reveal catalytically-active protein aggregates of *Escherichia coli* beta-glucuronidase. *PLoS One, 10*(6), e0130269.

Callanan, M. J., Russell, W. M., & Klaenhammer, T. R., (2007). Modification of Lactobacillus β-glucuronidase activity by random mutagenesis. *Gene, 389*(2), 122–127.

Cantarel, B. L., Coutinho, P. M., Rancurel, C., Bernard, T., Lombard, V., & Henrissat, B., (2009). The carbohydrate-active enzymes database (CAZy): An expert resource for glycogenomics. *Nucleic Acids Research, 37*(1), D233–D238.

Cha, S. K., Ortega, B., Kurosu, H., Rosenblatt, K. P., Kuro-o, M., & Huang, C. L., (2008). Removal of sialic acid involving klotho causes cell-surface retention of TRPV5 channel via binding to galectin–1. *Proc. Natl. Acad. Sci. USA, 105*(28), 9805–9810.

Cheng, C. M., Lu, Y. L., Chuang, K. H., Hung, W. C., Shiea, J., Su, Y. C., et al., (2008). Tumor-targeting prodrug-activating bacteria for cancer therapy. *Cancer Gene Ther., 15*(6), 393–401.

Cheng, T. C., Roffler, S. R., Tzou, S. C., Chuang, K. H., Su, Y. C., Chuang, C. H., et al., (2012). An activity-based near-infrared glucuronide trapping probe for imaging beta-glucuronidase expression in deep tissues. *J. Am. Chem. Soc., 134*(6), 3103–3110.

Chopra, A., (2004). N'-Fluorescein-N''-[4-O-(beta-d-glucopyranuronic acid)–3-difluoro-methylphenyl]-S-methylthiourea (FITC-TrapG) and N'-(p-aminophenyl) thioether of IR–820-N''-[4-O-(beta-d-glucopyranuronic acid)–3-difluoromethylphenyl]-S-methyl-thiourea (NIR-TrapG). In: *Molecular Imaging and Contrast Agent Database*. National Center for Biotechnology Information (US), Bethesda, MD.

Clairmont, C., Lee, K. C., Pike, J., Ittensohn, M., Low, K. B., Pawelek, J., et al., (2000). Biodistribution and genetic stability of the novel antitumor agent VNP20009, a genetically modified strain of *Salmonella typhimurium. J. Infect. Dis., 181*(6), 1996–2002.

Cummings, J., Ethell, B. T., Jardine, L., Boyd, G., Macpherson, J. S., Burchell, B., et al., (2003). Glucuronidation as a mechanism of intrinsic drug resistance in human colon cancer: Reversal of resistance by food additives. *Cancer Res., 63*(23), 8443–8450.

Dai, Y., Chen, S., Wang, L., Pei, X. Y., Kramer, L. B., Dent, P., & Grant, S., (2011). Bortezomib interacts synergistically with belinostat in human acute myeloid leukemia and acute lymphoblastic leukemia cells in association with perturbations in NF-κB and Bim. *British Journal of Haematology, 153*(2), 222–235.

Didier-Fichet, M. L., & Stoeber, F., (1968). On the glucuronidase and galacturonidase activities of *Escherichia coli* ML 30. *C R Acad. Sci. Hebd. Seances Acad. Sci. D., 266*(18), 1894–1897.

Dutton, G. J., (1980). *Glucuronidation of Drugs and Other Compounds*. CRC Press, Florida.

Ender, A., Goeppert, N., Grimmeisen, F., & Goldscheider, N., (2017). Evaluation of β-d-glucuronidase and particle-size distribution for microbiological water quality monitoring in Northern Vietnam. *Science of The Total Environment, 580*, 996–1006.

Eudes, A., Mouille, G., Thévenin, J., Goyallon, A., Minic, Z., & Jouanin, L., (2008). Purification, cloning and functional characterization of an endogenous beta-glucuronidase in *Arabidopsis thaliana*. *Plant and Cell Physiology, 49*(9), 1331–1341.

Farrell, L. B., & Beachy, R. N., (1990). Manipulation of beta-glucuronidase for use as a reporter in vacuolar targeting studies. *Plant Mol. Biol., 15*(6), 821–825.

Fincher, G. B., Stone, B. A., & Clarke, A. E., (1983). Arabinogalactan-proteins - structure, biosynthesis, and function. *Annual Review of Plant Physiology and Plant Molecular Biology, 34*, 47–70.

Fior, S., Vianelli, A., & Gerola, P. D., (2009). A novel method for fluorometric continuous measurement of β-glucuronidase (GUS) activity using 4-methyl-umbelliferyl-β-d-glucuronide (MUG) as substrate. *Plant Science, 176*(1), 130–135.

Fittkau, M., Voigt, W., Holzhausen, H. J., & Schmoll, H. J., (2004). Saccharic acid 1.4-lactone protects against CPT–11-induced mucosa damage in rats. *J. Cancer Res. Clin. Oncol., 130*(7), 388–394.

Flores, H., & Ellington, A. D., (2002). Increasing the thermal stability of an oligomeric protein, beta-glucuronidase. *Journal of Molecular Biology, 315*(3), 325–337.

Forbes, N. S., (2010). Engineering the perfect (bacterial) cancer therapy. *Nat. Rev. Cancer, 10*(11), 785–794.

Gallagher, P. M., D'Amore, M. A., Lund, S. D., & Ganschow, R. E., (1988). The complete nucleotide sequence of murine beta-glucuronidase mRNA and its deduced polypeptide. *Genomics, 2*(3), 215–219.

Gallegos, J. E., & Rose, A. B., (2015). The enduring mystery of intron-mediated enhancement. *Plant Sci., 237*, 8–15.

Gandhi, N. S., Freeman, C., Parish, C. R., & Mancera, R. L., (2012). Computational analyses of the catalytic and heparin-binding sites and their interactions with glycosaminoglycans in glycoside hydrolase family 79 endo-beta-D-glucuronidase (heparanase). *Glycobiology, 22*(1), 35–55.

Gehrmann, M. C., Opper, M., Sedlacek, H. H., Bosslet, K., & Czech, J., (1994). Biochemical properties of recombinant human β-glucuronidase synthesized in baby hamster kidney cells. *Biochemical Journal, 301*(3), 821–828.

George, J., (2008). Elevated serum beta-glucuronidase reflects hepatic lysosomal fragility following toxic liver injury in rats. *Biochem. Cell Biol., 86*(3), 235–243.

Gianì, S., Altana, A., Campanoni, P., Morello, L., & Breviario, D., (2009). In trangenic rice, α- and β-tubulin regulatory sequences control GUS amount and distribution through intron mediated enhancement and intron dependent spatial expression. *Transgenic Res., 18*(2), 151–162.

Gitzelmann, R., Bosshard, N. U., Superti-Furga, A., Spycher, M. A., Briner, J., Wiesmann, U., et al., (1994). Feline mucopolysaccharidosis VII due to beta-glucuronidase deficiency. *Vet. Pathol., 31*(4), 435–443.

Hänsch, R., Koprek, T., Mendel, R. R., & Schulze, J., (1995). An improved protocol for eliminating endogenous β-glucuronidase background in barley. *Plant Science, 105*(1), 63–69.

Harris, N. M., Anderson, W. R., Lwaleed, B. A., Cooper, A. J., Birch, B. R., & Solomon, L. Z., (2003). Epirubicin and meglumine γ-linolenic acid. *Cancer, 97*(1), 71–78.

Haskins, M. E., Aguirre, G. D., Jezyk, P. F., Schuchman, E. H., Desnick, R. J., & Patterson, D. F., (1991). Mucopolysaccharidosis type VII (Sly syndrome). Beta-glucuronidase-deficient mucopolysaccharidosis in the dog. *Am. J. Pathol., 138*(6), 1553–1555.

Hassan, M. I., Waheed, A., Grubb, J. H., Klei, H. E., Korolev, S., & Sly, W. S., (2013). High resolution crystal structure of human beta-glucuronidase reveals structural basis of lysosome targeting. *PLoS One, 8*(11), e79687, https: //doi.org/10.1371/journal. pone.0079687.

Hassan, M. I., Waheed, A., Grubb, J. H., Klei, H. E., Korolev, S., & Sly, W. S., (2015). Correction: High resolution crystal structure of human β-glucuronidase reveals structural basis of lysosome targeting. *PLoS One, 10*(9), e0138401, doi: 10.1371/journal. pone.0138401.

Hawksworth, G., Drasar, B. S., & Hill, M. J., (1971). Intestinal bacteria and the hydrolysis of glycosidic bonds. *J. Med. Microbiol., 4*(4), 451–459.

Heery, B., Briciu-Burghina, C., Zhang, D., Duffy, G., Brabazon, D., O'Connor, N., & Regan, F., (2016). ColiSense, today's sample today: A rapid on-site detection of β-D-Glucuronidase activity in surface water as a surrogate for *E. coli*. *Talanta, 148*, 75–83.

Hess, C., & Neri, D., (2014). Tumor-targeting properties of novel immunocytokines based on murine IL1beta and IL6. *Protein Eng. Des. Sel., 27*(6), 207–213.

Hill, M. J., Drasar, B. S., Hawksworth, G., Aries, V., Crowther, J. S., & Williams, R. E., (1971). Bacteria and aetiology of cancer of large bowel. *Lancet, 1*(7690), 95–100.

Himeno, M., Hashiguchi, Y., & Kato, K., (1974). β-Glucuronidase of bovine liver: Purification, properties, carbohydrate composition. *J. Biochem., 76*(6), 1243–1252.

Hsieh, Y. T., Chen, K. C., Cheng, C. M., Cheng, T. L., Tao, M. H., & Roffler, S. R., (2015). Impediments to enhancement of CPT–11 anticancer activity by *E. coli* directed beta-glucuronidase therapy. *PLoS One, 10*(2), e0118028, doi: 10.1371/journal.pone.0118028.

Hu, C. Y., Chee, P. P., Chesney, R. H., Zhou, J. H., Miller, P. D., & O'Brien, W. T., (1990). Intrinsic GUS-like activities in seed plants. *Plant Cell Rep., 9*(1), 1–5.

Islam, M. R., Grubb, J. H., & Sly, W. S., (1993). C-terminal processing of human beta-glucuronidase. The propeptide is required for full expression of catalytic activity, intracellular retention, and proper phosphorylation. *J. Biol. Chem., 268*(30), 22627–22633.

Islam, M. R., Tomatsu, S., Shah, G. N., Grubb, J. H., Jain, S., & Sly, W. S., (1999). Active site residues of human β-glucuronidase: Evidence for Glu540 as the nucleophile and Glu451 as the acid-base residue. *J. Biol. Chem., 274*(33), 23451–23455.

Iturriaga, G., Jefferson, R. A., & Bevan, M. W., (1989). Endoplasmic reticulum targeting and glycosylation of hybrid proteins in transgenic tobacco. *Plant Cell, 1*(3), 381–390.

Jain, S., Drendel, W. B., Chen, Z. W., Mathews, F. S., Sly, W. S., & Grubb, J. H., (1996). Structure of human beta-glucuronidase reveals candidate lysosomal targeting and active-site motifs. *Nat. Struct. Biol., 3*(4), 375–381.

Jancova, P., Anzenbacher, P., & Anzenbacherova, E., (2010). Phase II drug metabolizing enzymes. *Biomed. Pap. Med. Fac. Univ. Palacky Olomouc. Czech. Repub., 154*(2), 103–116.

Jefferson, R. A., (1989). The GUS reporter gene system. *Nature, 342*(6251), 837–838.

Jefferson, R. A., Burgess, S. M., & Hirsh, D., (1986). Beta-glucuronidase from *Escherichia coli* as a gene-fusion marker. *Proc. Natl. Acad. Sci. USA, 83*(22), 8447–8451.

Jefferson, R. A., Kavanagh, T. A., & Bevan, M. W., (1987). GUS fusions: Beta-glucuronidase as a sensitive and versatile gene fusion marker in higher plants. *EMBO J., 6*(13), 3901–3907.

Jefferson, R. A., Klass, M., Wolf, N., & Hirsh, D., (1987). Expression of chimeric genes in *Caenorhabditis elegans*. *J. Mol. Biol., 193*(1), 41–46.

Jeong, J. H., Kim, K., Lim, D., Jeong, K., Hong, Y., Nguyen, V. H., et al., (2014). Antitumoral effect of the mitochondrial target domain of noxa delivered by an engineered *Salmonella typhimurium*. *PLoS One, 9*(1), e80050, doi: 10.1371/journal.pone.0080050.

Jiang, S. N., Park, S. H., Lee, H. J., Zheng, J. H., Kim, H. S., Bom, H. S., et al., (2013). Engineering of bacteria for the visualization of targeted delivery of a cytolytic anticancer agent. *Mol. Ther., 21*(11), 1985–1995.

Jiang, S. N., Phan, T. X., Nam, T. K., Nguyen, V. H., Kim, H. S., Bom, H. S., et al., (2010). Inhibition of tumor growth and metastasis by a combination of *Escherichia coli*-mediated cytolytic therapy and radiotherapy. *Mol. Ther., 18*(3), 635–642.

Juan, T. Y., Roffler, S. R., Hou, H. S., Huang, S. M., Chen, K. C., Leu, Y. L., et al., (2009). Antiangiogenesis targeting tumor microenvironment synergizes glucuronide prodrug antitumor activity. *Clin. Cancer Res., 15*(14), 4600–4611.

Kaleem, I., Rasool, A., Lv, B., Riaz, N., Hassan, J. U., Manzoor, R., & Li, C., (2017). Immobilization of purified β-glucuronidase on ZnO nanoparticles for efficient biotransformation of glycyrrhizin in ionic liquid/buffer biphasic system. *Chemical Engineering Science, 162*, 332–340.

Keller, R. K., & Touster, O., (1975). Physical and chemical properties of beta-glucuronidase from the preputial gland of the female rat. *J. Biol. Chem., 250*(12), 4765–4769.

Khan, F. I., Shahbaaz, M., Bisetty, K., Waheed, A., Sly, W. S., Ahmad, F., & Hassan, M. I., (2016). Large scale analysis of the mutational landscape in β-glucuronidase: A major player of mucopolysaccharidosis type VII. *Gene, 576*(1), 36–44.

Khan, K., Rahim, F., Wadood, A., Taha, M., Khan, M., Naureen, S., et al., (2014). Evaluation of bisindole as potent beta-glucuronidase inhibitors: Synthesis and *in silico* based studies. *Bioorg. Med. Chem. Lett., 24*(7), 1825–1829.

Kim, H. S., Kim, J. Y., Park, M. S., Zheng, H., & Ji, G. E., (2009). Cloning and expression of beta-glucuronidase from *Lactobacillus brevis* in *E. coli* and application in the bioconversion of baicalin and wogonoside. *J. Microbiol. Biotechnol., 19*(12), 1650–1655.

Kimura, N. T., Taniguchi, S., Aoki, K., & Baba, T., (1980). Selective localization and growth of *Bifidobacterium bifidum* in mouse tumors following intravenous administration. *Cancer Res., 40*(6), 2061–2068.

Kolarova, H., Ambruzova, B., Svihalkov, L., Klinke, A., & Kubala, L., (2014). Modulation of endothelial glycocalyx structure under inflammatory conditions. *Mediators of Inflammation, 694312*, http://dx.doi.org/10.1155/2014/694312.

Konishi, T., Kotake, T., Soraya, D., Matsuoka, K., Koyama, T., Kaneko, S., et al., (2008). Properties of family 79 beta-glucuronidases that hydrolyze beta-glucuronosyl and 4-O-methyl-beta-glucuronosyl residues of arabinogalactan-protein. *Carbohydr. Res., 343*(7), 1191–1201.

Kornfeld, S., (1986). Trafficking of lysosomal enzymes in normal and disease states. *J. Clin. Invest., 77*(1), 1–6.

Koshland, D. E., (1953). Stereochemistry and the mechanism of enzymatic reactions. *Biological Reviews, 28*(4), 416–436.

Kosugi, S., Ohashi, Y., Nakajima, K., & Arai, Y., (1990). An improved assay for β-glucuronidase in transformed cells: Methanol almost completely suppresses a putative endogenous β-glucuronidase activity. *Plant Science, 70*(1), 133–140.

Kyle, J., Galvin, N., Vogler, C., & Grubb, J., (1992). GUS (β-glucuronidase) assay in animal tissue. In: Gallagher, S., (ed.), *GUS Protocols: Using β-Glucuronidase Gene as a Reporter Gene Expression* (pp. 189–203). Academic Press, San Diego.

Laxa, M., (2016). Intron-mediated enhancement: A tool for heterologous gene expression in plants? *Frontiers in Plant Science, 7*, 1977, doi: 10.3389/fpls.2016.01977.

Levvy, G. A., & Marsh, C. A., (1959). Preparation and properties of beta-glucuronidase. *Adv. Carbohydr. Chem., 14*, 381–428.

Levvy, G. A., (1954). Baicalinase, a plant β-glucuronidase. *Biochem. J., 58*(3), 462–469.

Liang, W. J., Wilson, K. J., Xie, H., Knol, J., Suzuki, S., Rutherford, N. G., et al., (2005). The gusBC genes of *Escherichia coli* encode a glucuronide transport system. *J. Bacteriol., 187*(7), 2377–2385.

Liu, S. C., Ahn, G. O., Kioi, M., Dorie, M. J., Patterson, A. V., & Brown, J. M., (2008). Optimized clostridium-directed enzyme prodrug therapy improves the antitumor activity of the novel DNA cross-linking agent PR–104. *Cancer Res., 68*(19), 7995–8003.

Loeffler, M., Le'Negrate, G., Krajewska, M., & Reed, J. C., (2007). Attenuated *Salmonella* engineered to produce human cytokine LIGHT inhibit tumor growth. *Proc. Natl. Acad. Sci., USA, 104*(31), 12879–12883.

Loeffler, M., Le'Negrate, G., Krajewska, M., & Reed, J. C., (2009). *Salmonella typhimurium* engineered to produce CCL21 inhibit tumor growth. *Cancer Immunol. Immunother., 58*(5), 769–775.

Matsuda, T., Hatano, K., Harioka, T., Taura, F., Tanaka, H., Tateishi, N., et al., (1995). Histochemical investigation of β-glucuronidase in culture cells and regenerated plants of *Scutellaria baicalensis* Georgi. *Physiol. Plant, 19*, 390–394.

Matsuura, T., Hosoda, K., Ichihashi, N., Kazuta, Y., & Yomo, T., (2011). Kinetic analysis of beta-galactosidase and beta-glucuronidase tetramerization coupled with protein translation. *J. Biol. Chem., 286*(25), 22028–22034.

Mazerska, Z., Mroz, A., Pawlowska, M., & Augustin, E., (2016). The role of glucuronidation in drug resistance. *Pharmacol. Ther., 159*, 35–55.

McBain, A. J., & Macfarlane, G. T., (1998). Ecological and physiological studies on large intestinal bacteria in relation to production of hydrolytic and reductive enzymes involved in formation of genotoxic metabolites. *J. Med. Microbiol., 47*(5), 407–416.

Meyer, J. E., & Schulz, G. E., (1997). Energy profile of maltooligosaccharide permeation through maltoporin as derived from the structure and from a statistical analysis of saccharide-protein interactions. *Protein. Sci., 6*(5), 1084–1091.

Miller, R. D., Hoffmann, J. W., Powell, P. P., Kyle, J. W., Shipley, J. M., Bachinsky, D. R., & Sly, W. S., (1990). Cloning and characterization of the human beta-glucuronidase gene. *Genomics, 7*(2), 280–283.

Miwa, T., (1932). *Acta Phytochim Tokyo, 6*, 154.

Moese, J. R., & Moese, G., (1964). Oncolysis by clostidia. I. activity of *Clostridium butyricum* (M–55) and other nonpathogenic Clostridia against the Ehrlich carcinoma. *Cancer Res., 24*, 212–216.

Morimoto, S., Tateishi, N., Matsuda, T., Tanaka, H., Taura, F., Furuya, N., et al., (1998). Novel hydrogen peroxide metabolism in suspension cells of *Scutellaria baicalensis* Georgi. *J. Biol. Chem., 273*(20), 12606–12611.

Muhitch, M. J., (1998). Characterization of pedicel β-glucuronidase activity in developing maize (*Zea mays*) kernels. *Physiologia Plantarum., 104*(3), 423–430.

Naleway, J. J., (1992). Histochemical, spectrophotometric, and fluorometric GUS substrates. In: Gallagher, S. R., (ed.), *GUS Protocols: Using the GUS Gene as a Reporter of Gene Expression* (pp. 61–76). Academic Press, San Diego, CA.

Nandan, D., Venkatesan, K., Katoch, K., & Dayal, R. S., (2007). Serum beta-glucuronidase levels in children with leprosy. *Lepr. Rev., 78*(3), 243–247.

Narita, M., Nagai, E., Hagiwara, H., Aburada, M., Yokoi, T., & Kamataki, T., (1993). Inhibition of beta-glucuronidase by natural glucuronides of kampo medicines using glucuronide of SN–38 (7-ethyl–10-hydroxycamptothecin) as a substrate. *Xenobiotica, 23*(1), 5–10.

Nelson, K. E., Clayton, R. A., Gill, S. R., Gwinn, M. L., Dodson, R. J., Haft, D. H., et al., (1999). Evidence for lateral gene transfer between Archaea and bacteria from genome sequence of *Thermotoga maritima*. *Nature, 399*(6734), 323–329.

Nishimura, Y., Rosenfeld, M. G., Kreibich, G., Gubler, U., Sabatini, D. D., Adesnik, M., & Andy, R., (1986). Nucleotide sequence of rat preputial gland beta-glucuronidase cDNA and *in vitro* insertion of its encoded polypeptide into microsomal membranes. *Proc. Natl. Acad. Sci., USA, 83*(19), 7292–7296.

Novel, G., Didier-Fichet, M. L., & Stoeber, F., (1974). Inducibility of β-glucuronidase in wild-type and hexuronate-negative mutants of *Escherichia coli* K–12. *J. Bacteriol., 120*(1), 89–95.

O'Dwyer, P. J., Leyland-Jones, B., Alonso, M. T., Marsoni, S., & Wittes, R. E., (1985). Etoposide (VP–16–213). *New England Journal of Medicine, 312*(11), 692–700.

Oshima, A., Kyle, J. W., Miller, R. D., Hoffmann, J. W., Powell, P. P., Grubb, J. H., et al., (1987). Cloning, sequencing, and expression of cDNA for human beta-glucuronidase. *Proc. Natl. Acad. Sci. USA, 84*(3), 685–689.

Paigen, K., (1979). Acid hydrolases as models of genetic control. *Annu. Rev. Genet., 13*, 417–466.

Paigen, K., (1989). Mammalian β-glucuronidase: Genetics, molecular biology, and cell biology. In: Waldo, E. C., & Klvle, M., (eds.), *Progress in Nucleic Acid Research and Molecular Biology* (Vol. 37, pp. 155–205). Academic Press, New York.

Pawlowska, M., Chu, R., Fedejko-Kap, B., Augustin, E., Mazerska, Z., Radominska-Pandya, A., & Chambers, T. C., (2013). Metabolic transformation of antitumor acridinone C–1305 but not C–1311 via selective cellular expression of UGT1A10 increases cytotoxic response: implications for clinical use. *Drug Metabolism and Disposition, 41*(2), 414–421.

Peterson, S. B., & Liu, J., (2013). Multi-faceted substrate specificity of heparanase. *Matrix Biology, 32*(5), 223–227.

Plegt, L., & Bino, R. J., (1989). β-Glucuronidase activity during development of the male gametophyte from transgenic and non-transgenic plants. *Mol. Gen. Genet., 216*, 321–327.

Powell, P. P., Kyle, J. W., Miller, R. D., Pantano, J., Grubb, J. H., & Sly, W. S., (1988). Rat liver beta-glucuronidase: cDNA cloning, sequence comparisons, and expression of a chimeric protein in COS cells. *Biochem. J., 250*(2), 547–555.

Ramadan, A. M., Eissa, H. F., El-Domyati, F. M., Saleh, O. M., Ibrahim, N. E., Salama, M., et al., (2011). Characterization of inhibitor(s) of β-glucuronidase enzyme activity in GUS-transgenic wheat. *Plant Cell Tissue and Organ Culture, 107*(3), 373–381.

Reboul, R., Geserick, C., Pabst, M., Frey, B., Wittmann, D., Lutz-Meindl, U., et al., (2011). Down-regulation of UDP-glucuronic acid biosynthesis leads to swollen plant cell walls and severe developmental defects associated with changes in pectic polysaccharides. *J. Biol. Chem., 286*(46), 39982–39992.

Reddy, S., Shaller, C. C., Doss, M., Shchaveleva, I., Marks, J. D., Yu, J. Q., & Robinson, M. K., (2011). Evaluation of the anti-HER2 C6.5 diabody as a PET radiotracer to monitor HER2 status and predict response to trastuzumab treatment. *Clin. Cancer Res., 17*(6), 1509–1520.

Reitsma, S., Slaaf, D. W., Vink, H., Van Zandvoort, M., & Oude, E. M. G. A., (2007). The endothelial glycocalyx: Composition, functions, and visualization. *Pflugers Arch., 454*(3), 345–359.

Roberts, A. P., Frampton, J., Karim, S. M. M., & Beard, R. W., (1967). Estimation of beta-glucuronidase activity in urinary-tract infection. *New England Journal of Medicine, 276*(26), 1468–1470.

Rosenfeld, M. G., Kreibich, G., Popov, D., Kato, K., & Sabatini, D. D., (1982). Biosynthesis of lysosomal hydrolases: Their synthesis in bound polysomes and the role of co- and post-translational processing in determining their subcellular distribution. *J. Cell. Biol., 93*(1), 135–143.

Russell, W. M., & Klaenhammer, T. R., (2001). Identification and cloning of gusA, encoding a new β-glucuronidase from *Lactobacillus gasseri* ADH. *Appl. Environ. Microbiol., 67*(3), 1253–1261.

Sakurama, H., Kishino, S., Uchibori, Y., Yonejima, Y., Ashida, H., et al., (2014). beta-Glucuronidase from *Lactobacillus brevis* useful for baicalin hydrolysis belongs to glycoside hydrolase family 30. *Appl. Microbiol. Biotechnol., 98*(9), 4021–4032.

Salar, U., Taha, M., Ismail, N. H., Khan, K. M., Imran, S., Perveen, S., et al., (2016). Thiadiazole derivatives as new class of beta-glucuronidase inhibitors. *Bioorg. Med. Chem., 24*(8), 1909–1918.

Salleh, H. M., Mullegger, J., Reid, S. P., Chan, W. Y., Hwang, J., Warren, R. A., & Withers, S. G., (2006). Cloning and characterization of *Thermotoga maritima* beta-glucuronidase. *Carbohydr. Res., 341*(1), 49–59.

Sands, M. S., & Birkenmeier, E. H., (1993). A single-base-pair deletion in the beta-glucuronidase gene accounts for the phenotype of murine mucopolysaccharidosis type VII. *Proc. Natl. Acad. Sci., USA, 90*(14), 6567–6571.

Sasaki, K., Taura, F., Shoyama, Y., & Morimoto, S., (2000). Molecular characterization of a novel β-glucuronidase from *Scutellaria baicalensis* Georgi. *J. Biol. Chem., 275*(35), 27466–27472.

Schoenbeck, M. A., Swanson, G. A., & Brommer, S. J., (2007). β-Glucuronidase activity in seedlings of the parasitic angiosperm *Cusctua pentagona*: Developmental impact of the β-glucuronidase inhibitor saccharic acid 1,4-lactone. *Functional Plant Biology, 34*(9), 811–821.

Schulz, M., & Weissenböck, G., (1987). Partial purification and characterization of a luteolin-triglucuronide-specific β-glucuronidase from rye primary leaves (*Secale cereale*). *Phytochemistry, 26*(4), 933–937.

Shimizu, T., Ohtani, K., Hirakawa, H., Ohshima, K., Yamashita, A., Shiba, T., et al., (2002). Complete genome sequence of *Clostridium perfringens*, an anaerobic flesh-eater. *Proc. Natl. Acad. Sci., USA, 99*(2), 996–1001.

Shipley, J. M., Grubb, J. H., & Sly, W. S., (1993). The role of glycosylation and phosphorylation in the expression of active human beta-glucuronidase. *J. Biol. Chem., 268*(16), 12193–12198.

Sly, W. S., Quinton, B. A., McAlister, W. H., & Rimoin, D. L., (1973). Beta glucuronidase deficiency: Report of clinical, radiologic, and biochemical features of a new mucopolysaccharidosis. *J. Pediatr., 82*(2), 249–257.

Solis-Ramos, L. Y., Gonzalez-Estrada, T., Andrade-Torres, A., Godoy-Hernandez, G., & De la Serna, E., (2010). Endogenous GUS-like activity in Capsicum chinense Jacq. *Electronic Journal of Biotechnology, 13*(4), doi: 10.2225/vol13-issue4-fulltext–3.

Speciale, G., Thompson, A. J., Davies, G. J., & Williams, S. J., (2014). Dissecting conformational contributions to glycosidase catalysis and inhibition. *Curr. Opin. Struct. Biol., 28*, 1–13.

Speleman, F., Vervoort, R., Van Roy, N., Liebaers, I., Sly, W. S., & Lissens, W., (1996). Localization by fluorescence *in situ* hybridization of the human functional beta-glucuronidase gene (GUSB) to 7q11.21 → q11.22 and two pseudogenes to 5p13 and 5q13. *Cytogenet Cell Genet., 72*(1), 53–55.

Sperker, B., Werner, U., Murdter, T. E., Tekkaya, C., Fritz, P., Wacke, R., et al., (2000). Expression and function of beta-glucuronidase in pancreatic cancer: Potential role in drug targeting. *Naunyn. Schmiedebergs Arch. Pharmacol., 362*(2), 110–115.

Sreenath, H., & Naveen, K., (2004). Survey of endogenous β-glucuronidase (GUS) activity in coffee tissues and development of an assay for specific elimination of this activity in transgenic coffee tissues. *20th International Conference on Coffee Science*, Bangalore, India.

Staker, B. L., Hjerrild, K., Feese, M. D., Behnke, C. A., Burgin, A. B., Jr., & Stewart, L., (2002). The mechanism of topoisomerase I poisoning by a camptothecin analog. *Proc. Natl. Acad. Sci., USA, 99*(24), 15387–15392.

Stoeber, F., (1957). Sur la biosynthése induite de la β-glucuronidase *chez Escherichia coli. C R. Acad. Sci., 244*, 950–952.

Stringer, A. M., Gibson, R. J., Logan, R. M., Bowen, J. M., Yeoh, A. S., & Keefe, D. M., (2008). Fecal microflora and beta-glucuronidase expression are altered in an irinotecan-induced diarrhea model in rats. *Cancer Biol. Ther., 7*(12), 1919–1925.

Stritzker, J., Weibel, S., Hill, P. J., Oelschlaeger, T. A., Goebel, W., & Szalay, A. A., (2007). Tumor-specific colonization, tissue distribution, and gene induction by probiotic *Escherichia coli* Nissle 1917 in live mice. *Int. J. Med. Microbiol., 297*(3), 151–162.

Su, Y. C., Cheng, T. C., Leu, Y. L., Roffler, S. R., Wang, J. Y., Chuang, C. H., et al., (2014). PET imaging of beta-glucuronidase activity by an activity-based 124I-trapping probe for the personalized glucuronide prodrug targeted therapy. *Mol. Cancer Ther., 13*(12), 2852–2863.

Sudan, C., Prakash, S., Bhomkar, P., Jain, S., & Bhalla-Sarin, N., (2006). Ubiquitous presence of β-glucuronidase (GUS) in plants and its regulation in some model plants. *Planta., 224*(4), 853–864.

Tabas, I., & Kornfeld, S., (1980). Biosynthetic intermediates of beta-glucuronidase contain high mannose oligosaccharides with blocked phosphate residues. *J. Biol. Chem., 255*(14), 6633–6639.

Taha, M., Ismail, N. H., Imran, S., Rashwan, H., Jamil, W., Ali, S., et al., (2016). Synthesis of 6-chloro–2-aryl–1H-imidazo[4,5-b]pyridine derivatives: Antidiabetic, antioxidant, β-glucuronidase inhibiton and their molecular docking studies. *Bioorganic Chemistry, 65*, 48–56.

Taha, M., Ismail, N. H., Imran, S., Selvaraj, M., Rashwan, H., Farhanah, F. U., et al., (2015). Synthesis of benzimidazole derivatives as potent β-glucuronidase inhibitors. *Bioorganic Chemistry, 61*, 36–44.

Tohyama, O., Imura, A., Iwano, A., Freund, J. N., Henrissat, B., Fujimori, T., & Nabeshima, Y. I., (2004). Klotho is a novel β-glucuronidase capable of hydrolyzing steroid β-glucuronides. *J. Biol. Chem., 279*(11), 9777–9784.

Tomatsu, S., Fukuda, S., Sukegawa, K., Ikedo, Y., Yamada, S., Yamada, Y., et al., (1991). Mucopolysaccharidosis type VII: Characterization of mutations and molecular heterogeneity. *Am. J. Hum. Genet., 48*(1), 89–96.

Tomatsu, S., Montano, A. M., Dung, V. C., Grubb, J. H., & Sly, W. S., (2009). Mutations and polymorphisms in GUSB gene in mucopolysaccharidosis VII (Sly Syndrome). *Hum. Mutat., 30*(4), 511–519.

Tomatsu, S., Sukegawa, K., Ikedo, Y., Fukuda, S., Yamada, Y., Sasaki, T., et al., (1990). Molecular basis of mucopolysaccharidosis type VII: Replacement of Ala619 in β-glucuronidase with Val. *Gene, 89*(2), 283–287.

Tranoy-Opalinski, I., Legigan, T., Barat, R., Clarhaut, J., Thomas, M., Renoux, B., & Papot, S., (2014). β-Glucuronidase-responsive prodrugs for selective cancer chemotherapy: An update. *Eur. J. Med. Chem., 74*, 302–313.

Tsui, L. C., Farrall, M., & Donis-Keller, H., (1988). Report of the committee on the genetic constitution of chromosomes 7 and 8. *Cytogenet Cell Genet, 49*(1–3), 60–70.

Tzou, S. C., Roffler, S., Chuang, K. H., Yeh, H. P., Kao, C. H., Su, Y. C., et al., (2009). Micro-PET imaging of beta-glucuronidase activity by the hydrophobic conversion of a glucuronide probe. *Radiology, 252*(3), 754–762.

Ueyama, J., Satoh, T., Kondo, T., Takagi, K., Shibata, E., Goto, M., et al., (2010). Beta-glucuronidase activity is a sensitive biomarker to assess low-level organophosphorus insecticide exposure. *Toxicol. Lett., 193*(1), 115–119.

Vancanneyt, G., Schmidt, R., O'Connor-Sanchez, A., Willmitzer, L., & Rocha-Sosa, M., (1990). Construction of an intron-containing marker gene: Splicing of the intron in transgenic plants and its use in monitoring early events in *Agrobacterium*-mediated plant transformation. *Mol. Gen. Genet, 220*(2), 245–250.

Vervoort, R., Islam, M. R., Sly, W. S., Zabot, M. T., Kleijer, W. J., Chabas, A., et al., (1996). Molecular analysis of patients with beta-glucuronidase deficiency presenting as hydrops fetalis or as early mucopolysaccharidosis VII. *Am. J. Hum. Genet., 58*(3), 457–471.

Vogler, C., Levy, B., Galvin, N., Lessard, M., Soper, B., & Barker, J., (2005). Early onset of lysosomal storage disease in a murine model of mucopolysaccharidosis type VII: Undegraded substrate accumulates in many tissues in the fetus and very young MPS VII mouse. *Pediatr. Dev. Pathol., 8*(4), 453–462.

Wakabayashi, M., (1970). β-glucuronidases in metabolic hydrolysis. In: Fishman, W. H., (ed.), *Metabolic Conjugation and Metabolic Hydrolysis* (Vol. 2, pp. 519–602). Academic Press, New York.

Wallace, B. D., Wang, H., Lane, K. T., Scott, J. E., Orans, J., Koo, J. S., et al., (2010). Alleviating cancer drug toxicity by inhibiting a bacterial enzyme. *Science, 330*(6005), 831–835.

Wang, C. C., & Touster, O., (1972). Studies of catalysis by β-glucuronidase: the effect of structure on the rate of hydrolysis of substituted phenyl-β-D-glucopyranosiduronic acids. *J. Biol. Chem., 247*(9), 2650–2656.

Weissmann, G., Zurier, R. B., Spieler, P. J., & Goldstein, I. M., (1971). Mechanisms of lysosomal enzyme release from leukocytes exposed to immune complexes and other particles. *J. Exp. Med., 134*(3), 149–165.

Westphal, K., Leschner, S., Jablonska, J., Loessner, H., & Weiss, S., (2008). Containment of tumor-colonizing bacteria by host neutrophils. *Cancer Res., 68*(8), 2952–2960.

White, I. N. H., (2003). Tamoxifen: Is it safe? Comparison of activation and detoxication mechanisms in rodents and in humans. *Curr. Drug Metab., 4*(3), 223–239.

Wilson, K. J., Hughes, S. G., & Jefferson, R. A., (1992). The *Escherichia coli* gus operon: Induction and expression of the gus operon in *E. coli* and the occurrence and use of GUS in other bacteria. In: Gallagher, S. R., (ed.), *GUS Protocols Using the GUS Gene as Reporter of Gene Expression* (pp. 7–22). Academic Press, San Diego.

Wilson, K. J., Sessitsch, A., Corbo, J. C., Giller, K. E., Akkermans, A. D., & Jefferson, R. A., (1995). beta-Glucuronidase (GUS) transposons for ecological and genetic studies of rhizobia and other gram-negative bacteria. *Microbiol., 141*(7), 1691–1705.

Wong, A. W., He, S., Grubb, J. H., Sly, W. S., & Withers, S. G., (1998). Identification of Glu–540 as the catalytic nucleophile of human β-glucuronidase using electrospray mass spectrometry. *J. Biol. Chem., 273*(51), 34057–34062.

Xiong, A. S., Peng, R. H., Zhuang, J., Chen, J. M., Zhang, B., Zhang, J., & Yao, Q. H., (2011). A thermostable beta-glucuronidase obtained by directed evolution as a reporter gene in transgenic plants. *PLoS One, 6*(11), e26773, https://doi.org/10.1371/journal.pone.0026773.

Xiong, A. S., Peng, R. H., Zhuang, J., Liu, J. G., Xu, F., Cai, B., et al., (2007). Directed evolution of beta-galactosidase from *Escherichia coli* into beta-glucuronidase. *J. Biochem. Mol. Biol., 40*(3), 419–425.

Xu, Y., Liu, Y., Rasool, A. E. W., & Li, C., (2017). Sequence editing strategy for improving performance of β-glucuronidase from *Aspergillus terreus*. *Chemical Engineering Science, 167*, 145–153.

Yazawa, K., Fujimori, M., Amano, J., Kano, Y., & Taniguchi, S., (2000). *Bifidobacterium longum* as a delivery system for cancer gene therapy: Selective localization and growth in hypoxic tumors. *Cancer Gene Ther., 7*(2), 269–274.

Yu, Y. A., Shabahang, S., Timiryasova, T. M., Zhang, Q., Beltz, R., Gentschev, I., et al., (2004). Visualization of tumors and metastases in live animals with bacteria and vaccinia virus encoding light-emitting proteins. *Nat. Biotechnol., 22*(3), 313–320.

Zawawi, N. K., Taha, M., Ahmat, N., Wadood, A., Ismail, N. H., Rahim, F., et al., (2015). Novel 2,5-disubstituted–1,3,4-oxadiazoles with benzimidazole backbone: A new class of beta-glucuronidase inhibitors and in silico studies. *Bioorg. Med. Chem., 23*(13), 3119–3125.

Zechel, D. L., & Withers, S. G., (2000). Glycosidase mechanisms: Anatomy of a finely tuned catalyst. *Acc. Chem. Res., 33*(1), 11–18.

Zhang, C., Zhang, Y., Chen, J., & Liang, X., (2005). Purification and characterization of baicalin-β-d-glucuronidase hydrolyzing baicalin to baicalein from fresh roots of *Scutellaria viscidula* Bge. *Process Biochem., 40*(5), 1911–1915.

Zhang, H., Man, J. H., Liang, B., Zhou, T., Wang, C. H., Li, T., et al., (2010). Tumor-targeted delivery of biologically active TRAIL protein. *Cancer Gene Ther., 17*(5), 334–343.

Zhang, L., Zhou, H., Belzile, O., Thorpe, P., & Zhao, D., (2014). Phosphatidylserine-targeted bimodal liposomal nanoparticles for *in vivo* imaging of breast cancer in mice. *J. Control Release, 183*, 114–123.

Zou, S., Huang, S., Kaleem, I., & Li, C., (2013). N-Glycosylation enhances functional and structural stability of recombinant beta-glucuronidase expressed in *Pichia pastoris*. *J. Biotechnol., 164*(1), 75–81.

CHAPTER 8

MICROBIAL COLLAGENASES: AN OVERVIEW OF THEIR PRODUCTION AND APPLICATIONS

JYOTSNA JOTSHI[1] and PRADNYA PRALHAD KANEKAR[2]

[1]*formerly Microbial Sciences Division, Agharkar Research Institute, Pune, Maharashtra 411004, India*

[2]*Department of Biotechnology, Modern College of Arts, Science, and Commerce, Shivajinagar, Pune–411005, Maharashtra, India*

8.1 INTRODUCTION

Collagen is an abundant structural protein forming one quarter or more of the body weight in all animals. In humans, collagen comprises one-third of the total protein, accounts for three-quarters of the dry weight of skin and is the most commonly occurring component of the extracellular matrix (ECM). Collagen is generally resistant to most proteinases except collagenases. These matrix metalloproteases (MMPs) are known to act on different matrix proteins including collagen (Woessner, 1991; Aimes and Quigley, 1995). Collagenases are synthesized by eukaryotes and prokaryotes. Of all the collagenolytic proteases, MMPs have been studied extensively than collagenolytic proteases from bacteria. Bacterial collagenases are flexible, as they are able to hydrolyze both water-insoluble native collagens and water-soluble denatured collagens (Mookhtiar et al., 1985) and hence have been grabbing most of the attention.

8.2 COLLAGEN

It was a great breakthrough in collagen research when Pauling and Corey (1951) along with their α helix and β sheet structure proposed a triple helical structure of collagen. The fiber diffraction data of collagen, studied by Ramachandran and Kartha (1954) illustrated right-handed triple helical structure comprising of three staggered, left-handed helices with all peptide bonds in the *trans* conformation and two hydrogen bonds within each triplet. This structure was refined further by Rich and Crick (1955) to the structure accepted today. From then onwards, researchers have been trying to study collagen and its structure more intricately.

8.2.1 STRUCTURE

Collagen is comprised of three parallel polypeptide strands in a left-handed, polyproline II-type helical conformation coil about each other which forms a right-handed triple helix. The tight packing of PPII helices requires that every third residue be Glycine (Gly), resulting in a repeating XaaYaaGly sequence, where Xaa and Yaa can be any amino acids. Proline (Pro, 28%) and 4-hydroxyproline (Hyp, 38%) commonly occur in the Xaa and Yaa positions in collagen. This repeat is common in all types of collagens, although it is disrupted at certain locations within the triple-helical domain of nonfibrillar collagens. The amino acid composition in collagen is unique in the sense that as many as 20% are post-translationally modified amino acids like hydroxyproline and hydroxylysine, which rarely occur in other proteins (Chattopadhyay, 1998).

One of the strikingly important properties par excellence of collagen fiber is its enormous physical strength. It has been reported that a one mm diameter collagen fiber bundle can have as high as 20 lbs tensile load (breaking load). This can be attributed to its highly arranged structure which makes the collagen a molecule with high mechanical strength. Hydrogen bonds between glycine of one chain and proline of the other chain along with the water-mediated bonds stabilize the triple helix. The inter-molecular covalent cross-links hold them close so that it is suited to the biological function of bearing loads. Thus, these fiber forming collagens are the major contributors to the mechanical strength of the protective and supporting tissues such as skins, hides, bones, tendons ligaments and cartilages (Chattopadhyay, 1998).

8.2.2 TYPES AND OCCURRENCE

Classification of collagens is based on the expression of different genes during tissue construction (Lozano et al., 1985). In vertebrates, twenty-nine different types of collagen, consisting of at least 46 distinct polypeptide chains have been identified (Brinckmann, 2005; Shoulders and Raines, 2009). The types of collagen include the classical fibrillar and network-forming collagens, the FACITs, MACITs, and MULTIPLEXINs. The most common type of collagen is the type I which is found in bone, tendon, skin, and ligaments, while the second commonly occurring is the collagen type III, found in elastic tissues such as blood vessels and various internal organs. The types V and XI are associated with the types I and II in bone and cartilage as well as in other tissues (Daboor et al., 2010).

8.3   COLLAGENASE

A rigid structure makes collagen susceptible to only a limited number of proteases which can decompose collagen. Additionally, due to the diversity of the structure of collagen, it becomes very difficult to differentiate true collagenases from other collagenolytic proteases and from gelatinases (Harrington, 1996). Vertebral collagenases play important roles in wound healing, growth regulation, tissue remodeling, and resorption. One of the functions of prokaryotic collagenases is to assist in the invasion of host tissue by rapid degradation of the collagen matrix. In addition to collagenases, bacteria also produce other enzymes, including hyaluronidases, phospholipases, and elastases. MMPs constitute a family of structurally related metal-dependent enzymes or subgroups which have been further classified according to their structure and substrate specificity. MMPs in each subgroup share several common structural and functional features. However, this systematic classification is not applicable to bacterial collagenolytic proteases as mammalian MMPs exhibit an essential difference in protein structure (Watanabe, 2004).

Until 1962, most of the interest in collagenases was centered on Clostridial collagenases. The discovery of first vertebrate collagenase came from Gross and Lapiere (1962) after they obtained evidence for a collagenase in bullfrog tadpole tissue culture media. Subsequent to this discovery, a large number of collagenases were found in mammals, marine life and other bacteria (Schoellmann and Fisher, 1966; Welton and Woods, 1973; Keil, 1979). Then onwards, collagenases from humans and other mammalian sources have been reported and are being actively studied to understand the pathology of human diseases and applicability of these enzymes (Daboor et al., 2010). The present knowledge of the characteristics of collagenase comes from the pioneering studies in the 1950s by Mandl et al. (1953), Evans and Wardlaw (1953) and further by Yoshida and Noda (1965), Kono (1968), Seifter and Harper (1970), and, Bond and Van Wart (1984a). Among the bacterial collagenolytic proteases, metalloproteases occur most frequently as compared to serine and other proteases. Hydrolysis by these metalloproteases mostly takes place between the peptide bond of residue X and Gly-Pro. Serine class of collagenolytic proteases shows a broader range of specificity and their products are hydrolyzed at various but specific peptide bonds (Watanabe, 2004). Vastly studied of these metalloproteases belong to the family of Clostridial collagenases which are known to be relatively large in size (116 kDa). Further studies indicated that class I and class II collagenases are products of two different genes and have distinct features in domain organization. The high number of different active forms detected was related, suggesting that truncated isoforms play important roles in the regulation of clostridial collagenases *in vivo* (Bond and Wart, 1984b; Yoshihara et al., 1994; Matsushita et al., 1999; Ducka et al., 2009). Structural studies have suggested that a catalytic domain (collagenase unit) and, at least, one collagen binding domain (recruitment domain) are needed to degrade native collagen (Matsushita et al., 1998).

Approximately 35% of the fish mass goes in a waste, that is typically disposed off at sea or in landfills (Shahidi, 1994). The newer sources of useful enzymes are being explored from marine environment because of their potential to make use of processing wastes (Shahidi and Kamil, 2001). Thus, the use of fish processing waste to obtain enzymes presents as a solution to the costly disposal of waste and serves as a value-added processing step, which will minimize environmental problems and improve the economics of the fish industry (Daboor et al., 2010).

8.3.1 CLASSIFICATION

Since long, enzymes have been classified on the basis of their catalytic activity. The approach of classifying peptidases by reaction catalyzed (as far as possible), and then by catalytic type, is followed in the comprehensive recommendations on enzyme nomenclature of the NC-IUBMB. This committee is the successor to the Enzyme Commission, and the index numbers that it assigns to enzymes are described as "EC" numbers. According to IUBMB classification, microbial collagenase has been classified as hydrolases group of enzymes which acts on peptide bonds and requires a metal for action, calcium, and zinc being more common. The site of cleavage is the amino-terminal end from a peptide hence an EC number of 3.4.24.3 is assigned to microbial collagenase.

Enzyme classification by MEROPS takes care of the inclusion of collagenases which do not belong to metalloprotease group by assembling them in a family with unknown catalytic type. Rawlings and Barrett (1993) introduced this modern system for classification of peptidases, which has been further developed since then. The system is now established worldwide for identifying peptidase families and forms the classification system in MEROPS database. Microbial collagenases belong to the MEROPS peptidase family M9. This comprises of zinc-dependent bacterial metalloproteinases from *Vibrio* and *Clostridium* and peptidases for which there is no information regarding their active site. These peptidases have been assigned to a family of unknown catalytic type – the family U32. Their proteolytic activity has been compared to the activity of the *prtC* gene product of *Porphyromonas gingivalis*, that, as stated by Kato et al. (1992) "degrades soluble and reconstituted fibrillar type I collagen, heat-denatured type I collagen, and azocoll, but not gelatin or the synthetic collagenase substrate 4-phenylazobenzyloxycarbonyl-Pro-Leu-Gly- Pro-D-Arg" (Duarte et al., 2016). The NC-IUBMB assigned these endopeptidases to EC 3.4.24.3 group (microbial collagenases) of the metalloendopeptidases sub-class (EC 3.4.24). The MEROPS M9 peptidase family is subdivided into subfamilies, M9A and M9B, based on differences in their amino acid sequence and catalytic function. The M9B subfamily includes the subfamily types M09.002 and M09.003 which includes class I and class II collagenases from *Clostridium* sp. (Rawlings et al., 2012). The multiple studies on the enzymes produced by *Clostridium histolyticum* resulted in creation of knowledgebase (Bond and Wart, 1984a,b; Van Wart and Steinbrink, 1985).

8.3.2 MECHANISM OF ACTION

The degradation mechanism of collagen by MMPs has been extensively investigated, and preliminary measurements have been carried out on the activation energy of the catalysis. Experiments on the cleavage of native and site-directed mutants of murine collagen by MMP–1 have shown the crucial aspects of recognition mechanism (Lin et al., 1999). Proteolysis of single collagen I molecules has been studied using atomic force microscopy which shows that simple Michaelis-Menten mechanism is followed in this process. Also, the results on bulk solution are compatible with those observed on single collagen molecules. Action of MMPs on collagen proceeds by a Ratchet mechanism whereby the enzyme moves along collagen fibrils in steps, alternately binding to and cleaving the available monomers (Saffarian et al., 2004). The authors have demonstrated that activated collagenase (MMP-1) moves progressively on the collagen fibril. The mechanism of movement is a biased diffusion with the bias component dependant on the proteolysis of its substrate, not ATP hydrolysis. The biological implications of MMP-1 acting as a molecular ratchet tethered to the cell surface suggest new mechanisms for its role in tissue remodeling and cell-matrix interaction. Another study carried out by Chung et al. (2004) explains the mechanism of breakdown of collagen by collagenase, typically taken as MMP-1.

According to this model, collagenase unwinds triple-helical collagen prior to peptide bond hydrolysis. This can be attributed to the structural dimensions of active site of collagenase and binding site cleft of the substrate, collagen. Thus it was suggested that the activity of collagenase is influenced by the conformation of the substrate. The strain on the fibrils could alter the conformational state of collagen triple helices and thus makes it difficult or impossible for collagenase to unwind and cleave the specific site. However, the exact mechanism and hence the related analytical information of unwinding is not yet fully understood.

8.4   MICROBES AS SOURCE OF COLLAGENASE

The first evidence for collagenase production by bacteria was reported by Weinberg and Randin (1932). In the late thirties of the twentieth century, Maschmann (1937) suggested the name "collagenase" for an enzyme from *C. perfringens* able to digest both gelatin and collagen. The term "true collagenase" was coined by virtue of their capacity to hydrolyze native bovine Achilles' tendon. The pioneering study of bacterial collagenases and its production thereof started in early 1950s with recognition of involvement of proteolytic *Clostridia* in tissue putrefaction and the subsequent isolation of an extracellular enzyme that was able to digest tendons (Mandl et al., 1953). Since then a number of other collagenases of both bacterial and mammalian origin have been identified and characterized (Harrington, 1996). Collagenases of mammalian origin cleave the triple helix of collagen at about three-quarters of the length from the N-terminus at Gly^{775}-Ile in the (I) chain, while those of the bacterial origin cleave at multiple sites at +Gly bonds (IUBMB classification). As compared to mammalian collagenases, bacterial collagenases are known to be more versatile and can act on different types of collagens. Also, bacterial collagenases are more potent as compared to their mammalian counterparts (Daboor et al., 2010). Microorganisms producing collagenases are listed in Table 8.1.

TABLE 8.1 Collagenase Producing Microorganisms

Microorganism	Enzyme type	References
Clostridium histolyticum	Metalloprotease	Mandl et al., 1953; De Bellis et al., 1954; Grant and Alburn, 1959; Yoshida and Noda, 1965; Kono, 1968; Bond and Van Wart, 1984a; Yoshihara et al., 1994
Clostridium novyi	Metalloprotease	Nezafat et al., 2015
Clostridium perfringens	Metalloprotease	Matsushita et al., 1994
Vibrio spp.	Metalloprotease	Merkel and Sipos, 1971; Takeuchi et al., 1992; Miyoshi et al., 2008
Achromobacter iophagus	Serine protease	Welton and Woods, 1975
Achromobacter lyticus	Serine protease	Ohara et al., 1989
Poryphyromonas gingivalis	Thiol protease, Serine protease	Kato et al., 1992; Lawson and Meyer, 1992; Bedi and Williams, 1994
Bacillus cereus	Metalloprotease	Makinen and Makinen, 1987; Sigal et al., 1998; Lund and Granum, 1999; Liu et al., 2010; Suphatharaprateep et al., 2011
Bacillus pumilus	Metalloprotease	Wu et al., 2010
Bacillus lichenformis	Metalloprotease	Baehaki et al., 2012
Streptococcus spp.	Metalloprotease, Serine protease	Jackson et al., 1997; Juarez and Stinson, 1999
Nocardiopsis	Serine protease	Abdel-Fatah, 2013
Streptomyces spp.	Metalloprotease	Petrova et al., 2006; Sakurai et al., 2009
Rhizoctonia solani	Metalloprotease	Hamdy, 2008
Treponema denticola	Serine protease	Sorsa et al., 1995
Trichophyton schoenlinii	Metalloprotease	Rippon, 1968

The most thoroughly studied bacterial collagenase is the enzyme from *C. histolyticum* which is frequently used for various applications. At least six distinct collagenases with molecular weights ranging from 68 to 12.5 kDa have been isolated from *C. histolyticum*. Based on their sequences, chromatographic profiles, collagen degradation sites, activities against different substrates, these enzymes have been subdivided into two classes. One of the collagenases that specifically degrade soluble or insoluble collagen and certain synthetic peptide sequences is inactive against elastin and hemoglobin (Lim et al., 1993).

Notably, over the last few years, an increasing number of pathogenic bacteria are being reported for collagenase production. In many cases, bacteria have been isolated from the site of infection itself (Harrington, 1996). To understand the etiology of a disease, studies like these are definitely of great significance, but exploring collagenases from other sources which are of some biotechnological significance is also of great importance. More recently, different collagenase producing bacteria have been isolated from different leather samples (Kate, 2015). With increasing applications of collagenase, it becomes important to draw attention towards bacterial sources for such large-scale production of this industrially important enzyme.

8.5   PRODUCTION OF COLLAGENASE

Bacterial collagenases are generally extracellular and secreted during cell growth. Mostly, the bacteria reported for the production of collagenase are strict or facultative anaerobes, making their cultivation and large-scale production of the enzyme a difficult and cost-intensive process. Though various bacteria have been reported to produce collagenase, *Clostridium histolyticum* and *Achromobacter iophagus* are the ones which have been studied extensively (Yoshida and Noda, 1965; Welton and Woods, 1975; Bond and Van Wart, 1984a,b). Most of the collagenase producing microorganisms are anaerobes, making the process a tedious one (Bedi and Williams, 1994; Barua et al., 1989) while many need long incubation time for optimal collagenase production (Rippon, 1968; Lawson and Meyer, 1992). Aerobic recombinants *E. coli* are being used to produce collagenase at industrial level. Mostly, authors have relied on the source of isolation for media optimization like Merkel et al. (1975) who used PYSW (containing peptone, yeast extract and 4 % Rila Marine Mix) medium for collagenase production from bacteria isolated from seawater. Bedi and Williams (1994) and Sakurai et al. (2009) used BHIB supplemented with hemin and menadione for collagenase production.

Peptone is an easy choice for collagenase production as it is used commercially for collagenase production by *Achromobacter iophagus* (Reid et al., 1978). Many bacteria require complex production media for their growth and production of collagenase enzyme. Kato et al. (1992) have reported the production of collagenase from *Porphyromonas gingivalis* using mycoplasma broth supplemented with sheep blood. Hare et al. (1983) in their studies have again used a complex medium designated as low-SNP medium containing succinate, nitrate, phospahte, and some other inorganic salts for the production of collagenase from *Vibrio alginolyticus*.

Collageanse production from other microorganisms has been in a medium rich of proteins most of which lack any carbohydrate supplementation. Reid et al. (1978) reported repression of collagenase synthesis by addition of glucose in the medium. It seems that the sugar supplied in the medium when utilized by the microorganism causes a slowdown in utilization of proteins from the medium, in turn, causing suppression in enzyme production.

Calcium is known to provide stability to the enzyme as well as enhance the enzyme activity (Matsushita et al., 1994). Miwa et al. (1982) have described the production of collagenase from *Bacillus* using a complex medium, where the production is dependent on the presence of at least one divalent metal ion. Also, Reid et al. (1978) have used 2 mM $CaCl_2$ as medium supplementation for optimal collagenase production. Almost all the collagenases are known to exhibit maximum collagenase activity in pH range of 7 to 7.5 (Kono et al., 1968; Juarez et al., 1999; Wu et al., 2010). Collagenase from *Clostridium histolyticum* was found to be active in a pH range of 4–9 with pH optima of 7.4 (Matsushita et al., 1998). Collagenases from *Porphyromonas gingivalis*, studied by Bedi and Williams (1994) exhibited pH optimum at 8 which was similar to the collagenase of *Bacteroides gingivalis* (Barua et al., 1989). Most of the collagenases are active and stable near neutral pH range. Some bacteria exhibit activity above pH 8 as well, but there are not many reports of collagenases active in pH below 5.

While developing a production process, it is customary to start with shake flask wherein mostly optimization is carried out on a working volume of 100 ml growth medium. The next step is bench scale where a tabletop fermenter is used to optimize the process parameters to scale it up to large-scale fermenter. Large-scale production involves optimization of various parameters at many levels to obtain best yields with minimum possible inputs. Since with every new process comes newer problems, there is still no standard protocol for optimum production of enzymes or for that matter any biomolecule in general.

Statistically designed experiments allow us to gain knowledge about the system being studied with a minimum number of experiments. By including replicate testing, experimental variation is achieved. Different types of designs like Taguchi Design of Experiments, Response Surface Methodology (RSM), Plackett Burman Design, Central Composite Design (CCD) are available. The choice of design is based upon the objectives of the experiment and the current state of knowledge.

Several methods have been developed for the downstream processing of enzymes, but their application is dependent on the limitations involved in the respective system. Enzymes whether intracellular or extracellular, are hardly found in pure form. Choice of fermentation process coupled with swift recovery program is of utmost importance in case of extracellular enzymes. Extracellular enzymes present in cell-free broth are difficult to store as crude preparation since large volumes of liquid are involved.

Production of collagenase at low temperatures (1–8°C) has been studied by Premratne et al. (1986) where the authors have studied enzyme production from psychrotrophic bacteria. Petrova et al. (2006) have used an incubation temperature of 28°C for enzyme production from *Streptomyces* sp. stain 3B. Although collagenase production has mostly been studied at 37°C by many authors (Bedi and Williams, 1994; Juarez et al., 1999; Supatharaprateep et al., 2011), production at lower temperatures is not unknown.

Collagenases are known to act at neutral or near neutral pH, and most of the production processes are carried out at that pH. Almost all the processes of collagenase production from either anaerobic or aerobic bacteria are carried out in a pH range of 7 to 8 (Dreisbach and Merkel, 1978; Bedi and Williams, 1994; Wu et al., 2010; Abdel-Fatah, 2013). The production of collagenase from recombinants is also carried out in this pH range. Table 8.2 enlists the nutritional media used for production of collagenase by different bacteria.

Bacterial collagenases are secreted extracellularly, and highest yields of the enzyme are mostly obtained from cultures grown to late exponential or stationary growth phase. Culture media are often supplemented with casein hydrolyzate, collagen or collagen hydrolyzates, gelatin, and polypeptones. Some studies of collagenase production indicate that the enzyme is induced by addition of denatured collagen or its hydrolyzate into the growth medium, although induction may depend on the growth phase in which the enzyme is secreted by different microorganisms (Lim et al., 1993).

TABLE 8.2   Production Media for Collagenase Production from Different Microorganisms

Microorganism	Nutritional medium	References
Achromobacter iophagus	Peptone	Reid et al., 1978
Bacillus tequilensis	Meat extract, peptone, gelatin, and wastes of meat industry	Kaur et al., 2013
Streptomyces sp.	Soy meal and starch with poultry wastes	Jain and Jain, 2010
	Complex mineral medium with peptone and starch	Petrova et al., 2006
Vibrio spp.	Seawater supplemented with casein and collagen, NZ-amine with acid extracted calf skin collagen, Rila Marine Mix with peptone and yeast extract	Merkel and Sipos, 1971
Vibrio alginolyticus	Succinate, nitrate, phosphate with inorganic salts	Hare et al., 1983
Nocardiopsis dassonvillei	Basal medium with Marine wastes and chitin waste	Abdel-Fattah, 2013
Porphyromonas gingivalis	Mycoplasma broth with sheep blood	Kato et al., 1992
Actinomycetes strains	Kosmachev medium with maize extract, peptone, and CaCO$_3$	Christov et al., 2000
Streptomyces sp.	Tryptone, yeast, malt extract, oatmeal, glycerine, asparagine, glucose, starch, and iron in combinations	Endo, 1986

8.6   APPLICATIONS OF COLLAGENASE

Bacterial collagenolytic proteases find direct as well as indirect applications. The applications can be classified as (1) those in which collagenases are used directly, and (2) those in which the reaction products produced by collagenases are used. Though collagenases are mostly used for therapeutics and tissue dissociation, its indirect application lags behind.

8.6.1 PHARMACOLOGY

Collagenase finds immense application in therapeutics wherein it is used for the treatment of skin ulcers and severe burns. Collagenase is able to digest collagen in necrotic tissue at physiological pH, thus contributing towards the formation of granulation tissue and subsequent epithelization of dermal ulcers and severely burnt areas. Ointments made of collagenase are being used to treat burns where scaring is reduced to minimum (Hansbrough et al., 1995). Collagenases are also used in experimental transplantation of pancreatic islet cells to alleviate diabetic symptoms (Barker, 1975). Collagenase also finds application in treatment of various conditions like Dupytren's disease, Peyronie's disease, Glaucoma (Herman, 1996). Earlier, the choice of treatment for these conditions was surgical methods, but controlled application of collagenase for their treatment is the better alternative possible today.

8.6.2 TISSUE DISSOCIATION

Collagenases find application in cell-cell separation in tissues. Industrially produced Clostridial collagenases have been identified as four types for the use in tissue dissociation. Sigma and Worthington which are the leading brands in collagenase production follow this classification: Type 1, which contains average amounts of assayed activities (collagenase, caseinase, clostripain, and tryptic activities). It is generally recommended for epithelial, liver, lung, fat, and adrenal tissue cell preparations. Type 2 which contains greater clostripain activity, is generally used for heart, bone, muscle, thyroid, and cartilage. Type 3 bears low proteolytic activity, and it is usually used for separation of mammary cells. Type 4 contains low tryptic activity, which is commonly used for islets and other applications where receptor integrity is crucial. This classification varies from one manufacturer to the other depending on the degree of purification, but the preparations are usually more or less the same. This process of cell-cell separation is extensively used in industries as well as in research to prepare cell lines.

8.6.3   FOOD

Collagenase is used in tenderization of meat. It finds applications in processing of meat and other food products to maintain its toughness or to make it tender. Based on the diverse biological functions of collagen, promising industrial and medical applications of collagenolytic proteases can be envisaged. The fact that most collagenolytic proteases are produced by pathogenic bacteria has resulted in low profile studies of this group of enzyme. However, the bacterial collagenolytic proteases have great utilities, and there is a need to develop effective applications for their use (Watanabe, 2004).

8.6.4   LEATHER INDUSTRY

Dyeing is an important process in leather industry and has constantly been under research for an effective and environmentally friendly process. Many good dyes suffer from incomplete exhaustion resulting in higher wastage in effluents. Kanth et al. (2008) have shown an uptake of about 99% dye after the hide was treated by collagenase. An increased dye penetration also resulted in shade brightness.

8.7   COLLAGEN RICH WASTES AND VALUE ADDED PRODUCTS

The United Nations Food and Agriculture Organization have estimated that by 2025, half of the world's seafood supply will come from global aquaculture, which at present is 35 percent to 40 percent. Industrial growth results in increased waste generation. Millions of pounds of aquatic-based food processing waste is generated by Louisiana seafood processors alone. Very similar is the scenario in Scotland, Japan, China, and Australia. Many processing facilities produce waste amounting to 1–2 million pounds per year. Several local landfills refuse fisheries wastes because of the nitrogenous runoff that may be produced. Seafood waste has also been traditionally used as low-value feeds and fertilizers. However, seafood processors are interested in obtaining higher value byproducts like health-enhancing products. Fishery waste is a mixture of many components like collagen that can be extracted and used.

Another quarter gaining attention is leather industry which forms one of the most important economic sectors in the world. Leather industry provides the necessities, such as leather shoes and garments, while using the by-products of the meat industry. It is known that only 20% of wet salted hides/skins are converted into commercial leather, while large percentage becomes leather shavings, trimmings, and splits which are mostly disposed off in landfills. Recently, local restrictions have caused the tanning industry to seek alternatives to dumping. Increasing criticism about the environmental and planning impacts of the industry has led the governments to undertake steps which include a strategic environmental assessment and framework for management of such wastes.

Large quantities of livestock waste such as skin, bone, and feathers, or fish waste such as bone and scales are produced during food and meat processing. This waste contains a large amount of the proteins such as keratin and collagen, and thus can be utilized as protein resources for novel food materials. Collagen is present at high levels in fish or animal bone and skin. Since protein in these wastes is hard-fiber protein, its treatment is very difficult. They are therefore presently only used as feed, and foaming agents for fire extinguishers (Hoshino, 1976; Eremeev et al., 2009) developed an apparatus, enzyme, and conditions for enzymatic hydrolysis to effectively utilize horn and hoof, and were able to produce enzymatic hydrolysates from them. Morimura et al. (2002) developed an extraction process for collagen from pig skin or fish bone, for use in cosmetic materials. Jotshi (2011) has extensively studied production of bacterial collagenases and their use in preparation of low molecular weight peptides having various applications.

Recently, many studies have been performed on enzymatic hydrolysates of proteins. Physiological functions were found to be present in peptides produced from proteins by enzymatic hydrolysis. The collagen molecule is rich in amino acids like hydroxyproline thus making its hydrolysate a potentially valuable dietary supplement. Peptides also find applications in skin creams as well as cosmetic procedures. Therefore, the design of efficient approaches to collagen containing stock bioconversion allowing maximal preservation of components with high biological value is in great demand.

The main methods of bioconversion of collagen containing stock are thermal, acid, and alkaline hydrolysis. Sometimes, these procedures are used in combination to obtain desirable results. Few researchers have used a combination of acid/alkaline procedure followed by enzymatic hydrolysis. These processes are mostly unreliable and might produce unambiguous results. Enzymatic hydrolysis like with collagenases, allowing the preparation of hydrolysates with the maximum amount of labile amino acids and highly reproducible results is the most attractive from the point of view of ecological safety and preservation of functional properties of the hydrolysis products.

Collagenases are also used for treatment of various conditions like Peyronies disease, Dupytrens disease, glaucoma, burns, and wounds. The use of this enzyme in controlled conditions has avoided the need of surgery in various conditions. The list of biologically active peptides resulting from the hydrolysis of various proteins in mammalian organisms has been growing rapidly. Recently, many studies have been performed on enzymatic hydrolysates of proteins. These products which are formulated from proteins and perform physiological functions are value-added products since they come from a source of no or very less value (Li et al., 2005; Zhongkai et al., 2005). Chemical treatments like acid, alkali, and heat treatment have been used to generate collagen hydrolysates. Berg et al. (1980) have described a method for extraction of collagen and collagen degradation products using hot water extraction method followed by treatment with protease from 4 to 24h. Another study by Ermakova et al. (1998) describes a method for treating collagen material with acid, alkali, and surfactants wherein tanned collagen containing waste has been used to prepare material for decorating wall panels, leather accessories and shoes. Morimura et al. (2002) developed a process for extraction of protein and production of peptides by enzymatic hydrolysis from bone and skin wastes. Here, the authors employed a pretreatment step of extraction of collagen in acidic conditions for enzymatic treatment to produce peptides to be used in cosmetic materials Ohba et al. (2003) used fish bone and meat meal to generate collagen hydrolysate by treating it with acetone, 0.6 M HCl and finally enzymatic treatment with protease. They have reported high antioxidative effect of their collagen hydrolysate.

Another study by Gousterova et al. (2005) gave a comparative analysis of collagen and keratin degradation by thermoactinomycetes against alkaline hydrolysis method. Vaslleva et al. (2007) suggested the enzymatic treatment of extracted collagen for utilization in microbiologic practice in place of nitrogen sources like peptone. Aftab et al. (2006) studied the effects of various parameters on biodegradation of leather waste to generate hydroxyproline. An elaborate study by Eremeev et al. (2009) on generation of keratin hydrolysate throws light on the importance of optimization of parameters like stock: water, enzyme, and incubation time for generation of products of interest.

8.8   APPLICATIONS OF COLLAGEN PEPTIDES AND HYDROLYSATES

The collagenolytic proteases generate peptides from collagen, although their sequences are variable. Collagen peptides find use as effective agents in the treatment of several diseases, including suppression of collagen-induced arthritis and protection against bacteria (Watanabe, 2004). Following are some of the uses of collagen peptides.

8.8.1   MEDICAL USES

Collagen peptides have the important biological characters, e.g., high mechanics performance, promoting growth of cells, biological compatibility. They have been orally administered as pretreatment against osteoporosis (Ku et al., 1993). Collagen peptides are also known to protect against gastric ulceration, relax hypertension and stimulate metabolism (Khare et al., 1995). Many biomaterials made of collagen and its peptides are being used in the form of injection solutions, gels, films or shields used for ophthalmology and dentistry. They also find uses in surgical beauty treatments like glycolic peeling, laser reducing wrinkles.

8.8.2   HEALTHY FOODS AND FOOD PROCESSING

Collagen and its peptides are used as health foods as they help in repairing functions in dementia, heart, and blood vessel diseases. The food supplements are marketed in the form of powder, tablet, capsule, oral liquid. Collagen has the effect of dispersion and emulsion as well as gelatinizing function, hence is used as fine food additive in food processing.

8.8.3   COSMETICS

Low molecular weight collagen peptides are cosmetic ingredients which help in anti-aging, wrinkle softening, skin whitening, and moisture retention. Many skin creams in market like Verisol (Gelita Co., Germany) use these collagen peptides. Other than these, collagen peptides are used as ingredients in toning solutions, skin masks, hair packs, shampoos, conditioners, and several other products. Interestingly, most of these are peptides generated using fish collagen. Fish proteins are known for their activity in enhancing the beauty of skin; hence peptides of fish origin are gaining popularity (http://www.biotechlearn.org.nz).

8.8.4   INDUSTRIES

Though lesser known, collagen hydrolysate is used as paint additive and for bioengineering. A copolymer of collagen with plastic has also been reported (Chattopadhyay, 1998). With so many applications, the process of generation of collagen peptides has to be defined in the strictest terms. Controlled use of collagenolytic proteases for generation of collagen peptides will most likely become important as the chemical cleavage method requires highly reactive cyanogen bromide and can yield ambiguous results (Watanabe, 2004).

8.9   CONCLUSION AND FUTURE PERSPECTIVES

Collagen is a structural protein found in abundance in extra-cellular matrices of connective tissues. This structure is formed of three helices which come together to form a triple helix, tropocollagen. The triple helix is stabilized by interchain hydrogen bonding and inter and intramolecular cross-links and thus forms the compact collagen fibril. This structure makes the collagen a molecule with high mechanical strength. Collagen is generally resistant to most proteinases, except collagenases.

Collagenases are known to degrade collagen by unwinding triple-helical collagen prior to peptide bond hydrolysis. There have been several reports of collagenase producing bacteria like *Clostridium histolyticum*, *Bacillus* sp., *Achromobacter iophagus*, *Porphyromonas gingivalis*,

Pseudomonas aeruginosa, Streptomyces. Recently, researchers have been focusing on collagenase production from bacteria inherent to collagen source, other than the site of infections. Production of collagenase from safer bacteria, relatively less pathogenic than the reported ones is being explored.

Collagenases find uses in cell-cell separation in tissues, therapeutics. Ointments made of collagenase are used in the treatment of dermal ulcers and severely burnt areas. Peptides and amino acids generated from keratin and collagen stocks and wastes have been gaining attention lately hence collagen-rich wastes like tannery, fish, and poultry are being used to produce collagenase.

Collagenases have many applications, and newer ones are being envisaged. Pathogenic bacteria, isolated from sites of infections have been primary source of the collagenases. Production using a potent pathogen always poses a threat to the ones handling the process. Hence, in many industries recombinants have been developed to produce collagenase on large scale.

Newer bacteria producing collagenase have been reported from a long time and are being reported till today. Though there are numerous reports of collagenolytic bacteria, many of these lacks a detailed investigation. Again, significance of collagenase production by bacteria as etiological agent is the prime focus of many authors. As a result, industrial significance of this enzyme has been less studied.

Collagen-rich wastes like fish and tannery wastes are posing as big environmental problems. Efforts are being diverted to management of this waste and at the same time utilizing it to improve the economics of fish industry. Marine products are also gaining attention because of the rise in number of mad cow disease cases and other disease associated with collagen products of bovine and porcine origin. Fish collagen, which is safer to use, abundantly available and also easy to hydrolyze, is fast becoming popular as food additive.

In this scenario, it seemed appropriate to study fish collagen and particularly, an enzyme specific for such type of collagen. To ensure the specificity of such an enzyme, fish collagen was looked for the presence of collagenase producing bacteria. Collagenases of these bacteria could be useful in treating the fish waste rich in collagen to generate value-added products.

KEYWORDS

- cell-cell separation
- clostridial collagenase
- collagen
- collagenase
- collagen peptides
- collagenolytic protease
- fish waste
- food additives
- gelatinase
- matrix metalloproteases
- microbial production
- peptidase
- serine protease
- tannery waste

REFERENCES

Abdel-Fattah, A. M., (2013). Production and partial characterization of collagenase from marine *Nocardiopsis dassonvillei* NRC2aza using chitin wastes. *Egypt. Pharma. J., 12,* 109–114.

Aftab, M. N., Hameed, A., Ikram-ul-Haq., & Cheng, C. R., (2006). Biodegradation of leather waste by enzymatic treatment. *Chinese J. Proc. Eng., 6,* 462–465.

Aimes, R. T., & Quigley, J. P., (1995). Matrix metalloproteinase–2 is an interstitial collagenase. *J. Biol. Chem., 270,* 5872–5876.

Baehaki, A., Maggy, T. S., Sukarno, S. D., Azis, B. S., Setyahadi, S., & Meinhardt, F., (2012). Purification and characterization of collagenase from *Bacillus licheniformis* F11. 4. *Afr. J. Microbiol. Res., 6*(10), 2373–2379.

Barker, C. F., (1975). Transplantation of the islets of Langerhans and the histocompatibility of endocrine tissue. *Diabetes, 24,* 766–775.

Barua, P. K., Neiders, M. E., Topolnycky, A., Zambon, J. J., & Birkedal-Hansen, H., (1989). Purification of an 80,000-Mr glycylprolyl peptidase from *Bacteroides gingivalis. Infec. Immun., 57,* 2522–2528.

Bedi, G. S., & Williams, T., (1994). Purification and characterization of a collagen degrading protease from *Porphyromonas gingivalis. J. Biol. Chem., 269,* 599–606.

Berg, A., Eckmayer, Z., Monsheimer, R., & Pfleiderer, E., (1980). Methods for treating raw materials containing collagen. *US Patent*, 4220724.

Bond, M. D., & Van Wart, H. E., (1984a). Purification and separation of individual collagenases of *Clostridium histolyticum* using red dye ligand chromatography. *Biochem., 23*, 3077–3085.

Bond, M. D., & Van Wart, H. E., (1984b). Characterization of the individual collagenases from *Clostridium histolyticum*. *Biochem., 23*, 3085–3091.

Brinckmann, J., (2005). Collagens at a glance. *Top Curr. Chem., 247*, 1–6.

Chattopadhyay, B., (1998). Molecular biology of collagen-Part I. *J. Indian Leather Technol. Assoc., 48*, 473–492.

Christov, P., Gousterova, A., Goshev, I., Tzvetkova, R., & Nedkov, P., (2000). Optimization of the biosynthetic conditions for collagenase production by certain thermophilic actinomycete strains. *Comptes Rendus de l'Academie Bulgare des Sciencesi., 53*(3), 115–118.

Chung, L., Dinakarpandian, D., Yoshida, N., Lauer-Fields, J. L., Fields, G. B., Visse, R., & Nagase, H., (2004). Collagenase unwinds triple-helical collagen prior to peptide bond hydrolysis. *EMBO J., 23*, 3020–3030.

Daboor, S. M., Budge, S. M., Ghaly, A. E., Brooks, S. L., & Dave, D., (2010). Extraction and purification of collagenase enzymes: A critical review. *Am. J. Biochem. Biotechnol., 6*, 239–263.

De Bellis, R., Mandl, I., MacLennan, J., & Howes, E., (1954). Separation of proteolytic enzymes of *Clostridium histolyticum*. *Nature, 174*, 1191–1192.

Dreisbach, J. H., & Merkel, J. R., (1978). Induction of collagenase production in *Vibrio* B–30. *J. Bacteriol., 135*, 521–527.

Duarte, A. S., Correia, A. C. M., & Esterres, A. C., (2016). Bacterial collagenases - a review. *Crit. Rev. Microbiol., 42*(1), 106–126.

Ducka, P., Eckhard, U., & Schonauer, U. E., (2009). A universal strategy for high-yield production of soluble and functional clostridial collagenases in *E. coli*. *Appl. Microbiol. Biotechnnol., 83*, 1055–1065.

Endo, A., (1986). Novel collagenase discolysin and production method thereof. *US Patent*, 4624924.

Eremeev, N. L., Nikolaev, I. V., Keruchenko, I. D., Stepanova, E. V., Satrutdinov, A. D., Zinovev, S. V., et al., (2009). Enzymatic hydrolysis of keratin containing stock for obtaining protein hydrolyzates. *Appl. Biochem. Microbiol., 45*, 648–655.

Ermakova, I. R., Khaytin, B. S., & Beleske, A. L., (1998). Method of producing high molecular products from collagen-containing materials, and product produced by the same. *US Patent, 5720778*.

Evans, D. G., & Wardlaw, A. C., (1953). Gelatinase and collagenase production by certain species of *Bacillus*. *J. Gen. Microbiol., 8*, 481–487.

Gousterova, A., Braikova, D., Goshev, I., Christov, P., Tishinov, K., Vaslleva, E. T., et al., (2005). Degradation of keratin and collagen containing wastes by newly isolated thermoactinomycetes or by alkaline hydrolysis. *Lett. Appl. Microbiol., 40*, 335–340.

Grant, N., & Alburn, H., (1959). Studies on the collagenase of *Clostridium histolyticum*. *Arch. Biochem. Biophys., 82*, 245–249.

Gross, J., & Lapiere, C. M., (1962). Collagenolytic activity in amphibian tissues: A tissue culture assay. *Proc. Natl. Acad. Sci., USA, 48*, 1014–1022.

Hamdy, H. S., (2008). Extracellular collagenase from *Rhizoctonia solani*: Production, purification and characterization. *Ind. J. Biotechnol., 7*, 333–340.

Hansbrough, J. F., Achauer, B., Dawson, J., Himel, H., Luterman, A. H., Slater, H., et al., (1995). Wound healing in partial-thickness burn wounds treated with collagenase ointment versus silver sulfadiazine cream. *J. Burn Care Rehab., 16*, 241–247.

Hare, P., Scott-Burden, T., & Woods, D. R., (1983). Characterization of extracellular alkaline protease and collagenase induction in *Vibrio alginolyticus*. *J. Gen. Microbiol., 129*, 1141–1147.

Harrington, D. J., (1996). Bacterial collagenases and collagen degrading enzymes and their potential role in human disease. *Infect. Immun., 64*, 1885–1891.

Herman, I. M., (1996). Stimulation of human keratinocyte migration and proliferation *in vitro*: Insight into the cellular responses to injury and wound healing. *Wounds, 8*, 33–41.

Hoshino, M., (1976). Foaming property of the hydrolysis products of keratin. I. Molecular weight distribution and average molecular weights at maximum in foaming property. *Yukagaku, 25*(11), 793–795.

Jackson, R. J., Lim, D. V., & Dao, M. L., (1997). Identification and analysis of a collagenolytic activity in *Streptococcus mutans*. *Curr. Microbiol., 34*, 49–54.

Jain, R., & Jain, P. C., (2010). Production and partial characterization of collagenase of *Streptomyces exfoliatus* CFS 1068 using poultry feather. *Ind. J. Exp. Biol., 48*, 174–178.

Jotshi, J., (2011). *Microbial Production of Collagenolytic Protease and its Application in Recovery of Value Added Products* (pp. 1–184). PhD Thesis, Savitribai Phule Pune University, Pune, India.

Juarez, Z., & Stinson, M. W., (1999). An extracellular protease of *Streptococcus gordonii* hydrolyzes type IV collagen and collagen analoges. *Infect. Immun., 67*, 271–278.

Kanth, S. V., Venba, R., Madhan, B., Chandrababu, N. K., & Sadulla, S., (2008). Studies in the influence of bacterial collagenase in leather dyeing. *Dyes and Pigments, 76*(2), 338–347.

Kate, S., & Pethe, A., (2015). Production of collagenase by *Bacillus* KM369985 isolated from leather sample. *Int. J. Res. Biosciences, 4*(4), 81–87.

Kato, T., Takahashi, N., & Kuramitsu, H. K., (1992). Sequence analysis and characterization of the *Porphyromonas gingivalis* prtC gene, which expresses a novel collagenase activity. *J. Bacteriol., 174*, 3889–3895.

Kaur, P. S., & Azmi, W., (2013). Cost-effective production of a novel collagenase from a non-pathogenic isolate *Bacillus tequilensis*. *Current Biotechnol., 2*(1), 17–22.

Keil, B., (1979). Some newly characterized collagenase from prokaryotes and lower eukaryotes. *Mol. Cell Biochem., 23*, 87–108.

Khare, S. D., Krco, C. J., Griffiths, M. M., Luthra, H. S., & David, C. S., (1995). Oral administration of an immunodominant human collagen peptide modulates collagen-induced arthritis. *J. Immunol., 155*, 3653–3659.

Kono, T., (1968). Purification and characterization of collagenolytic enzymes from *Clostridium histolyticum*. *Biochem., 7*, 1106–1114.

Ku, G., Kronenberg, M., Peacock, D. J., Tempst, P., Banquerigo, M. L., Braun, B. S., et al., (1993). Prevention of experimental autoimmune arthritis with a peptide fragment of type II collagen. *Eur. J. Immunol., 23*, 591–599.

Lawson, D. A., & Meyer, T. F., (1992). Biochemical characterization of *Porphyromons (Bacteroides) gingivalis* collagenase. *Inf. Immun., 60*, 1524–1529.

Li, Y. C., Zhu, D. Y., & Jin, L. Q., (2005). Preparation and analysis of collagen polypeptide from hides by enzymes. *J. Soc. Leather Technol. Chem., 89*, 103–106.

Lim, D. V., Jackson, R. J., & Pull-Von, G. C. M., (1993). Purification and assay of bacterial collagenases. *J. Microbiol. Methods, 18*, 241–253.

Lin, H., Clegg, D. O., & Lal, R., (1999). Imaging real-time proteolysis of single collagen I molecules with an atomic force microscope. *Biochem., 38*, 9956–9963.

Liu, L., Ma, M., Yu, X., & Wang, W., (2010). Screening of collagenase producing strain and purification of *Bacillus cereus* collagenase. *Sheng Wu Gong Cheng Xue Bao, 26*, 194–200.

Lozano, G., Ninomiya, Y., Thompson, H., & Olsen, B. R., (1985). A distinct class of vertebrate collagen genes encodes chicken type IX collagen polypeptides. *Proc. Natl. Acad. Sci., USA, 82*, 4050–4054.

Lund, T., & Granum, P. E., (1999). The 105 kDa protein component of *Bacillus cereus* non-hemolytic enterotoxin (Nhe) is a metalloprotease with gelatinolytic and collagenolytic activity. *FEMS Microbiol. Lett., 178*, 355–361.

Makinen, K. K., & Makinen, P. L., (1987). Purification and properties of an extracellular collagenolytic protease produced by the human oil bacterium *Bacillus cereus* (Strain Soc 67). *J. Biol. Chem., 262*, 12488–12495.

Mandl, I., MacLennan, J., Howes, E., Debellis, R., & Sohler, A., (1953). Isolation and characterization of proteinase and collagenase from *Clostridium histolyticum. J. Clin. Invest., 32*, 1323–1327.

Maschmann, E., (1937). Uber Bakterienproteasen II. *Biochem. Ztschr., 295*, 1–10.

Matsushita, O., Jung, C. M., Katayama, S., Minami, J., Takahashi, Y., & Okabe, A., (1999). Gene duplication and multiplicity of collagenases in *Clostridium histolyticum. J. Bacteriol., 181*, 923–933.

Matsushita, O., Jung, C. M., Minami, J., Katayama, S., Nishi, N., & Okabe, A., (1998). A study of the collagen-binding domain of a 116-kDa *Clostridium histolyticum* collagenase. *J. Biol. Chem., 273*, 3643–3648.

Matsushita, O., Yoshihara, K., Katayama, S., Minami, J., & Okabe, A., (1994). Purification and characterization of a *Clostridium perfringens* 120 kDa collagenase and nucleotide sequence of the corresponding gene. *J. Bacteriol., 176*, 149–156.

Merkel, J. R., & Sipos, T., (1971). Marine bacterial proteases. I. characterization of an endopeptidase produced by *Vibrio* B–30. *Arch. Biochem. Biophys., 145*, 126–136.

Merkel, J. R., Dreisbach, J. H., & Ziegler, H. B., (1975). Collagenolytic activity of some marine bacteria. *Appl. Microbiol., 29*, 145–151.

Miwa, N., Masuda, Y., Kawarabuki, S., Sai, T., & Saito, T., (1982). Thermophilic collagenases - thermophilic bacteria capable of producing thermophilic collagenases, and process for producing said collagenases. *US Patent, 4315988*.

Miyoshi, S., Nitanda, Y., Fujii, K., Kawahara, K., Li, T., Maehara, Y., et al., (2008). Differential gene expression and extracellular secretion of the collagenolytic enzymes by the pathogen *Vibrio parahaemolyticus. FEMS Microbiol. Lett., 283*, 176–181.

Mookhtiar, K. A., Steinbrink, D. R., & Van Wart, H. E., (1985). Mode of hydrolysis of collagen-like peptides by class I and class II *Clostridium histolyticum* collagenases: Evidence for both endopeptidase and tripeptidylcarboxypeptidase activities. *Biochem., 24*, 6527–6533.

Morimura, S., Nagata, H., Uemura, Y., Fahmi, A., Shigematsu, T., & Kida, K., (2002). Development of an effective process and evaluation for utilization of collagen contained in livestock and fish waste. *Proc. Biochem., 37*, 1403–1412.

Nezafat, N., Negahdaripour, M., Gholami, A., & Ghasemi, Y., (2015). Computational analysis of collagenase from different *Vibrio, Clostridium* and *Bacillus* strains to find new enzyme sources. *Trends in Pharma Sciences, 1,* 213–222

Ohara, T., Makino, K., Shinagawa, H., Nakata, A., Norioka, S., & Sakiyama, F., (1989). Cloning, nucleotide sequence, and expression of *Achromobacter* protease I gene. *J. Biol. Chem., 264,* 20625–20631.

Ohba, R., Deguchi, T., Kishikawa, M., Arsyad, F., Morimura, S., & Kida, K., (2003). Physiological functions of enzymatic hydrolysates of collagen or keratin containing livestock and fishwaste. *Food Sci. Technol. Res., 9,* 91–93.

Pauling, L., & Corey, R. B., (1951). The structure of fibrous proteins of the collagen-gelatin group. *Proc. Natl. Acad. Sci., USA, 37,* 272–281.

Petrova, D., Derekova, A., & Vlahov, S., (2006). Purification and properties of individual collagenases from *Streptomyces* sp. strain 3B. *Folia Microbiol., 51,* 93–98.

Premaratne, R. J., Nip, W. K., & Moy, J. H., (1986). Characterization of proteolytic and collagenolytic psychrotrophic bacteria of ice stored freshwater prawn, *Macrobrachium rosenbergii. Marine Fisheries Rev., 48,* 44–47.

Ramachandran, G. N., & Kartha, G., (1954). Structure of collagen. *Nature, 174,* 269–270.

Rawlings, N. D., & Barrett, A. J., (1993). Evolutionary families of peptidase. *Biochem. J., 290,* 205–218.

Rawlings, N. D., Barrett, A. J., & Bateman, A., (2012). MEROPS: The database of proteolytic enzymes, their substrates, and inhibitors. *Nucleic. Acids Res., 40,* 43–50.

Reid, G. C., Robb, F. T., & Woods, D. R., (1978). Regulation of extracellular collagenase production in *Achromobacter iophagus. J. Gen. Microbiol., 109,* 149–154.

Rich, A., & Crick, F. H., (1955). The structure of collagen. *Nature, 176,* 915–916.

Rippon, J. W., (1968). Extracellular collagenase from *Trichophyton schoenleinii. J. Bacteriol., 95,* 43–46.

Saffarian, S., Collier, I. E., Marmer, B. L., Elson, E. L., & Goldberg, G., (2004). Interstitial collagenase is an ATP - independent molecular motor driven by proteolysis of collagen. *Science, 306,* 108–111.

Sakurai, Y., Inoue, H., Nishii, W., Takahashi, T., Iino, Y., Yamamoto, M., & Takahashi, K., (2009). Purification and characterization of a major collagenase from *Streptomyces parvulus. Biosci. Biotechnol. Biochem., 73,* 21–28.

Schoellmann, G., & Fischer, E., (1966). A collagenase from *Pseudomonas aeruginosa. Biochem. Biophys. Acta., 122,* 557–561.

Seifter, S., & Harper, E., (1970). The collagenase. In: Boye, P. D., (ed.), *The Enzymes* (Vol. 2, pp. 649–667). Academic Press, New York.

Shahidi, F., & Kamil, Y. V. A. J., (2001). Enzymes from fish and aquatic invertebrates and their application in the food industry. *Trends Food Sci. Technol., 12,* 435–464.

Shahidi, F., (1994). Proteins from seafood processing discards. In: Sikorski, Z. E., Pan, B. S., & Shahidi, F., (eds.), *Seafoods Proteins* (pp. 226–240). Chapman and Hall, New York.

Shoulders, M. D., & Raines, R. T., (2009). Collagen structure and stability. *Ann. Rev. Biochem., 78,* 929–958.

Sigal S., Schickler, H., Chet, I., & Spiegel, Y., (1998). Purification and characterization of a *Bacillus cereus* collagenolytic/proteolytic enzyme and its effect on *Melo idogyne javanica* cuticular proteins. *European Journal of Plant Pathology, 104*, 59–67.

Sorsa, T., Ding, Y. L., Ingman, T., Salo, T., Westerlund, U., Haapasalo, M., Tschesche, H., & Konttinen, Y. T., (1995). Cellular source, activation, and inhibition of dental plaque collagenase. *J. Clin. Periodontol., 22*, 709–717.

Suphatharaprateep, W., Cheirsilp, B., & Jongjareonrak, A., (2011). Production and properties of two collagenases from bacteria and their application for collagen extraction. *Nat. Biotechnol., 28*, 649–655.

Takeuchi, H., Shibano, Y., Morihara, K., Fukushima, J., Inami, S., Keil, B., et al., (1992). Structural gene and complete amino acid sequence of *Vibrio alginolyticus* collagenase. *Biochem. J., 281*, 703–708.

Van Wart, H. E., & Steinbrink, D. R., (1985). Complementary substrate specificities of class-I and class-II collagenases from *Clostridium histolyticum. Biochem., 24*, 6520–65206.

Vaslleva, T. E., Nustorova, M., & Gousterova, A., (2007). New protein hydrolyzates from collagen wastes used as peptone for bacterial growth. *Curr. Microbiol., 54*, 54–57.

Watanabe, K., (2004). Collagenolytic proteases from bacteria. *Appl. Microbiol. Biotechnol., 63*, 520–526.

Weinberg, M., & Randin, A., (1932). Proprietes physicochimiques du ferment fibrolytique d'origine microbienne . *Compt. Rend. Soc. Biol., 110*, 352.

Welton, R., & Woods, D., (1973). Halotolerant collagenolytic activity of *Achromobacter iophagus. J. Gen. Microbiol., 75*, 191–196.

Welton, R., & Woods, D., (1975). Collagenase production by *Achromobacter iophagus. Biochem. Biophys. Acta., 384*, 228–234.

Woessner, J. F., (1991). Matrix metalloproteinase and their inhibitors in connective tissue remodeling. *FASEB J., 5*, 2145–2154.

Wu, Q., Li, C., Chen, H., & Shuliang, L., (2010). Purification and characterization of a novel collagenase from *Bacillus pumilus* Col-J. *Appl. Biochem. Biotechnol., 160*, 129–139.

Yoshida, E., & Noda, H., (1965). Isolation and characterization of collagenase I and II from *Clostridium histolyticum. Biochem. Biophys. Acta., 105*, 562–574.

Yoshihara, K., Matsushita, O., Minami, J., & Okabe, A. J., (1994). Cloning and nucleotide sequence analysis of the *colH* gene from *Clostridium histolyticum* encoding a collagenase and a gelatinase. *J. Bacteriol., 176*, 6489–6496.

Zhongkai, Z., Li, G., & Shi, B., (2005). Physiochemical properties of collagen, gelatin and collagen hydrolyzate derived from bovine limed split wastes. *J. Soc. Leather Technol. Chem., 90*, 23–28.

CHAPTER 9

CHARACTERIZATION, PRODUCTION, AND APPLICATION OF MICROBIAL LIPASES FOR FOOD, AGRICULTURE, HEALTH, AND ENVIRONMENT

JEYABALAN SANGEETHA[1], RAVICHANDRA HOSPET[2], DEVARAJAN THANGADURAI[2], ABHISHEK CHANNAYYA MUNDARAGI[2], PURUSHOTHAM PRATHIMA[2], SHIVASHARANA CHANDRABANDA THIMMAPPA[3], and MUNISWAMY DAVID[4]

[1]Department of Environmental Science, Central University of Kerala, Kasaragod 671316, Kerala, India

[2]Department of Botany, Karnatak University, Dharwad, 580003, Karnataka, India

[3]Department of Microbiology and Biotechnology, Karnatak University, Dharwad, 580003, Karnataka, India

[4]Department of Zoology, Karnatak University, Dharwad, 580003, Karnataka, India

9.1 INTRODUCTION

Microbial enzymes considered more useful than the enzymes from animals or plants because of distinct catalytic activity, ease of genetic manipulation, the rapid growth of microbial cells on inexpensive media and the high yield makes microbial enzymes more stable than any other plant and animal enzymes (Wiseman, 1995). Lipases found widely in yeasts, molds, and bacteria (Gaoa et al., 2000). The enzymes from microbial sources are currently receiving key attention because of their catalytic performance and potential applications in industries mainly in detergents, oils, pharma-ceutical, and dairy (Cardenas et al., 2001; Momsia and Momsia, 2013). Generally, bacterial enzymes are more preferred over the fungal enzymes because of their higher catalytic performance and for to withstand at higher pH conditions. To boost the cellular yields and the enzymatic activities or to produce new altered enzymes, environmental, and genetic improvement techniques can be performed more suitably on bacterial cells than any other organisms (Hasan et al., 2006; Rasmey et al., 2017).

Lipases are one of the main biocatalysts with potentially wide appli-cations in bio-industries and have been utilized as a part of *in situ* lipid metabolism and *ex situ* industrial and agricultural applications. Lipases are triacylglycerol acyl hydrolases (EC3.1.1.3) that catalyze the hydrolysis of triacylglycerol to diglycerides, monoglycerides, and unsaturated fatty acids. They frequently express different activities such as cholesterol esterase, phospholipase, amidase, lysophospholipase, cutinase (Svendsen, 2000). For the first time, microbial lipase structure studied was of *Rhizomucor miehei* lipase (RML) from X-ray crystallographic analysis. The study reveals that enzyme had an active site triad as that of serine proteases. X-ray crystallography is now appeared to be a more powerful tool for structural analysis of biological macromolecules.

Lipases are abundant in plants, animals, fungi, and bacteria; where they play a significant role in lipid metabolism. For the past few years, lipases have gained much consideration due to their versatile activities towards various chemical and physical standards. Lipases from various sources have examined for their synthetic and hydrolytic properties. The most important characteristics of lipase are its ability to use all monoglycerides, diglycerides, and triglycerides as well as free fatty acids in transesterifica-tion, high yield in non-aqueous media, low product inhibition, low reac-tion time, resistance to altered pH, temperature, alcohol, and reusability of immobilized enzyme (Kumar et al., 2012). Lipases are essential for the bioconversion of lipids from one organism to other and within the organ-isms, and they exhibit unique feature of acting at an interface between the aqueous and non-aqueous phase; this novel feature distinguishes them from esterase (Pandey et al., 1999; Ray, 2015).

Most enzymes today are produced by the fermentation of bio-based substrates (Louwrier, 1998). Lipids constitute a large part of earth's biomass, and lipolytic enzymes play a vital role in emphasizing these water-insoluble compounds. Lipolytic enzymes are involved in the breakdown of cells and thus helps in the mobilization of lipids within the cells of organisms as well in the transfer of lipids from one organism to other (Beisson et al., 2000; Hasan et al., 2006). Several thousand enzymes exhibiting different substrate specificities are known, but comparatively very few enzymes have been isolated in pure and crystallized form, and less has been known about their structure and function. The advancement in protein engineering and genetic modification makes their application most suitable to bio-industries (Cheetham, 1995; Hasan et al., 2006). The existence of microbial lipases has been noticed as early as in 1901 in *Bacillus pyocyaneus*, *B. prodigiosus* and *B. fluorescens* by Sir Eijkman. Enzymes catalyzing triglycerides have been studied for over 300 years, and the ability of the lipases to catalyze the hydrolysis and synthesis of esters has been observed nearly 70 years ago (Hossain et al., 2010; Momsia and Momsia, 2013). In recent years, there has been a progressive increase in the application of extracellular microbial lipases and these are one of the main focuses of research because of its potential micro-biotechnological usage over the decades (Rasmey et al., 2017).

Microbial lipases got much attention in the modern food industry instead of traditional chemical processes and widely used in the production of various fruit beverages, baked foods, and fermented vegetables. Ethyl, isobutyl, amyl, and isoamyl acetates are widely used flavor esters (Hasan et al., 2006; Verma and Kanwar, 2008). Several kinds of lipases are synthesized from Gram-positive and Gram-negative bacteria, but Gram-negative bacterial enzyme production got much importance and those bacteria belongs to genus *Pseudomonas* where more than seven lipase producing species are categorized such as *Pseudomonas fragi*, *Pseudomonas glumae*, *Pseudomonas aeruginosa*, *Pseudomonas alcaligenes*, *Pseudomonas cepacia*, *Pseudomonas fluorescens* and *Pseudomonas putida* (Jaeger et al., 1994; Kojima et al., 2003). The fungal lipases of industrial importance have spread out in last few years, replacing the time consuming and expensive biochemical processes. This is the main reason for increased production of new alternative strains and enzymes of industrial importance. There are many important fungal groups known for lipase production such as *Mucoromycotina*, *Rhizomucor*, *Rhizopus*, and *Mucor* species (Sharma et al., 2011; Singh and Mukhopadhyay, 2012; Papp et al., 2016; Miklós et al., 2017).

The 3-dimensional structure of lipase helps in understanding, designing, and developing lipases of specific importance. More than 12 microbial lipases from several sources have been categorized, and comprehension information on lipase engineering has been documented (Sandana Mala and Takeuchi, 2008; Verma et al., 2012). Considering high versatility with respect to unique properties such as chemoselectivity, regioselectivity, stereoselectivity, lipases holds 5 to 10% of the enzyme market among hydrolysis. In the near future, it would not be surprising if lipases become high priority enzyme in enzyme application area.

9.2 SOURCES OF LIPASES

In wake of current encroachment in microbiology and biotechnology, lipases have emerged as key enzyme due to its versatile nature and used in various industrial applications. Lipase enzyme is used in various food, leather, cosmetics, detergents, and pharmaceutical industries and industrial waste management (Park et al., 2005; Gupta et al., 2007). Lipases are extensively present in nature, whereas, microbial lipases are commercially utilized. Several challenges have been made so far for the separation of lipase-producing microbes from various habitats suitable to its high industrial substances. However, various additional sources of lipase-like plant or animal are available, but the bacterial lipases are extensively used due to its specificity of the reaction, stereospecificity, and also less energy consumption than any other conservative methods (Gilham and Lehner, 2000; Wang et al., 2009). Microbes including bacteria, yeast, and fungi are reported to generate lipase enzyme from different locality such as manufacturing wastes, vegetable oil dispensation unit, and dairies. In fact, the environment polluted with oil enhances the growth of oil-degrading microorganisms being a standard carbon source. Some significant lipase produced by bacteria includes *Bacillus*, *Pseudomonas*, and *Burkholderia* (Gupta et al., 2007; Wang et al., 2009).

Lipases mainly present in the fore-stomach tissues of calves and pancreatic tissues of hogs and pigs. The difficulty in utilizing animal lipase due to the incidence of trypsin in pancreases of pig where lipases were extracted, which escorts the pungent flavor amino acids, in which remains the animal hormones or viruses as well as its detrimental effects in the handing out of vegans (Vakhlu and Kour, 2006). Plant lipases are also accessible but are not utilized hence the yield is low and the method involved. Therefore, microbial lipases are considered because it is having industrial and cost-effective advantages, in which the organisms are refined in media consisting of suitable nutrient concentration beneath appropriate conditions (Srivastava,

2008). Microbes also produce lipase enzyme, but it differs in accordance with the strains, deliberation of growth medium, farming environment, pH, temperature, and the nature of carbon and nitrogen sources (Gupta et al., 2004; Souissi et al., 2009; Treichel et al., 2010). Usually, bacteria, fungi, and actinomycetes are known to produce extracellular lipases, assist the enzyme improvement as of the culture broth; thus *Candida, Pseudomonas, Mucor,* and *Geotrichum* sp. are considered as the major commercially feasible strains (Ertugrul et al., 2007).

9.2.1 ANIMAL SOURCES

For industrial processes, pancreatic, and pregastric animal lipases were used. Hence the lipase synthesized from pancreatic juice is the first commercialized lipase for industrial process. Claude Bernard (1856) revealed a lipase action in the pancreatic juice which catalyzed impenetrable oil droplets to soluble products. Lipases of pancreatic along with pregastric removed from calf, goat, and animal protein gullets used in hastening of flavor production, cheese maturity and also for lipolysis of butter, plump, and emulsion (El-Hofi et al., 2011). The animal phospholipases are used for the industrial purpose are PLA2s (Phospholipases A2) (De Maria et al., 2007). Thus, PLA2 is formed by porcine pancreas, commercially manufactured for the synthesis of lysolecithin, further, used for food as an emulsifier and antifungal agent. Degumming of edible oil can be done from porcine PLA2 (Casado et al., 2012). In recent times, *Asterina pectinifera* and *Dasyatis pastinaca* are recommended as new resources of PLA2, potentially applicable as biocatalysts; thus it is one of the substitution causes of porcine pancreatic PLA2.

9.2.2 PLANT SOURCES

A lipase oilseed is the predominant form used as a biocatalyst against several organic solvents when the entreaty in the environment is very low (Barros et al., 2010; Mounguengui et al., 2013). Most of the muesli seeds, the lipases from rice, wheat, barley, oat, and maize were hardened to be used as a catalyst, but the majority of them possessed defiant to maximum temperatures and alkali condition (Mohamed et al., 2000; Bhardwaj et al., 2001). Chong et al. (2007) reported that acylglycerols were synthesized using decayed lipase of rice bran which esterifies palm oil FAs through glycerol. Hence, the rubber trees were exploited to extracted lipase from the latex for industrial applications (Paques and Macedo, 2006). Recently, the lipase from *Carica papaya* has emerged as a phenomenal multipurpose biocatalyst and for most of the biotechnological applications was recorded using this enzyme, such as modification of fats and oils, the formation of organic means through esterification and transesterification activity, asymmetric resolution of diverse chiral acids (De María et al., 2006). Phospholipase D (PLD) is one of the important plant phospholipases used as biocatalyst, hence act as suitable to the catalyst for transphosphatidylation activity (Ulbrich-Hofmann et al., 2005). Cabbage is one of the most important traditional resources of PLD. Thus cabbage PLD used to modify polar head-groups and for the construction of synthetic phospholipids for nutraceutical and medicinal purposes (Mukherjee, 2005).

9.2.3   MICROBIAL SOURCES

Yeast and fungi are generally used commercial lipases and phospholipases. Commonly exploited microbes are *Aspegillus, Candida, Penicillium, Rhizomucor, Thermomyces,* and *Yarrowia. Candida antartica* lipase B (CALB) is most repeatedly oppressed enzyme for various biocatalytic procedure, and it has several patents, owing to its 1,3-specificity, high enantio-selectivity beside secondary alcohols and primary amines, exhibits astonishing stability towards organic solvents at excessive temperatures (Lock et al., 2007). *Candida rugosa* lipase is also one of the commercially used lipases the mixture is available in different isoforms, and the production is generally certified and preferred for food industry. For biodiesel production the fungal lipases namely, *Rhizopus oryzae* lipase (ROL), RML, *Thermomyces lanuginosus* lipase (TLL) and *Fusarium heterosporum* lipase (FHL) were artificially synthesized (Hwang et al., 2014), while it is intended for food dispensation and other nutraceutical inventions of lipases from *Aspergillus niger, Penicillium roqueforti, Penicilium cyclopium,* and *Rhizomucor javanicus,* where ROL and RML was also synthesized (Ferreira-Dias et al., 2013). For detergent additives, TLL is commercially used for extraction of lipase (Singh and Mukhopadhyay, 2012). Thus, for degumming of vegetable oils the microbial phospholipases, PLA1s and PLA2s extracted by *Fusarium oxysporum, T. lanuginosus, A. niger,* and *Trichoderma reesei* respectively (De Maria et al., 2007; Casado et al., 2012). Whereas PLDs synthesized from Actinomycetes are widely used in industrial purposes due to its high transphosphatidylation and hydrolytic activities (Casado et al., 2012). Bacteria are predominant genera investigated and approved for the manufacture of lipases as well as phospholipases such as *Pseudomonas, Bacillus,* and *Streptomyces.* Bacterial lipases and phospholipases were commercially processed and used for various purposes, such as fabrication of biodiesel, degumming of oil, nutraceutical synthesis and in formulations of detergents, but they are least applicable for food production. However, biodiesel is manufactured by extracting lipases from *Pseudomonas fluorescens, Burkholderia cepacia* and *Bacillus thermocatenulatus* (Hwang et al., 2014), although lipases from *Pseudomonas mendocina, Pseudomonas alcaligenes, Pseudomonas glumae* and *B. cepacia* used for preparation of detergent additives (Gupta et al., 2004). For degumming of vegetable oil the bacterial phospholipases like PLA2, PLB, and PLC extracted from *Streptomyces violaceroruber, P. fluorescens* and *Bacillus anthracis* strains are used (De Maria et al., 2007). Streptomyces PLDs are useful in food and nutraceutical industries (Casado et al., 2012).

9.3   TYPES OF MICROBIAL LIPASES

When compared to plant and microbial lipases only certain bacterial lipases were studied. The majority of the bacterial lipases are known to be glycoproteins, whereas extracellular bacterial lipases are recognized to be lipoproteins. Bacterial lipases identified till to date are constitutive; otherwise, they are imprecise with respect to its substrate specificity, and few are thermostable in nature. Several species of bacteria such as *Achromobacter, Alcaligenes, Pseudomonas, Staphylococcus,* and *Chromobacterium* were utilized for the mining of lipases. In most of the Staphylococcal, lipoprotein exhibits the lipase property. Whereas, lipases obtained from *S. aureus* and *S. hyicus* possess molecular weight of 34 and 46KDa. They are enhanced by Ca^{++} and repressed by ethylene diamine tetraacetic acid (EDTA). Their maximum pH lies between 7.5 and 9.0. That kind of lipases has two preserved domains separated by 100 amino acids which are most likely to form active sites (Godtfredsen, 1990).

Brockerhoff and Jensen (1974) have studied and published about the uses of fungal lipases in detail. Such lipases are utilized more as it exhibits low cost for production, thermal, and pH stability, substrate specificity and movement in organic solvents. For commercial purpose, several fungal species are exploited such as *Aspergillus niger, Humicola lanuginosa, Mucor miehei, Rhizopus arrhizus, R. delemar, R. japonicus, R. niveus,* and *R. oryzae.*

Vakhlu and Kour (2006) reported globally there are several yeast species documented and they are capable of producing lipases such as *Candida rugosa, Candida parapsilopsis, Candida deformans, Candida curvata, Candida valida, Yarrowia lipolytica, Rhodotorula glutinis, Rhodotorula pilimornae, Pichia burtonii, Saccharomycopsis crataegenesis,* and *Trichosporon asteroids. Candida* sp., *Geotrichum* sp., and *Y. lipolytica* the gene which encodes the lipase was cloned and hyper-expressed (Wang et al., 2007). Lipases extracted from *C. rugosa* and *C. antarctica* were highly used in various fields, and various microorganisms were also used nowadays for production of lipases.

9.4 MASS PRODUCTION OF MICROBIAL LIPASES

Several microorganisms have been exploited for the production of lipases as most commercially useful lipases are of microbial origin (Kumar and Kanwar, 2012; Muthumari et al., 2016). Several studies on isolation and screening of lipase-producing microorganisms have been reported from diverse habitats which include household waste to industrial waste (Kumar and Kanwar, 2012; Muthumari et al., 2016). Further, some studies have also been reported on the isolation of lipase-producing microorganisms from extreme environments such as hot springs and alkali-rich soils. Several classes of lipase-producing microorganisms have been reported until recent past, which includes bacteria, fungi, and actinomycetes (Kumar and Kanwar, 2012; Muthumari et al., 2016). Among bacteria, major producers are species belonging to genus *Bacillus, Pseudomonas, Staphylococcus,* and *Burkholderia* and these bacterial enzymes are extracellular in nature (Gupta et al., 2004). Extensive studies on fungal lipases have been reported from species belonging to *Zygomycetes* followed by *Hyphomycetes* and yeasts (Godtfredson, 1990; Subash et al., 2013).

Currently, lipases of fungal origin are widely exploited for commercial production. Fungi such as *Aspergillus* spp., *Penicillium* spp., and

Thermomyces spp. are the few among many others that are used in commercial production of lipases. These lipases find numerous industrial applications ranging from dairy- to pharmaceutical industry (Hasan et al., 2006). Numerous attempts are made in past and still being made in the isolation of high yielding lipase microorganisms. In this context newer approaches derived by modern molecular techniques has played instrumental role in elucidating the diversity hidden among microbes (Lee et al., 2006; López-López et al., 2014; Almaabadi et al., 2015). The advanced techniques such as metagenomics have revolutionized the current understandings of microbial diversity which is quite rapid and easier way to detect the unculturable microbes from different origins including extreme environments (Rondon et al., 2000; Lorenz and Eck, 2005; Yun and Ryu, 2005; Lämmle et al., 2007; Kennedy et al., 2008; Steele et al., 2009; Kakirde et al., 2010; Sahoo et al., 2017). Literature suggests that microbes isolated from the extreme environment possess unique traits with enhanced features when compared to the enzyme of normal origin. Among some of the unique features exhibited by these enzymes the first and foremost is the temperature followed by pH. It has been shown that at higher temperatures most of the enzymes become unstable, to overcome these issues several enzymes of temperature tolerant extremophiles have been screened from diverse habitats (Hasan et al., 2006).

As most of microbial lipases are extracellular, the yield of lipase produced is dependent upon multiple intrinsic and extrinsic factors, viz., temperature, pH, nitrogen-, carbon-, and lipid sources (Jaeger and Eggert, 2002; Treichel et al., 2010). Numerous studies have been carried out with respect to optimization of important fermentation factors, majority of them are on optimization of media compositions (Sharma et al., 2001). Lipase production has been found maximum in the media comprising higher lipid content, and reports suggest that lipids especially triglycerides, free fatty acids and glycerol act as potential inducers and have profound effect on lipase yield (Gupta et al., 2004).

During the last few decades, several technological upgrades have been made in the fermentation techniques which have revolutionized the industrial production of enzymes (Pandey et al., 2000; Cavalcanti et al., 2005). Fermentation techniques such as solid-state fermentation (SSF) and submerged fermentation (SMF) have been extensively studied for the commercial production of enzymes (Subramaniyam and Vimala, 2012). Brief information on both the techniques is detailed. SMF involves the use of liquid medium for the production of products (Teng and Xu, 2008; Renge et al., 2012). It generally involves the growth of selected microorganisms in a closed container (usually fermentation tanks or bioreactors) under controlled conditions comprising liquid media enriched with nutrients and oxygen. Most preferably bacterial lipases are produced by SMF. SSF is the oldest technique used in fermentation (Pandey, 2003). It is the most followed technique for the production of several valuable products including enzymes. This technique offers multiple benefits when compared to its counterpart, viz., high productivity, stability, and lower catabolic repression (Hölker et al., 2004; Kumar and Kanwar, 2012; Muthumari et al., 2016). SSF provide natural environment for the growth of microorganisms and usually involves a cheap substrate such as agricultural waste which offer greater economic feasibility when compared to SMF (Rivera-Muñoz et al., 1991; Castilho et al., 2000; Viniegra-González et al., 2009; Kumar and Kanwar, 2012; Muthumari et al., 2016). Filamentous fungi are best suited for this kind of fermentation (Colla at al., 2009). Several studies have been carried out in recent past for the production of lipase using this technique (Pandey, 1992; Pandey et al., 2000; Suryanarayan and Mazumdar, 2001; Couto and Sanromán, 2006). Comparative investigations for the production of lipases using both the fermentation techniques have also been demonstrated by many authors (Mateos Diaz et al., 2006; Azeredo et al., 2007; Sun and Xu, 2009).

Production of lipase has been reported from several sources which can be broadly classified into synthetic medium and agriculture residues (Kumar and Kanwar, 2012; Muthumari et al., 2016). Optimization studies with respect to fermentation medium have been reported by several authors. Extensive studies on optimization of fermentation substrates using RSM have been carried out in recent times (Treichel et al., 2010). Gupta et al. (2007) reported optimization of fermentation medium for the enhanced production of lipase using bacteria *Burkholderia multivorans*. Wherein, they applied statistical approach response surface methodology coupled with Plackett-Burman experimental design, with the optimized conditions of glucose, olive oil, NH_4Cl, yeast extract, K_2HPO_4, $MgCl_2$ and $CaCl_2$ they found 12 fold increased lipase. Wang et al. (2008) optimized the fermentation parameters for membrane-bound lipase production from fungi *Rhizopus chinensis*. Their study indicated that lipase activity increased substantially with the optimized peptone, olive oil, maltose, K_2HPO_4, and $MgSO_4 \cdot 7H_2O$. Ruchi et al. (2008) reported optimization studies using *Pseudomonas aeruginosa* for enhanced productivity, and similarly, use of *Aspergillus carneus* has also been reported earlier. Fungus *Antrodia cinnamomea* under SMF resulted in 54 U mL^{-1} of lipase productivity with incubation duration of 17 days following screening and optimization of different substrate sources, viz., glycerol (0.5%), sodium nitrate (0.5%), and thiamine (0.1%). He and Tan (2006) reported 9,600 U mL^{-1} of lipase activity using immobilized *Candida* sp. Immobilization technique is largely accepted technique in the industrial production of several chemical compounds, as it minimizes loss of inoculum/enzymes during down-stream process and greatly reduces the cost incurred further increasing the stability and productivity (Johri et al., 1990; Frusaki et al., 1992; Datta et al., 2013). During the past three decades, there has been a notable innovation in immobilization technologies, viz., adsorption, and encapsulation, and greater emphasis is given to overcome the technological errors such as stability and productivity (Elibol and Ozer, 2000; Yang et al., 2005). In this context, numerous enzymes have been entitled to wide array of chemical production through immobilization technique including lipases. A reliable with increased efficiency, stability, and cheaper supporting material is a research of interest. In search of novel support materials, several attempts have been made. Supporting materials such as calcium alginate, κ-carrageenan, and polyacrylamide gel are few among the many supporting materials proposed. Immobilized cells of *Aspergillus niger* (Ellaiah et al., 2004), *Candida rugosa* (Ferrer and Carles, 1992; Benjamin and Pandey, 1997), *Sporotrichum thermophile* (Johri et al., 1990) and *Ralstonia pickettii* (Hemachander et al., 2001) have been used in the production of lipases (Muthumari et al., 2016). Pencreach et al. (2007) demonstrated that immobilization of *Pseudomonas cepacia* on microporous polypropylene support (commercial grade) resulted in enhanced enzyme stability and temperature resistance with a Half-lives at 80°C of 11 and 4 min for the immobilized and free enzymes, respectively. The immobilized whole cell is another alternative technique which has received a considerable recognition in recent times (Kennedy et al., 1990; Ramkrishna and Prakasham, 1999). Nanotechnology coupled with immobilization techniques is a future thrust area which offers potential advantages over conventional techniques (Datta et al., 2013; Li et al., 2017). A recent statistical report indicates that more than 1800 patents have been filed towards the commercial application of lipases in various streams and more than 2200 scientific reports have been published globally (Daiha et al., 2015).

9.5 INDUSTRIAL APPLICATIONS OF MICROBIAL LIPASES

In recent years, lipases attracted much interest in enzyme technology. This is because of physicochemical modification in substrates of lipases and its unique activity towards various biological conditions. The main concept is that microbial lipases often exhibit versatile substrate particularity and hence can be utilized for the transformation of various unnatural substrates (Ray, 2013). Lipases are triacylglycerol (EC 3.1.1.3), which hydrolyze ester linkages of glyceride at water-oil interface. At the time of hydrolysis, lipases catalyze acyl groups from glycerides constituting lipase-acyl complex and then transferring its acyl group to OH groups of water. In non-aqueous conditions, these hydrolytic enzymes transfer acyl group of carboxylic acid to nucleophile other than water (Martinelle and Hult, 1995). Hence, lipases can be sugars, acylate alcohols, thiols, and amine producing several of stereospecific esters and amides (Singh et al., 2003; Zouaoui et al., 2012). Generally, lipases synthesized by animals, plants, and microorganism. Microbial lipases got importance for industrial application due to their versatile features such as potential stability, selectivity, and broad specificity of substrate (Dutra et al., 2009; Griebeler et al., 2011). The various microorganisms are known for producers of extra-cellular lipases such as fungi, bacteria, and yeast (Abada, 2008). Fungal strains are mainly isolated in solid-state fermentation conditions, while yeast and bacteria are isolated in SMF (Kumar and Ray, 2014). Potential of lipases to make specific biotransformation has made them popular in food, organic synthesis, detergents, and pharmaceutical industries (Chi et al., 2005; Grbavcic, 2007; Gupta et al., 2007; Veerapagu et al., 2013). The importance of exiting lipases has gradually enhanced since 1980s. This actually results in the huge success achieved by cloning techniques, enzyme expression from microbe and increasing demand for such catalysts with potential and key properties make lipases more important in enzyme technology (Bornscheuer, 2002; Menoncin et al., 2010).

Advancement in lipase-based technologies for production of bio-compounds is readily expanding application of these enzymes (Azim et al., 2001; Andualema and Gessesse, 2012). Lipases plays potential major role in the processing of astaxanthin a food colorant; methyl ketones, flavor compounds of characteristic of blue cheese; 4-hydroxydecanoic acid utilized as precursor of γ-decalactone, a fruit flavor; dicarboxylic acids for prepolymers; inter-esterification of cheaper glycerides to more important forms such as cocoa butter substitutes for application in choco-late production (Undurraga et al., 2001). The value of chirality with respect to biological activity has got enhanced demand for sustainable methods for biosynthesis of pure enantiomers comprising chiral anti-inflammatory drugs like ibuprofen and naproxen (Kim et al., 1995; Ducret, 1998; Xie, 1998; Arroyo, 1999; Chen and Tsai, 2000) antihypertensive compounds such as angiotensin-permuting enzyme inhibitors (e.g., captopril, enala-pril, zofenopril, lisinopril, and ceranopril) and calcium channel obstructing drugs such as metoprolol and diltiazem. Lipases for detergents manufacture needs to be thermotolerant and remain volatile in alkaline condition. The industrial potential of microbial lipase is still prohibited to their utilization as detergents to clear oils and fats stains (Jaeger and Reetz, 1998; Hasan et al., 2010). Recently, their potential usage has been widening to catalysis of trans-esterification of plants seed oils for biodiesel production (Bajaj, 2010; Ghaly et al., 2010).

Emerging fields for potential use of microbial lipases also include resolution of racemic elements to synthesize optically active compounds and plastics biodegradation (Kashmiri et al., 2006). Such versatility of microbial lipases owes to their broad applicability for wide range of spectrum substrate and organic solvents stability (Snellman and Colwell, 2004; Kiran et al., 2008). Microbial lipases are one of leading biocatalysts with improved ability for contributing to multibillion dollar. During the synthetic application of lipases, as soluble lipases lose their potentiality in non-aqueous media, it is necessary to perform immobilization (Bruno, 2004; Kumar et al., 2016). Such synthetic characteristics make widespread usage in several areas including biochemical and organic conversions (Agu et al., 2013; Sharma and Kanwar, 2014). Advancement in technology of 3D structures of microbial lipases performs a specific role in designing lipases for particular purposes (Table 9.1).

9.5.1 MICROBIAL LIPASES IN FOOD, PAPER AND PULP INDUSTRY

Oils and fats are the main constituents of foods. Nutritional and sensory importance and physical factors of triglyceride are influenced by facts like position of fatty acid compounds in glycerol backbone, length of fatty acid and their degree of instaurations. Lipases have the ability to

TABLE 9.1 Industrial Uses of Microbial Lipases

Industry	Importance	Applications	References
Detergents	Hydrolysis of fats	Removal of oil stains from fabrics	Jaeger and Reetz, 1998; Ito et al., 2001
Dairy foods	Hydrolysis of milk fat, cheese ripening, modification of butterfat	Improving flavor agents in milk, cheese, and butter	Kaffarnik et al., 2014
Bakery foods	Improvement of flavor compounds	Shelf-life prolongation	Verma et al., 2012
Beverages	Improved flavor and aroma	Alcoholic and non-alcoholic beverages	Andualema and Gessesse, 2012
Food dressings	Quality enhancement	Mayonnaise, whippings, and dressings	Macedo et al., 2003
Meat and fish	Flavor improvement	Meat and fish, fat removal	Seitz, 1974; Kazlauskas and Bornscheur, 1998
Fats and oils	Transesterification	Cocoa butter, fatty acids, margarine, glycerol	Nakajima et al., 2000; Undurraga et al., 2001
Chemicals	Enantioselectivity, synthesis	Chiral blocks, chemicals	Vulfson, 1994; Patil et al., 2011
Pharmaceuticals	Hydrolysis	Specialty lipids and digestive aids	Vulfson, 1994
Cosmetics	Synthesis enhancement	Emulsifiers, moisturizers	Nishat and Rathod, 2015
Leather and Paper	Hydrolysis	Leather materials and paper with improved quality	Fukuda et al., 1990; Kobayashi et al., 2015

change the characteristics of lipids by relocating fatty acids chain in glycerides and altering the position of one or more of fatty acid with new one. In this way, relatively expensive and less potential lipid can be altered to higher grade fat (Undurraga et al., 2001; Rajendran et al., 2009). A high-value fat, cocoa butter contains stearic and palmitic acids and has melting point of nearly 370°C. Lipase-based methods involving hydrolysis and production is well utilized commercially to improve few of undesirable fats to cocoa butter substitutes (Undurraga et al., 2001; Pinyaphong and Phutrakul, 2009). Lipases used to get PUFAs from plant and animal lipids like tuna oil, menhaden, and borage oil. PUFAs and their mono and diglycerides are used to produce several types of pharmaceuticals including anti-inflammatory, thrombolytic, and anticholesterolemic (Gill and Valivety, 1997; Belarbi, 2000). Microbial lipases have been utilized for improving factors responsible for flavors in bakery products, cheese ripening, and beverages and utilized to aid removal of fats from fish and meat products (Muthumari, 2016). The hydrophobic elements of wood known as pitch causes severe drawbacks in paper manufacture (Hasan et al., 2010). Lipases are utilized successfully to remove it from pulp produced during paper manufacture. Japanese Nippon Paper Industries have come out with a pitch control technique that uses fungal lipase *Candida rugosa* to hydrolyze about 90% of wood triglycerides (Sharma et al., 2001).

9.5.2 MICROBIAL LIPASES APPLICATION IN ORGANIC SYNTHESIS

Application of Microbial lipases in chemical synthesis is becoming more interesting area of research. Lipases are subjected to catalyze variety of stereo-, chemo-, regio, and selective transformations (Saxena et al., 1999; Yadav et al., 2000; Muthumari, 2016). Many of lipases utilized as biocatalysts in organic chemical synthesis are of microbial origin. Such enzymes perform at hydrophilic-lipophilic interface and withstand organic solvents in reaction mixtures. Applications of lipase in production of enantiopure chiral molecules have been discussed by Yadav et al. (2000). At water-liquid interface amount of water will determine the results of lipase-catalyzed reactions. Lack of adequate water results esterification and transesterification (Rantakyla et al., 1996). Hydrolysis is favored phenomena when excess of water is present during enzyme catalytic process (Klibanov, 1997). Lipase-catalyzed reaction in supercritical solvents has been analyzed (Rantakyla et al., 1996; King et al., 2001).

Hydrolysis of ester is usually performed by lipase in the two-phase aqueous substrate (Vaysse et al., 1997; Xu et al., 2002). Pencreac and Baratti (1996) observed hydrolysis of p-nitro phenyl palmitate (pNPP) in n-heptane by a lipase preparation of *P. cepacia*. Hasan et al. (2010) utilized lipase trapped in hydrophobic sol-gel matrix for several kinds of transformations. Mutagenesis has been employed to enhance enantioselectivity of lipases (Bornscheuer, 2000; Liebeton et al., 2000). The lipase-acyl transferase from *C. parapsilosis* has tended to catalyze fatty hydroxamic acids synthesis in bi-phasic aqueous medium (Therisod and Klibanov, 1987; Yeo, 1998).

9.5.3 APPLICATION IN RESOLUTION OF RACEMIC ACIDS AND ALCOHOLS

Stereoselectivity of microbial lipases was utilized to resolve several racemic organic mixtures in immiscible bi-phasic systems (Klibanov, 1990; Tsai and Dordick, 1996; Kiran and Divakar, 2001). Racemic alcohols could resolve into enantiomerically pure form by the lipase catalyzed transesterification. Bhushan et al. (2011) reported that esterification reactions in non-aqueous media by lipase-B from *C. antarctica* was stereoselective towards R-isomer of ketoprofen in achiral solvents like isobutyl methyl ketone and (S+)-carvone. In a study, purified lipase from *C. rugosa* was observed to its crude equivalent in hydrated and anhydrous hydrophobic organic solvents. Profens (2-aryl propinoic acids), important group of non-steroidal anti-inflammatory drug is pharmacologically active in (S)-enantiomer form (Landoni and Soraci, 2001). (S)-Ibuprofen ((S)–2(4-isobutylphenyl) propionic acid) is 160 times more vigorous than its antipode in preventing prostaglandin synthesis. Subsequently, notable results were being made to improve optically refined profens through asymmetric chemical and catalytic kinetic resolutions (Lee et al., 1995; Van Dyck et al., 2001; Xin et al., 2001). Microbial enzymes are helpful in improving racemic mixtures. Miyazawa et al. (1998) and Weber et al. (1999) observed solvent-free thio-esterification fatty acids with long-chain thiols by lipases from *R. miehei* and *C. antarctica*.

9.5.4   MICROBIAL LIPASES IN REGIOSELECTIVE ACYLATIONS AND ESTER SYNTHESIS

Chen et al. (1995) and Kodera et al. (1998) analyzed lipase from *A. niger* to catalyze regioselective deacylation of preacylated methyl-β-D-glucopyranoside. From et al. (1997) observed esterification of lactic acid and alcohols by a lipase of *C. antarctica* in hexane. Esterification of five positional isomers of acetylenic fatty acids with n-butanol was analyzed by Lie et al. (1998), using eight different lipases. Janssen et al. (1999) studied on esterification of fatty acids, and sulcatol in toluene, catalyzed by *C. rugosa* lipase and Krishnakant and Madamwar (2001) studied using immobilized lipase on silica-based organogels for synthesis of esters.

9.5.5   MICROBIAL LIPASES IN OLEOCHEMICAL AND DETERGENT INDUSTRY

Lipases application in oleochemical process lessens energy and thermal degradation during hydrolysis, alcoholysis, acidolysis, and glycerolysis (Vulfson, 1994; Bornscheuer, 2000). Even though lipases formulated by the nature of hydrolytic separation of ester bonds of triacylglycerols, lipases can perform the synthesis of ester in low water conditions. Esterification and hydrolysis can form randomly in activity known as inter-esterification. Based on type of substrates, lipases can catalyze acidolysis, alcoholysis, and transesterification (Balcao et al., 1996). Due to their capability to hydrolyze fats, lipase plays major role as additives in laundry and detergent industry. Detergent lipases are mainly characterized by some key requirements such as low substrate specificity and high catalytic properties. Lipases with novel activity are found through combination of protein engineering and continuous screening (Cardenas et al., 2001; Muthumari, 2016).

9.5.6   MICROBIAL LIPASES AS BIOSENSORS AND THEIR PHARMACEUTICAL APPLICATION

Microbial lipase application as biosensors is a promising new emerging field of research. The biosensors are the analytical tool that uses bio-compounds as sensing system materials and transverse bio response into electrical signal. A self-contained integrated biosensor device that has capacity to provide required quantitative analytical results by biological recognition particles those are in direct linkage with transduction particles (Thévenot et al., 2001). The polymer-enzyme group poly succinate was integrated into sensor, which in terms degraded by lipases. The potential areas of such usage of sensor system include detection of enzyme concentrations and the formation of disposable enzyme-linked immune sensors, which employ polymer-degrading enzymes as enzyme labels (Sumner, 2001). Lipases immobilized onto pH/oxygen electrodes in correlation with glucose oxidase and this act as lipid biosensors, and it can utilize in triglycerides and blood cholesterol determinations also (Arya, 2006; Ray, 2012a,b). The fundamental concept of utilizing lipase as biosensors is to produce glycerol from the triacylglycerols in analytical sample and to quantify synthesized glycerol by chemical or enzymatic method. Linko and Wu (1996) improved a protocol for characterization of organo phosphorous pesticides of surface acoustic wave impedance sensors by the lipase hydrolysis. Lipases are utilized to improve PUFAs (polyunsaturated fatty acids) from animals and plants lipids, and their mono and diacylglycerides compounds are utilized to produce several kinds of pharmaceuticals (Dong, 1999). Due to their metabolic benefits, increasingly PUFAs are utilized as food substitutes, nutraceuticals, and pharmaceuticals. PUFAs are important for production of prostaglandins and lipid membranes. The notable effort is being made to get optically pure compounds, having pharmacologically more sustainable than its antipodes. Profens, a non-steroidal anti-inflammatory drug is active in (s)-enantiomer form. Microbial lipases also used in the biosynthesis of artificial sweetener sucralose by hydrolysis of octaacetylsucrose (Linko and Wu, 1996; Huang et al., 2001; Hasan et al., 2006; Demontis et al., 2007).

9.6 CONCLUSION AND FUTURE PERSPECTIVES

Microbial lipases considered as versatile biocatalysts due to its wide application in bio-industries. Most of the microbial populations such as yeast, bacteria, and molds have ability to synthesize hydrolytic lipase. The potential utilization of microbial lipases in food technology exhibits the need of development of cost-effective methodologies for quality production, scaling up and purification. However diverse substrate usage, enantioselectivity, and stereoselective biotransformation prove its novel properties. Immobilized enzymes influence mainly on processes of catalytic transformations by microbial lipases. Gene cloning, mutagenesis, and biochemical characterization appear to be one of the emerging fields of research for developing or improving enzyme properties to the desired objectives. Simultaneously, advanced techniques are being developed in bioreactor technologies by effectively utilizing microbial lipases. Hence, the rapid advancement of lipase technology will improve the novel application in various fields of science in future.

KEYWORDS

- acidolysis
- alcoholysis
- *Aspergillus* spp.
- biosensors
- *Candida antarctica*
- chemoselectivity
- esterification
- microbial lipases
- *Penicillium* spp.

- phospholipases
- polyunsaturated fatty acids
- regioselectivity
- *Rhizomucor miehei*
- *Rhizopus oryzae*
- Stereoselectivity
- *Thermomyces lanuginosus*
- triacylglycerol acyl hydrolase
- x-ray crystallography

REFERENCES

Abada, E. A. E., (2008). Production and characterization of a mesophilic lipase isolated from *Bacillus stearothermophilus* AB–1. *Pak. J. Biol. Sci., 11*, 1100–1106.

Agu, K. C., Ogbue, M. O., Abuchi, H. U., Onunkwo, A. U., Chidi-Onuorah, L. C., & Awah, N. S., (2013). Lipase production by fungal isolates from palm oil-contaminated soil in Awka Anambra State, Nigeria. *Int. J. Agri. Biosci., 2*(6) 386–390.

Almaabadi, A. D., Gojobori, T., & Mineta, K., (2015). Marine metagenome as a resource for novel enzymes. *Genomics Proteomics Bioinformatics, 13*, 290–295.

Andualema, B., & Gessesse, A., (2012). Microbial lipases and their industrial applications: Review. *Biotechnol., 11*, 100–118.

Arroyo, M., Sanchez-Montero, J. M., & Sinisterra, J. V., (1999). Thermal stabilization of immobilized lipase B from *Candida antarctica* on different supports: Effect of water activity on enzymatic activity in organic media. *Enzyme Microb. Technol., 24*, 3–12.

Arya, S. K., Chaubey, A., & Malhotra, B. D., (2006). Fundamentals and applications of biosensors. *Proceedings of the Indian National Science Academy, 72*(4), 249.

Azeredo, L. A., Gomes, P. M., Santanna, Jr, G. L., Castilho, L. R., & Freire, D. M., (2007). Production and regulation of lipase activity from *Penicillium restrictum* in submerged and solid-state fermentations. *Curr. Microbiol., 54,* 361–365.

Azim, A., Sharma, S. K., Olsen, C. E., & Parmer, V. S., (2001). Lipase-catalyzed synthesis of optically enriched α-haloamides. *Bioorg. Med. Chem., 9,* 1345–1348.

Bajaj, A., Lohan, P., Jha, P. N., & Mehrotra, R., (2010). Biodiesel production through lipase-catalyzed transesterification: An overview. *J. Mol. Catal. B: Enzym., 62*(1), 9–14.

Balcao, V. M., Paiva, A. L., & Malcata, F. X., (1996). Bioreactors with immobilized lipases: State of the art. *Enzyme Microbiol. Technol., 18,* 392–416.

Barros, M., Fleuri, L. F., & Macedo, G., (2010). Seed lipases: Sources, applications, and properties - a review. *Braz. J. Chem. Eng., 27,* 15–29.

Beisson, F., Arondel, V., & Verger, R., (2000). Assaying *Arabidopsis* lipase activity. *Biochem. Soc. Trans., 28,* 773–775.

Belarbi, E. H., Molina, E., & Chisti, Y., (2000). A process for high yield and scalable recovery of high purity eicosapentaenoic acid esters from microalgae and fish oil. *Enzyme Microbiol. Technol., 26,* 516–529.

Benjamin, S., & Pandey, A., (1997). Enhancement of lipase production during repeated batch culture using immobilized *Candida rugosa. Process Biochem., 32,* 437–440.

Bhardwaj, K., Raju, A., & Rajasekharan, R., (2001). Identification, purification, and characterization of a thermally stable lipase from rice bran, a new member of the (phospho) lipase family. *Plant Physiol., 127,* 1728–1738.

Bhushan, I., Kumar, A., Modi, G., & Jamwal, S., (2011). Chiral resolution of differently substituted racemic acetyl–1-phenyl ethanol using lipase from *Bacillus subtilis. J. Chem. Technol. Biotechnol., 86*(2), 315–318.

Bornscheuer, U. T., (2000). *Enzymes in Lipid Modification* (pp. 1–394). Wiley-VCH Verlag, Weinheim, Germany.

Bornscheuer, U. T., Bessler, C., Srinivas, R., & Krishna, S. H., (2002). Optimizing lipases and related enzymes for efficient application. *Trends Biotechnol., 20,* 433–437.

Brockerhoff, H., & Jensen, R. G., (1974). *Lipolytic Enzymes* (pp. 1–327). Academic Press, New York.

Bruno, M. L., Filho, J. L. L., Melo, E. H. M., & Castro, F. H., (2004). Ester synthesis catalyzed by *Mucor miehei* lipase immobilized on magnetic polysiloxane-polyvinyl alcohol particles. *Appl. Biochem. Biotechnol., 113,* 189–199.

Cardenas, F., Alvarez, E., De Castro-Alvarez, M. S., Sanchez-Montero, J. M., Valmaseda, M., Elson, S., & Sinisterra, J. V., (2001). Screening and catalytic activity in organic synthesis of novel fungal and yeast lipases. *J. Mol. Catal. B: Enzym., 14,* 111–123.

Casado, V., Martín, D., Torres, C., & Reglero, G., (2012). Phospholipases in food industry: a review. In: Sandoval, G., (ed.), *Lipases and Phospholipases: Methods and Protocols* (pp. 495–523). Springer, New York, USA.

Castilho, L. R., Polato, C. M., Baruque, E. A., Santanna, Jr, G. L., & Freire, D. M., (2000). Economic analysis of lipase production by *Penicillium restrictum* in solid-state and submerged fermentations. *Biochem. Eng. J., 4*(3), 239–247.

Cavalcanti, E. A. C., Gutarra, M. L. E., Freire, D. M. G., Castilho, L. R., & Júnior, G. L. S., (2005). Lipase production by solid-state fermentation in fixed-bed bioreactors. *Braz. Arch. Biol. Technol., 48,* 79–84.

Cheetham, P. S., (1995). Principles of industrial biocatalysis and bioprocessing. In: Wiseman, A., (ed.), *Handbook of Enzyme Biotechnology* (pp. 83–234). Ellis Horwood Pvt. Ltd., UK.

Chen, H. P., Hsiao, K. F., Wu, S. H., & Wang, K. T., (1995). Regioselectivity enhancement by partial purification of lipase from *Aspergillus niger*. *Biotechnol. Lett., 17*, 305–308.

Chen, J. C., & Tsai, S. W., (2000). Enantioselective synthesis of (S)-ibuprofen ester prod rug in cyclohexane by *Candida rugosa* lipase immobilized on accurel MP1000. *Biotechnol. Prog., 16*, 986–992.

Chi, Y., Jeong, S., Lee, K., & Park, H., (2005). Effects of methanol on the catalytic properties of porcine pancreatic lipase. *J. Microbiol. Biotechnol., 15*(2), 296–301.

Chong, F. C., Tey, B. T., Dom, Z. M., Cheong, K. H., Satiawihardja, B., Ibrahim, M. N., et al., (2007). Rice bran lipase-catalyzed esterification of palm oil fatty acid distillate and glycerol in organic solvent. *Biotechnol. Bioprocess Eng., 12*, 250–256.

Colla, L. M., Rezzadori, K., Câmara, S. K., Debon, J., Tibolla, M., Bertolin, T. E., & Costa, J. A. V., (2009). A solid-state bioprocess for selecting lipase-producing filamentous fungi. *Z. Naturforsch., 64*, 131–137.

Couto, S. R., & Sanromán, M. A., (2006). Application of solid-state fermentation to food industry - a review. *J. Food Eng., 76*(3), 291–302.

Daiha, K. G., Angeli, R., Oliveira, S. D., & Almeida, R. V., (2015). Are lipases still important biocatalysts? A study of scientific publications and patents for technological forecasting. *PLoS One, 10*(6), e0131624, doi: 10.1371/journal.pone.0131624.

Datta, S., Christena, L. R., & Rajaram, Y. R. S., (2013). Enzyme immobilization: An overview on techniques and support materials. *3 Biotech., 3*(1), 1–9.

De Maria, L., Vind, J., Oxenbøll, K. M., Svendsen, A., & Patkar, S., (2007). Phospholipases and their industrial applications. *Appl. Microbiol. Biotechnol., 74*, 290–300.

De María, P. D., Sinisterra, J. V., Tsai, S. W., & Alcántara, A. R., (2006). *Carica papaya* lipase (CPL): An emerging and versatile biocatalyst. *Biotechnol. Adv., 24*, 493–499.

Demontis, V., Monduzzi, M., Mula, G., Setzu, S., Salis, S., & Salis, A., (2007). Porous silicon-based potentiometric biosensor for triglycerides. *Phys. Status Solidi., 204*(5), 1434–1438.

Dong, H., Gao, S., Han, S. P., & Cao, S. G., (1999). Purification and characterization of a *Pseudomonas* sp. lipase and its properties in non-aqueous media. *Biotechnol. Appl. Biochem., 30*, 251–256.

Ducret, A., Lortie, R., & Trani, M., (1998). Lipase-catalyzed enantioselective esterification of ibuprofen in organic solvent under controlled water activity. *Enzyme Microb. Technol., 22*, 212–216.

Dutra, J. C. V., Terzi, S. C., Bevilaqua, J. V., Damaso, M. C. T., Couri, S., & Langone, M. A. P., (2009). Lipase production in solid-state fermentation monitoring biomass growth of *Aspergillus niger* using digital image processing. *Appl. Biochem. Biotechnol., 147*, 63–75.

El-Hofi, M., El-Tanboly, E. S., & Abd-Rabou, N. S., (2011). Industrial application of lipases in cheese making: A review. *Int. J. Food Saf., 13*, 293–302.

Elibol, M., & Ozer, D., (2000). Lipase production by immobilized *Rhizopus arrhizus*. *Process Biochem., 36*, 219–233.

Ellaiah, P., Prabhakar, T., Ramakrishna, B., Taleb, A. T., & Adinarayana, K., (2004). Production of lipase by immobilized cells of *Aspergillus niger*. *Process Biochem., 39*, 525–528.

Ertugrul, S., Donmez, G., & Takac, S., (2007). Isolation of lipase-producing *Bacillus* sp. from olive mill wastewater and improving its enzyme activity. *J. Hazard Mat., 149,* 720–724.

Ferreira-Dias, S., Sandoval, G., Plou, F., & Valero, F., (2013). The potential use of lipases in the production of fatty acid derivatives for the food and nutraceutical industries. *Elect. J. Biotechnol., 16,* 1–34.

Ferrer, P., & Carles, S., (1992). Lipase production by immobilized *Candida rugosa* cells. *Appl. Microbiol. Biotechnol., 37,* 737–741.

From, M., Adlercreutz, P., & Mattiasson, B., (1997). Lipase-catalyzed esterification of lactic acid. *Biotechnol. Lett., 19,* 315–317.

Frusaki, S., & Seki, M., (1992). Use and engineering aspects of immobilized cells in biotechnology. *Adv. Biochem. Eng. Biotechnol., 23,* 161–185.

Fukuda, S., Hayashi, S., Ochiai, H., Liizumi, T., & Nakamura, K., (1990). Improvers for deinking of wastepaper. *Japanese Patent, 2,* 229–290.

Gaoa, X. G., Cao, S. G., & Zhang, K. C., (2000). Production, properties and application to nonaqueous enzymatic catalysis on lipase from a newly isolated *Pseudomonas* strain. *Enzyme Microb. Technol., 27,* 74–82.

Ghaly, A. E., Dave, D., Brooks, M. S., & Budge, S., (2010). Production of biodiesel by enzymatic transesterification: Review. *Am. J. Biochem. Biotechnol., 6*(2), 54–76.

Gilham, D., & Lehner, R., (2000). Techniques to measure lipase and esterase activity *in vitro. Methods, 36,* 139–147.

Gill, I., & Valivety, R., (1997). Biotransformation and biotechnological applications. *Trends Biotechnol., 15,* 470–478.

Godtfredsen, S. E., (1990). Microbial lipases. In: Fogarty, W. M., & Kelly, E. T., (eds.), *Microbial Enzymes and Biotechnology* (pp. 255–274). Elsevier Applied Sciences, The Netherlands.

Grbavcic, S. Z., Dimitrijevic-Brankovic, S. I., Bezbradica, D. I., & Siler-Marinkovic, S. S., (2007). Effect of fermentation conditions on by *Candida utilis. J. Serb. Chem. Soc., 72,* 757–765.

Griebeler, N., Polloni, A. E., Remonatto, D., Arbter, F., Vardanega, R., & Cechet, J. L., (2011). Isolation and screening of lipase-producing fungi with hydrolytic activity. *Food Bioprocess Technol., 4,* 578–586.

Gupta, N., Shai, V., & Gupta, R., (2007). Alkaline lipase from a novel strain *Burkholderia multivorans*: Statistical medium optimization and production in a bioreactor. *Process Biochem., 42*(2), 518–526.

Gupta, R., Gupta, N., & Rathi, P., (2004). Bacterial lipases: An overview of production, purification, and biochemical properties. *Appl. Microbiol. Biotechnol., 64,* 763–781.

Hasan, F., Shah A. A., Javed, S., & Hameed, A., (2010). Enzymes used in detergents: Lipases. *Afr. J. Biotechnol., 9*(31), 4836–4844.

Hasan, F., Shah, A. A., & Hameed, A., (2006). Industrial applications of microbial lipases. *Enzyme Microb. Technol., 39*(2), 235–251.

He, Y. Q., & Tan, T. W., (2006). Use of response surface methodology to optimize culture medium for lipase for production of lipase with *Candida* sp. 99–125. *J. Mol. Catal. B Enzym., 43,* 9–14.

Hemachander, C., Bose, N., & Puvanakrishnan, R., (2001). Whole cell immobilization of *Ralstonia pickettii* for lipase production. *Process Biochem., 36,* 629–633.

Hölker, U., Höfer, M., & Lenz, J., (2004). Biotechnology advantages of laboratory-scale solid-state fermentation with fungi. *Appl. Microbiol. Biotechnol., 64*, 175–186.

Hossain, M. Z., Shrestha, D. S., & Kleve, M. G., (2010). Biosensors for biodiesel quality sensing. *J. Ark. Acad. Sci., 64*, 80–85.

Huang, X. R., Li, Y. Z., Liu, L. L., Yang, G. L., Qu, Y. B., & Zhang, W. J., (2001). A novel method for fabrication of a glass-electrode-based lipase sensor. *Chinese Chem. Lett., 12*(5), 453–456.

Hwang, H. T., Qi, F., Yuan, C., Zhao, X., Ramkrishna, D., Liu, D., & Varma, A., (2014). Lipase-catalyzed process for biodiesel production: protein engineering and lipase production. *Biotechnol. Bioeng., 111*, 639–653.

Ito, T., Kikuta, H., Nagamori, E., Honda, H., Ogino, H., Ishikawa, H., & Kobayashi, T., (2001). Lipase production in two-step fed-batch culture of organic solvent-tolerant *Pseudomonas aeruginosa* LST–03. *J. Biosci. Bioeng., 91*, 245–250.

Jaeger, K. E, Ransac, S., Dijkstra, B. W, Colson, C., Van Heuvel, M., & Misset, O., (1994). Bacterial lipases. *FEMS Microbiol. Rev., 15*, 29–63.

Jaeger, K. E., & Eggert, T., (2002). Lipases for biotechnology. *Curr. Opin. Biotechnol., 13*(4), 390–397.

Jaeger, K. E., & Reetz, M. T., (1998). Microbial lipases from versatile tools for biotechnology. *Trends Biotechnol., 16*, 396–403.

Janssen, E. M., Sjurenes, J. B., Vakurov, A. V., & Halling, P. J., (1999). Kinetics of lipase-catalyzed esterification in organic media: Correct model and solvent effects on parameters. *Enzyme Microbiol. Technol., 24*, 463–470.

Johri, B. N., Alurralde, J. D., & Klein, J., (1990). Lipase production by free and immobilized protoplasts *Sporotrichum thermophile* Apinis. *Appl. Microbiol. Biotechnol., 33*, 367–371.

Kaffarnik, S., Kayademir, Y., Heid, C., & Vetter, W., (2014). Concentrations of volatile 4-alkyl branched fatty acids in sheep and goat milk and dairy products. *J. Food Sci., 79*(11), 2209–2214.

Kakirde, K. S., Parsley, L. C., & Liles, M. R., (2010). Size does matter: Application-driven approaches for soil metagenomics. *Soil Biol. Biochem., 42*, 1911–1923.

Kashmiri, M. A., Adnan, A., & Butt, B. W., (2006). Production, purification and partial characterization of lipase from *Trichoderma viride*. *Afr. J. Biotechnol., 5*(10), 878–882.

Kazlauskas, R. J., & Bornscheur, U. T., (1998). Biotransformations with lipases. In: Rehm, H. J., Pihler, G., Stadler, A., & Kelly, P. J. W., (eds.), *Biotechnology* (pp. 37–192). Wiley-VCH, New York.

Kennedy, J. F., Melo, E. H. M., & Jumel, K., (1990). Immobilized enzymes end cells. *Chem. Eng. Porg., 86*(7), 81–89.

Kennedy, J., Marchesi, J. R., & Dobson, A. D. W., (2008). Marine metagenomics: Strategies for the discovery of novel enzymes with biotechnological applications from marine environments. *Microb. Cell. Fact, 21*, 7–27.

Kim, K. J., Kim, M. G., Lee, S. B., & Lee, W. M., (1995). Enzymatic resolution of racemic ibuprofen esters: Effects of organic solvents and temperature. *J. Ferment. Bioeng., 6*, 613–615.

King, J. W., Snyder, J. M., Frykman, H., & Neese, A., (2001). Sterol ester production using lipase-catalyzed reactions in supercritical carbon dioxide. *Food Res. Technol., 212*, 566–569.

Kiran, G. S., Shanmughapriya, S., Jayalakshmi J., Selvin, J., Gandhimathi, R., Sivaramakrishnan, S., et al., (2008). Optimization of extracellular psychrophilic alkaline lipase produced by marine *Pseudomonas* sp. (MSI057). *Bioprocess Biosyst. Eng., 31*(5), 483–492.

Kiran, K. R., & Divakar, S., (2001). Lipase catalyzed synthesis of organic acid esters of lactic acid in non-aqueous media. *J. Biotechnol., 87*, 109–121.

Klibanov, A. M., (1990). Asymmetric transformations catalyzed by enzymes in organic solvents. *Acta. Chem. Res., 23*, 114–120.

Klibanov, A. M., (1997). Why are enzymes less active in organic solvents than in water? *Trends Biotechnol., 15*, 97–101.

Kobayashi, S., (2015). Enzymatic ring-opening polymerization and polycondensation for the green synthesis of polyesters. *Polym. Adv. Tech., 26*(7), 677–686.

Kodera, Y., Sakurai, K., & Satoh, Y., (1998). Regioselective deacetylation of peracetylated monosaccharide derivatives by polyethylene glycol modified lipase for oligosaccharide synthesis. *Biotechnol. Lett., 20*, 177–180.

Kojima, Y., Kobayashi, M., & Shimizu, S. A., (2003). Novel lipase from *Pseudomonas fluorescens* HU380: Gene cloning, overproduction, denaturation-activation, two-step purification, and characterization. *J. Biosci. Bioeng., 96*(3), 242–249.

Krishnakant, S., & Madamwar, D., (2001). Synthesis by lipase immobilized on silica and microemulsion based organogels (MBGs). *Process Biochem., 36*, 607–611.

Kumar, A., & Kanwar, S. S., (2012). Lipase production in solid-state fermentation (SSF): Recent developments and biotechnological applications. *Dyn. Biochem. Process Biotechnol. Mol. Biol., 6*(S1), 13–27.

Kumar, A., Dhar, K., Kanwar S. S., & Arora, P. K., (2016). Lipase catalysis in organic solvents: Advantages and applications. *Biological Procedures Online, 18*(2), doi: 10.1186/s12575–016–0033–2.

Kumar, A., Sharma, P., & Kanwar, S. S., (2012). Lipase catalyzed esters syntheses in organic media: A review. *Int. J. Insti. Pharma. Life Sci., 2*(2), 91–119.

Kumar, D. S., & Ray, S., (2014). Fungal lipase production by solid state fermentation – an overview. *J. Anal. Bioanal. Tech., 5*, 230. doi: 10.4172/2155-9872.1000230.

Lämmle, K., Zipper, H., Breuer, M., Hauer, B., Buta, C., Brunner, H., & Rupp, S., (2007). Identification of novel enzymes with different hydrolytic activities by metagenome expression cloning. *J. Biotechnol., 127*, 575–592.

Landoni, M. F., & Soraci, A., (2001). Pharmacology of chiral compounds 2-arylpropionic acid derivatives. *Curr. Drug Metabol., 2*(1), 37–51.

Lee, M. H., Lee, C. H., Oh, T. H., Song, J. K., & Yoon, J. H., (2006). Isolation and characterization of a novel lipase from a metagenomic library of tidal flat sediments: Evidence for a new family of bacterial lipases. *Appl. Environ. Microbiol., 72*(11), 7406–7409.

Lee, W. M., Kim, K. J., Kim, M. G., & Lee, S. B., (1995). Enzymatic resolution of racemic ibuprofen esters: Effects of organic cosolvents and temperature. *J. Ferment. Bioeng., 6*, 613–615.

Li, C., Jiang, S., Zhao, X., & Liang, H., (2017). Co-immobilization of enzymes and magnetic nanoparticles by metal-nucleotide hydrogel nanofibers for improving stability and recycling. *Molecules, 22*, 179, doi: 10.3390/molecules22010179.

Lie, K., Jie, M. S., & Xun, F., (1998). Studies of lipase catalyzed esterification reactions of some acetylenic fatty acids. *Lipids, 33*, 71–75.

Liebeton, K., Zonta, A., Schimossek, K., Nardini, M., Lang, D., Dijkstra, B. W., Reetz, M. T., & Jaeger, K. E., (2000). Directed evolution of an enantioselective lipase. *Chem. Biol.,* *7*(9) 709–718.

Linko, Y. Y., & Wu, X. Y., (1996). Biocatalytic production of useful esters by two forms of lipase from *Candida rugosa. J. Chem. Technol. Biotechnol., 65,* 163–170.

Lock, L. L., Corbellini, V. A., & Valente, P., (2007). Lipases produced by yeasts: Powerful biocatalysts for industrial purposes. *Tecno-Logica, 11,* 18–25.

López-López, O., Cerdán, M. E., & González, S. M. I., (2014). New extremophilic lipases and esterases from metagenomics. *Curr. Protein Pept. Sci., 15*(5), 445–455.

Lorenz, P., & Eck, J., (2005). Metagenomics and industrial applications. *Nat. Rev. Microbiol., 3,* 510–516.

Louwrier, A., (1998). Industrial products: The return to carbohydrate-based industries. *Biotechnol. Appl. Biochem., 27,* 1–8.

Macedo, G. A., Lozano, M. M. S., & Pastore, G. M., (2003). Enzymatic synthesis of short chain citronellyl esters by a new lipase from *Rhizopus* sp. *J. Biotechnol., 6,* 72–75.

Martinelle, M., & Hult, K., (1995). Kinetics of acyl transfer reactions in organic media catalyzed by *Candida antarctica* lipase B. *Biochim. Biophys. Acta, 1251*(2), 191–197.

Mateos Diaz, J. C., Rodriguez, J. A., Roussos, S., Cordovac, J., Abousalhamd, A., Carriered, F., & Barattia, J., (2006). Lipase from thermotolerant fungus *Rhizopus homothallicus* is more thermostable when produced using solid state fermentation than liquid fermentation procedures. *Enzyme Microb. Technol., 39,* 1042–1050.

Menoncin, S., Domingues, N. M., Freire, D. M. G., Toniazzo, G., Cansian, R. L., & Oliveira, J. V., (2010). Study of the extraction, concentration, and partial characterization of lipases obtained from *Penicillium verrucosum* using solid-state fermentation of soybean bran. *Food Bioprocess Technol., 3,* 537–544.

Miklós, T., Alexandra, K., Tamás, P., Shine, K., Naiyf, S. A., & Csaba, V., (2017). Purification and properties of extracellular lipases with transesterification activity and 1,3-regioselectivity from *Rhizomucor miehei* and *Rhizopus oryzae. J. Microbiol. Biotechnol., 27*(2), 277–288.

Miyazawa, T., Yukawa, T., Chezi, S., Yanagihara, R., & Yamada, T., (1998). Resolution of 2-phenoxy–1-propanols by *Pseudomonas* species lipase catalyzed highly enentioselective transesterification: Influence of reaction conditions on the enantioselectivity toward primary alcohol. *Biotechnol. Lett., 20,* 235–238.

Mohamed, M., Mohamed, T. M., Mohamed, S. A., & Fahmy, A. S., (2000). Distribution of lipases in the gramineae, partial purification and characterization of esterase from *Avena fatua. Bioresour. Technol., 73,* 227–234.

Momsia, T., & Momsia, P., (2013). A review on microbial lipase-versatile tool for industrial applications. *Int. J. Life Sci. Pharm. Res., 2*(4), 1–3.

Mounguengui, R. W. M., Brunschwig, C., Baréa, B., Villeneuve, P., & Blin, J., (2013). Are plant lipases a promising alternative to catalyze transesterification for biodiesel production? *Prog. Energy Combust. Sci., 39,* 441–456.

Mukherjee, K. D., (2005). Plant lipases as biocatalysts. In: Kuo, T. M., & Gardner, H. W., (eds.), *Lipid Biotechnology* (pp. 399–415). Marcel Dekker Inc., New York, USA.

Muthumari, G. M., Thilagavathi, S., & Hariram, N., (2016). Industrial enzymes: Lipase producing microbes from waste volatile substances. *Int. J. Pharm. Sci. Res., 7*(5), 2201–2208.

Nakajima, M., Snape, J., & Khare, S. K., (2000). Method in non-aqueous enzymology. In: Gupta, M. N., (ed.), *Biochemistry* (pp. 52–69). Birkhauser Verlag, Basel, Switzerland.

Nishat, R. T., & Rathod, V. K., (2015). Enzyme catalyzed synthesis of cosmetic esters and its intensification: A review. *Process Biochem., 50*(11), 1793–1806.

Pandey, A., (1992). Recent developments in solid state fermentation. *Process Biochem., 27*, 109–117.

Pandey, A., (2003). Solid-state fermentation. *Biochem. Eng. J., 13*, 81–84.

Pandey, A., Benjamin, S., Soccol C. R., Nigam, P., Krieger, N., & Soccol, V. T., (1999). The realm of microbial lipases in biotechnology. *Biotechnol. Appl. Biochem., 29*, 119–131.

Pandey, A., Soccol, C. R., & Mitchell, D., (2000). New developments in solid state fermentation: I - bioprocesses and products. *Process Biochem., 35*, 1153–1169.

Papp, T., Nyilasi, I., Csernetics, Á., Nagy, G., Takó, M., & Vágvölgyi, C., (2016). Improvement of industrially relevant biological activities in Mucoromycotina fungi. In: Schmoll, M., & Dattenböck, C., (eds.), *Gene Expression Systems in Fungi: Advancements and Applications* (pp. 97–118). Springer International Publishers, Switzerland.

Paques, F. W., & Macedo, G. A., (2006). Plant lipases from latex: Properties and industrial applications. *Quim. Nova., 29*, 93–99.

Park, H., Lee, K., Chi, Y., & Jeong, S., (2005). Effects of methanol on the catalytic properties of porcine pancreatic lipase. *J. Microbiol. Biotech., 15*(2), 296–301.

Patil, K. J., Chopda, M. Z., & Mahajan, R. T., (2011). Lipase biodiversity. *Indian J. Sci. Tech., 4*(8), 971–982.

Pencreac, H. G., & Baratti, J. C., (1996). Hydrolysis of *p*-nitrophenyl palmitate in *n*-heptane by *Pseudomonas cepacia* lipase: A simple test for the determination of lipase activity in organic media. *Enzyme Microbiol. Technol., 18*, 417–422.

Pinyaphong, P., & Phutrakul, S., (2009). Synthesis of cocoa butter equivalent from palm oil by *Carica papaya* lipase-catalyzed interesterification. *Chiang Mai. J. Sci., 36*(3), 359–368.

Rajendran, A., Palanisamy, A., & Thangavelu, V., (2009). Lipase catalyzed ester synthesis for food processing industries. *Braz. Arch. Biol. Technol., 52*, 207–219.

Ramkrishna, S. V., & Prakasham, R. S., (1999). Microbial fermentation with immobilized cells. *Curr. Sci., 77*, 87–100.

Rantakyla, M., Alkio, M., & Paltonen, O., (1996). Stereospecific hydrolysis of 3-(4-methoxyphenyl) glycosidic ester in supercritical carbon dioxide by immobilized lipase. *Biotechnol. Lett., 18*, 1089–1094.

Rasmey, A. M., Aboseidah, A. A., Gaber, S., & Mahran, F., (2017). Characterization and optimization of lipase activity produced by *Pseudomonas monteilli* 2403-KY120354 isolated from ground beef. *Afr. J. Biotechnol., 16*(2), 96–105.

Ray, A., (2012a). Application of lipase in industry. *Asian J. Pharm. Technol., 2*(2), 33–37.

Ray, S., (2012b). Fermentative production of an alkaline extracellular lipase using an isolated bacterial strain of *Serratia* sp (C4). *J. Microbiol. Biotechnol. Res., 2*(4), 545–557.

Ray, S., (2013). Extracellular microbial lipase production, purification, recovery, and immobilization: A review. *J. Bioprocess Technol., 98*, 240–259.

Ray, S., (2015). Applications of extracellular microbial lipase: a review. *Int. J. Res. Biotechnol. Biochem., 5*(1), 6–12.

Renge, V. C., Khedkar, S. V., & Nandurkar, N. R., (2012). Enzyme synthesis by fermentation method: A review. *Sci. Revs. Chem. Commun., 2*(4), 585–590.

Rivera-Muñoz, G., Tinoco-Valência, J. R., Sánchez, S., & Farrés, A., (1991). Production of microbial lipases in a solid state fermentation system. *Biotechnol. Lett., 13*, 277–280.

Rondon, M. R., August, P. R., Bettermann, A. D., Brady, S. F., Grossman, T. H., Liles, M. R., et al., (2000). Cloning the soil metagenome: A strategy for accessing the genetic and functional diversity of uncultured microorganisms. *Appl. Environ. Microbiol., 66*, 2541–2547.

Ruchi, G., Anshu, G., & Khare, S. K., (2008). Lipase from solvent tolerant *Pseudomonas aeruginosa* strain: Production optimization by response surface methodology and application. *Biores. Technol., 99*, 4796–4802.

Sahoo, R. K., Kumar, M., Sukla, L. B., & Subudhi, E., (2017). Bioprospecting hot spring metagenome: Lipase for the production of biodiesel. *Environ. Sci. Pollut. Res., 24*, 3802–3809.

Sandana, M. J. G., & Takeuchi, S., (2008). Understanding structural features of microbial lipases: an overview. *Anal. Chem. Insights, 3*, 9–19.

Saxena, R. K., Ghosh P. K., Gupta, R., Davidson, W. S., Bradoo, S., & Gulati, R., (1999). Microbial lipases: Potential biocatalysts for the future industry. *Curr. Sci., 77*(1), 101–115.

Seitz, E. W., (1974). Industrial applications of microbial lipases: A review. *J. Am. Oil Chem. Soc., 51*, 12–16.

Sharma, D., Sharma, B., & Shukla, A. K., (2011). Biotechnological approach of microbial lipase: A review. *Biotechnol., 10*, 23–40.

Sharma, R., Chisti, Y., & Banerjee, U. C., (2001). Production, purification, characterization, and applications of lipases. *Biotechnol. Adv., 19*, 627–662.

Sharma, S., & Kanwar, S. S., (2014). Organic solvent tolerant lipases and applications. *Scientific World J., 625258*, http: //dx.doi.org/10.1155/2014/625258.

Singh, A. K., & Mukhopadhyay, M., (2012). Overview of fungal lipase: A review. *Appl. Biochem. Biotechnol., 166*, 486–520.

Singh, S. K., Felse, A. P., Nunez, A., Foglia, T. A., & Gross, R. A., (2003). Regioselective enzyme-catalyzed synthesis of phospholipid esters, amides, and multifunctional monomers. *J. Organ. Chem., 68*(14), 5466–5477.

Snellman, E. A., & Colwell, R. R., (2004). *Acinetobacter* lipase: Molecular biology, biochemical properties and biotechnological potential. *J. Ind. Microbiol. Biotechnol., 31*(9), 391–400.

Souissi, N., Bougatef, A., Triki-Ellouz, Y., & Nasri, M., (2009). Production of lipase and biomass by *Staphylococcus simulans* grown on sardinella (*Sardinella aurita*) hydrolysates and peptone. *Afr. J. Biotechnol., 8*(3), 451–457.

Srivastava, M. L., (2008). *Fermentation Technology* (pp. 1–404). Alpha Science, Oxford, UK.

Steele, H. L., Jaeger, K. E., Daniel, R., & Streit, W. R., (2009). Advances in recovery of novel biocatalysts from metagenomes. *J. Mol. Microbiol. Biotechnol., 16*, 25–37.

Subash, C. B. G., Periasamy, A., Thangavel, L., & Azariah, H., (2013). Strategies to characterize fungal lipases for applications in medicine and dairy industry. *Bio. Med. Research International*, doi: 10.1155/2013/154549.

Subramaniyam, R., & Vimala, R., (2012). Solid state and submerged fermentation for the production of bioactive substances: A comparative study. *Int. J. Sci. Nat., 3*, 480–486.

Sumner, C., Krause, S., Sabot, A., & McNeil, C. J., (2001). Biosensor based on enzyme catalyzed degradation of thin polymer films. *Biosens. Bioelectron., 16*, 709–714.

Sun, S. Y., & Xu, Y., (2009). Membrane-bound synthetic lipase specifically cultured under solid-state fermentation and submerged fermentation by *Rhizopus chinensis*: A comparative investigation. *Biores. Technol., 100*(3), 1336–1342.

Suryanarayan, S., & Mazumdar, K., (2001). *Solid State Fermentation.* US Patent 6197573.

Svendsen, A., (2000). Lipase protein engineering. *Biochem. Biophys. Acta., 1543*, 223–238.

Teng, Y., & Xu, Y., (2008). Culture condition improvement for whole cell lipase production in submerged fermentation by *Rhizopus chinensis* using statistical method. *Biores. Technol., 99*, 3900–3907.

Therisod, M., & Klibanov, A. M., (1987). Regioselective acylation of secondary hydroxyl groups in sugars catalyzed by lipases in organic solvents. *J. Am. Chem. Soc., 109*, 3977–3981.

Thévenot, D. R., Toth, K., Durst, R. A., & Wilson, G. S., (2001). Electrochemical biosensors: Recommended definitions and classification. *Biosens. Bioelectron., 16*(1-2), 121–131.

Treichel, H., De Oliveira, D., Mazutti, M. A., Luccio, M. D., & Oliveira, J. V., (2010). A review on microbial lipases production. *Food Bioprocess Technol., 3*, 182–196.

Tsai, S. W., & Dordick, J. S., (1996). Extraordinary enantio specificity of lipase catalysis in organic media induced by purification and catalyst engineering. *Biotechnol. Bioeng., 52*, 296–300.

Ulbrich-Hofmann, R., Lerchner, A., Oblozinsky, M., & Bezakova, L., (2005). Phospholipase D and its application in biocatalysis. *Biotechnol. Lett., 27*, 535–543.

Undurraga, D., Markovits, A., & Erazo, S., (2001). Cocoa butter equivalent through enzymic interesterification of palm oil midfraction. *Process Biochem., 36*, 933–939.

Vakhlu, J., & Kour, A., (2006). Yeast lipases: Enzyme purification, biochemical properties, and gene cloning. *Elect. J. Biotechnol., 9*(1), 69–81.

Van Dyck, S. M. O., Lemiere, G. L. F., Jonckers, T. H. M., Dommisse, R., Pieters, L., & Buss, V., (2001). Kinetic resolution of a dihydrobenzofuran-type neolignan by lipase-catalyzed acetylation. *Tetrahedron Asymmetry, 12*, 785–789.

Vaysse, L., Dubreucq, E., Pirat, J. L., & Galzy, P., (1997). Fatty hydroxamic acid biosynthesis in aqueous medium in the presence of the lipase-acyl transferase from *Candida parasilosis. J. Biotechnol., 53*, 41–46.

Veerapagu, M., Sankara, N. A., Ponmurugan, K., & Jeya, K. R., (2013). Screening selection identification production and optimization of bacterial lipase from oil spilled soil. *Asian J. Pharma. Clin. Res., 6*, 62–67.

Verma, M. L., & Kanwar, S. S., (2008). Properties and application of poly (methacrylic acid-cododecyl methacrylate-cl-N,N-methylene bisacrylamide) hydrogel immobilized *Bacillus cereus* MTCC 8372 lipase for the synthesis of geranyl acetate. *J. Appl. Polymer. Sci., 110*, 837–846.

Verma, N., Thakur, S., & Bhatt, A. K., (2012). Microbial lipases: Industrial applications and properties, a review. *Int. Res. J. Biol. Sci., 1*(8), 88–92.

Viniegra-González, G., Favela-Torres, E., Aguilar, C. N., Rómero-Gomez, S. J., Díaz-Godínez, G., & Augur, C., (2009). Advantages of fungal enzyme production in solid state over liquid fermentation systems. *Biochem. Eng. J., 13*(2), 157–167.

Vulfson, E. N., (1994). Industrial applications of lipases. In: Woolley, P., & Peterson, S. B., (eds.), *Lipases - Their Structure, Biochemistry, and Application* (pp. 271–288). Cambridge University Press, UK.

Wang, D., Xu, Y., & Shan, T., (2008). Effects of oils and oil-related substrates on the synthetic activity of membrane-bound lipase from *Rhizopus chinensis* and optimization of the lipase fermentation media. *Biochem. Eng. J., 41*, 30–37.

Wang, L., Chi, Z. M., Wang, X. H., Liu, Z. Q., & Li, J., (2007). Diversity of lipase-producing yeasts from marine environments and oil hydrolysis by their crude enzymes. *Annal. Microbiol., 4*, 2–7.

Wang, S. L., Lin, Y. T., Liang, T. W., & Chio, S. H., (2009). Purification and characterization of extracellular lipases from *Pseudomonas monteilii* TKU009 by the use of soybeans as the substrate. *J. Ind. Microbiol. Biotechnol., 36*, 65–73.

Weber, N., Klein, E., & Mukerjee, K. D., (1999). Long chain acyl thioesters prepared by solvent free thioesterification and transesterification catalyzed by microbial lipases. *Appl. Microbiol. Biotechnol., 51*, 401–404.

Wiseman, A., (1995). Introduction to principles. In: Wiseman, A., (ed.), *Handbook of Enzyme Biotechnology* (pp. 3–8). Ellis Horwood Ltd., Padstow, Cornwall, UK.

Xie, Y. C., Liu, H. Z., & Chen, J. Y., (1998). *Candida rugosa* lipase catalyzed esterification of racemic ibuprofen and chemical hydrolysis of S-ester formed. *Biotechnol. Lett., 20*, 455–458.

Xin, J. Y., Li, S. B., Xu, Y., Chui, J. R., & Xia, C. G., (2001). Dynamic enzymatic resolution of naproxen methyl ester in a membrane bioreactor. *J. Chem. Technol. Biotechnol., 76*, 579–585.

Xu, Y., Wang, D., Mu, X. Q., Zhao, G. A., & Zhang, K. C., (2002). Biosynthesis of ethyl esters of short-chain fatty acids using whole-cell lipase from *Rhizopus chinesis* CCTCC M201021 in non-aqueous phase. *J. Mol. Catal. B: Enzym., 18*(1), 29–37.

Yadav, R. P., Agarwal, P., & Upadhyay, S. N., (2000). Microbial lipases: Tool for drug discovery. *J. Sci. Ind. Res., 59*, 977–987.

Yang, X., Wang, B., Cui, F., & Tan, T., (2005). Production of lipase by repeated batch fermentation with immobilized *Rhizopus arrhizus*. *Process Biochem., 40*(6) 2095–2103.

Yeo, S. H., Nihira, T., & Yamada, Y., (1998). Screening and identification of a novel lipase from *Burkholderia* sp. YY62 which hydrolyzes t-butyl esters effectively. *J. Gen. Appl. Microbiol., 44*, 147–152.

Yun, J., & Ryu, S., (2005). Screening for novel enzymes from metagenome and SIGEX, as a way to improve it. *Microb. Cell Fact, 4*, 8–10.

Zouaoui, B., & Abbouni, B., (2012). Isolation, purification and properties of lipase from *Pseudomonas aeruginosa*. *Afr. J. Biotechnol., 11*, 12415–12421.

CHAPTER 10

MICROORGANISMS AS POTENTIAL BIOSORBENTS OF HEAVY METALS

ANTONY ALEX KENNEDY AJILDA[1], NAGARAJAN PADMINI[1], NATESAN SIVAKUMAR[2], and GOPAL SELVAKUMAR[1]

[1]Department of Microbiology, Science Campus, Alagappa University, Karaikudi, Tamil Nadu–630003, India

[2]School of Biotechnology, Madurai Kamaraj University, Madurai, Tamil Nadu–625021, India

10.1 INTRODUCTION

The term heavy metals have been used in many literatures without any clear-cut definition in IUPAC classification system. Heavy implicit high density and metal point out the pure element or an alloy of metallic elements (Duffus, 2002). Heavy metals contain specific gravity that is at least five times greater than the specific gravity of water; metals have high atomic weight, an atomic number greater than 20 (Hale and Margham, 1988) and also lethal to nature, environment, and all living organisms collectively called heavy metals. There are 23 metals belongs to this group which includes antimony (Sb), arsenic (As), bismuth (Bi), cadmium (Cd), cerium (Ce), chromium (Cr), cobalt (Co), copper (Cu), gallium (Ga), gold (Au), iron (Fe), lead (Pb), manganese (Mn), mercury (Hg), nickel (Ni), platinum (Pt), silver (Ag), tellurium (Te), thallium (Tl), tin (Sn), uranium (U), vanadium (V) and zinc (Zn) (Glanze, 1996). They play a crucial role in function of living organisms and accomplish their physiological part. Humans obtain their allocation of trace elements from water and food, an indispensable connection in the food chain being plant existence, which also supports animal life. The superfluity of this elements cause acute or chronic toxicity (Volesky et al., 1983).

Source of heavy metals comprises acid mine drainage metals with uranium containing sulfide minerals and low-level radioactive waste material. In leather tanning process, common tanning agents are alum, glutaraldehyde, heavy oils, formaldehyde, and chromium. Ferrous metal industries involved in primary iron and steel production and coal-fired power generation containing low levels of uranium, thorium, and other naturally occurring radioactive isotopes and trace amounts of mercury exist in coal and other fossil fuels (Volesky and Naja, 2007). Accumulation of these heavy metals in the soil and marine ecosystem leads to health and ecological concerns (Fendorf et al., 2004; Zaki and Hammam, 2014). For instance, Su et al. (2014) reviewed the heavy metal content in urban and agricultural soil worldwide. Presence of Cr, Cu, Pb, Zn, Ni, and Cd content in urban soils are 66.08, 49.60, 1733.94, 289.78, 29.14, and 1.52 mg/kg, respectively. Besides Cr, Cu, Pb, Zn, Ni, Cd, Hg, and As occurrence in agricultural soils are 46.69, 38.03, 51.19, 117.35, 26.12, 1.50, 0.28, and 21.19 mg/kg, respectively. According to Wei and Yang (2009), around 65% of cities in China are polluted by heavy metals especially in urban soils, and urban road dusts through anthropogenic activities. Similarly, in Nigeria, Fe, Zn, Cu, and Cr concentration were average compared to that of standards by World Health Organization (WHO) and European Union (EU). At the same time, concentration of Pb, Mn, Cd, and Ni were extremely higher than the standard (Mafuyai et al., 2014). Contaminated leaf samples from tree in Syria have high risk of Cd content more than plants susceptibility by main environmental pollution by industry (Mansour, 2014; Tchounwou et al., 2014). The levels of heavy metals in the eatable vegetables (parsley plant, tomato, cabbage, and crop plants) through pesticide and chemical fertilizers have been reviewed by Al-Jaboobi et al. (2014). On the other hand, in India, changes in the metal contamination levels in estuarine sediments have been assessed by Chakraborty et al. (2014). Also, Cochin and Ulhas estuarine sediments were contains high level of metal loads compared to the East Coast estuarine sediments.

The elimination of heavy metals has done by conventional methods like chemical precipitation, lime coagulation, ion exchange, reverse osmosis, electro-winning, cementation, and solvent extraction (Rich and Cherry, 1987; Ahalya et al., 2003; Ahluwalia and Goyal, 2007). This conventional methods are nonselective and ineffective in the removal of low concentration of metals. For alternative of conventional methods, live or dead cells of bacteria, fungi, algae, and seaweeds are used for heavy metal removal (Volesky and Holant, 1995; Tchounwou et al., 2014). Biosorption is the method used to eliminate the heavy metal in the environment by dead biomass (Volesky, 1990). Hence, this chapter discusses the effect of heavy metal in human and aquatic animals, and also deals with the biosorption mechanism, techniques, and recent trends. In addition, the metal binding mechanism of bacteria, fungi, algae, yeast, and exopolysaccharides were discussed.

10.2 IMPACT OF HEAVY METALS

Water sources can be contaminated due to industrial and domestic wastes. Acid rain can make this process worse by heavy metals trapped into soils, groundwater, lakes, streams, and rivers. Plants are exposed to heavy metals through the uptake of water. Cultivation of crops for human or animal consumption on contaminated soil can potentially leads to uptake and accumulation of trace metals in edible plant parts with a possible risk to human and animal health (Bryan and Langston, 1992; Bubb and Lestar, 1994). Irrigation water may transport dissolved metals to agricultural field; as a result, heavy metals accumulation in soils. These metals can bind to the cellular components, such as structural proteins, enzymes, and nucleic acids, which interfere with their functioning of plants, animals, and humans (Rajeshwari and Sailaja, 2014).

10.2.1 IMPACT OF HEAVY METALS IN THE ENVIRONMENT

Rainwater dissolves rocks and ores and physically transports material to streams and rivers. Depositing and stripping materials from adjacent soil, eventually transporting these substances to the ocean to be precipitated as sediment or taken up in rainwater (Goyer, 2001). These metals are potent to affect crop production due to the risk of bioaccumulation and biomagnifications (Rajeswari and Sailaja, 2014). Cu, Zn, Fe, Mn, Mo, and B are essential metals to the normal growth of plants, whereas Cu, Zn, Fe, Mn, Mo, Co, and Se are essential to the growth and health of animals and human beings. Cu, Zn, Pb, and Cd are the most environmentally concerning elements that have been often reported to cause contamination of soil, water, and food chains (He et al., 2005). Heavy metals at the low concentration, plays a major role as nutrient for plants, animals, and human health. Excess amount of these metals leads to impact on life forms (Phipps, 1981; Crounse et al., 1983).

Environmental contamination of these metals naturally occurring through anthropogenic effects like industrial production and use, mining, and smelting operations, agricultural, and domestic use of metals and metal-containing compounds (Shallari et al., 1998; He et al., 2005). Soil erosion, metal corrosion, atmospheric deposition of metal ions, leaching of heavy metals, sediment re-suspension and metal evaporation from water resources to soil and groundwater were one of the major sources of environmental contamination (Nriagu, 1989). Natural calamities like volcanic eruptions and weathering have also been contributed to the environmental contamination (Shallari et al., 1998; Bradl, 2002). Industrial sources include metal processing in refineries, petroleum combustion, coal burning in power plants, nuclear power stations and high tension lines, wood preservation, textiles, microelectronics, plastics, and paper processing plants (Pacyna, 1996). Apart from these sources, in China e-waste recycling is the major cause of heavy metal contamination from processing unit. Accumulation of these metals in the surrounding air, dust, soils, sediments, and plants leads to negative impact on the environment (Song and Li, 2014). Metals can affect various food chains in the total environment which includes mammals, birds, microorganisms, nematodes, and plants. In Table 10.1 listed the heavy metals impact on terrestrial organisms, microorganisms, mammals, and birds.

10.2.2   IMPACT OF HEAVY METALS IN THE AQUATIC ENVIRONMENT

Water is one of the fundamental sources for all life forms like plants and animals (Vanloon and Duffy, 2005). We can get water from two different natural resources, one is surface water which includes freshwater lakes, rivers, streams, and another one is groundwater which includes water and well water (Mendie, 2005; Tchounwou et al., 2014). Water is capable to absorb, dissolve, and suspend many different compounds due to the presence of polarity and hydrogen bonds. Hence, naturally, water can able to attain the contaminants from humans and animals as well as other biological activities (Momodu and Anyakora, 2010).

TABLE 10.1 Impact of Heavy Metals in Environment

Metal	Impact on mammals and birds	Impact on microorganisms	Impact on terrestrial organisms	References
Lead	Affect the blood system, central nervous system, the kidney, reproductive, and immune systems	Not very toxic	Visible toxic effects on photosynthesis and growth, impaired reproduction by nematode and caterpillars	Tchounwou et al., 2014
Mercury	Neurological impairment, reproductive effects, liver damage and significant decreases in intestinal absorption in animals, reduction in food intake and consequent poor growth, effects on enzyme systems, cardiovascular function, blood parameters, immune response, kidney function and structure, and behavior	Mercury affects the soil microbiological activity and also plays an important role in nutrient imbalance leads to influence the soil organisms and trees	Plants are insensitive to mercury toxic effects	Boening, 2000; Sharma et al., 2014; Saha et al., 2017
Cadmium	Acute and chronic effects in mammals, cause kidney damage and lung emphysema in mammals and kidney damage in birds	Effect the growth and replication in microorganisms; some species are eliminated after cadmium exposure	Stomatal opening, transpiration, and photosynthesis have been affected by cadmium exposure; invertebrates are insensitive to cadmium	Khan et al., 2014
Arsenic	In chicken, causes depression, ataxia, lameness, and stunted growth, body weight loss, less feed consumption, and neurological disorders, in mammals, cause increased frequency of defecation, excessive salivation and keratosis	Arsenic is toxic to almost all bacteria, by inhibiting basic cellular functions, which are linked with energy metabolism	Eliminate the shrub and small plants	Tchounwou et al., 2014
Chromium	Severe progressive proteinuria in mammals	Inhibition of growth and various metabolic processes such as photosynthesis or protein synthesis in microorganisms	In plants, accumulation of chromium cause chlorosis	Ghosh, 2004

Owing to industrialization, raising the population and the number of factories has been increased. Sewage, urbanization, agriculture, and industrial disposal has significantly increased the discharge of heavy metals into aquatic ecosystem like estuaries, rivers, streams, ocean, and lakes (Ezemonye and Kadiri, 2000; Alkarkhi et al., 2009). Metals were dissolved into water and are easily absorbed by fish and other aquatic organisms. Smaller concentrations can be toxic because metals undergo bio-concentration, which means that their concentration in an organism is higher than in water. Metal toxicity produces adverse biological effects on an organism's survival, activity, growth, and metabolism or reproduction. Metals can be lethal or harm to the organism without assassination. If metals were accumulated in an aquatic organism, it may pass through the upper class of the carnivores (humans) via food chain. Probably humans were obtained heavy metals through food chain by consumption of fish. This process leads to cause bioaccumulation and biomagnifications in the environment and also cause lethal effect to organisms (Ayandiran et al., 2009; Jiwan and Ajay, 2011; Govind and Madhuri, 2014).

Metals can also affect the locomotion of fish in many ways: (1) fish may attracted or away from the contaminated area; (2) alter sensory perception and reduce responses to normal olfactory cues associated with such activities as feeding, mate selection or homing (all having a component of locomotor behavior); (3) cause alterations in free locomotor activity, manifested as hypoactivity or hyperactivity; (4) alter locomotor components such as turning frequency or angular orientation; and (5) reduce swimming performance or the endurance of a fish swimming against water current (Atchison et al., 1987). The hematological effects of fish were surveyed by Vosyliene (1999). The exposure and time of heavy metals can change hemostatic system of fish. A short-term exposure causes osmotic dis-balance and increased volume of erythrocytes. Under the osmotic stress secreted epinephrine causes the contraction of spleen and erythrocytes. Their higher count correspondingly increases the share of hematocrit. A long-term exposure causes decrease in erythrocyte count. A decrease in the concentration of hemoglobin (Soundararajan and Veeraiyan, 2014), by the effect of toxic metals on gills, furthermore decrease in oxygen also indicates anemia or confirms negative changes.

Vinodhini and Narayanan et al. (2009) was studied the exposure of Cd, Ni, Pb, Cr in common crop (*Cyprinus carpio* L.). The results showed the impairment in respiratory function, alteration of hemoglobin function, elevated level of cholesterol, glucose iron and copper, because prolonged metallic stress in fish makes adaptation difficult and creates weakness in fish. Cellular disintegration and cytopathological changes in the cornea of fish lead to impairment of the visual system (Baatrup, 1991; Tchounwou et al., 2014).

During embryo formation, heavy metals reduced the swelling of egg leads to reduced space for developing embryo. Abnormal cleavage and blastula formation causes malformation and death of embryo. At the organogenesis, the metabolic and development rate was reduced. At the time of hatching, it will slow down the gland development and alter the hatching rate causes death of unhatched and newly hatched larvae (Jezierska et al., 2009). Biomonitoring is a technique to access the changes in the environment due to anthropogenic activities. It also used for the assessment of the pollution levels of water bodies. Normally fish and algae used for the heavy metal bioindicator to access the acute toxicity. A good bioindicator must have following characters: (1) it can accumulate high levels of pollutants without death; (2) it lives in a sessile style, thus definitely representing the local pollution; (3) it has enough abundance and wide distribution for the repetitious sampling and comparison; (4) its life is long enough for the comparison between various ages; (5) it can afford suitable target tissue or cell for the further research at microcosmic level; (6) easy sampling and easy raising in the laboratory; (7) it keeps alive in water; (8) it occupy the important position in food chain; and (9) well dose-effect relationship can be observed in it (Zhou et al., 2008).

In this view, bivalves are most adopted mollusk to monitor the environmental pollution. Two gastropod species (*Rapana venosa* and *Neverita didyma*) and three bivalve species (*Mytilus edulis, Crassostrea talienwhanensis,* and *Ruditapes philippinarum*) were collected from Chinese Bohai Sea for the investigation of heavy metal contaminants. *R. venosa* and *R. philippinarum* were hopeful bioindicators for monitoring Cd and Ni pollution in waters (Liang et al., 2004). Fish also used for the assessment of the quality of aquatic environment and serves as a bioindicator. The results were confirmed that the enzymic and non-enzymic biomarkers of oxidative stress can be sensitive indicators of aquatic pollution (Farombi et al., 2007). Siebert et al. (1996) used the aquatic moss *Fontinalis antipyretica* L. ex Hedw. as a bioindicator in aquatic systems for heavy metals assessment.

10.2.3 IMPACT OF HEAVY METALS ON HUMAN HEALTH

Heavy metals are essential to maintain the chemical and physiological role in all living systems. It exceed beyond the threshold level leads to the health impact on human beings (Chronopoulos, 1997; Ernst, 2002; Jaishankar et al., 2014). These metals were entering into our body through food, drinking water and air (Prasher, 2009). Accumulation of these heavy metals causes serious damage to hematopoietic, central nervous, cardiovascular, peripheral nervous system, renal, and gastrointestinal (GI) systems (Ibrahim et al., 2006).

Long-term irrigation of contaminated sewage water leads to accumulation of heavy metals in environment (Morira et al., 2002; Singh and Agarwal, 2008). Canned foods containing relatively high amount of heavy metals compared to the fresh-marketed foods (Al-Thagafi et al., 2014). Wang et al. (2015) analyzed the health risk of consumption of crop plants cultivated by using sewage irrigation in China. The concentration of Pb, Cd, Zn, and As high in the wheat sample and As, Cu, and Cd concentration high in rice sample compared to the Chinese standard level. In Pakistan and China concentration of heavy metals present in the vegetables were also high (Wang et al., 2005; Khan, 2009; Guerra et al., 2012; Chang et al., 2014).

Apart from air, water, food, and heavy metals may also enter into the body via cosmetics. Hexavalent chromium present in the lipstick was a carcinogen and increases the risk of lung cancer and may also damage the small capillaries in kidneys and intestine. Nickel was present in the lipstick and eyeshadows. About 1ppm of nickel may trigger the pre-existing allergy. Fairness creams containing mercury may cause nephritic syndrome. Table 10.2 summarized about some important health effects caused by heavy metal accumulation in human beings. Effects of heavy metal poisoning during pregnancy were reviewed by Neeti and Prakash (2013). Chronic arsenic exposure may increase the risk of fetal and infant death. Exposure of lead causes miscarriage, premature birth, low birth weight and it affects development of fatuous brain and growth of newborn baby (Altmann et al., 1993). Copper increases the chance of miscarriage, complication during delivery, low birth weight, and muscular weakness in newborn baby and leads to neurological problems in growing children. Mercury affects the mental growth, nervous system and cognitive thinking in newborn baby. Cadmium at lower concentration in pregnant woman

TABLE 10.2 Sources and Their Health Effects of Important Heavy Metals

Heavy metals	Sources	Health effects
Arsenic	Smelting industry, wood preservatives, pesticides, herbicides, fungicides, and paints	Bronchitis, dermatitis, poisoning, cancers of the liver, lung, bladder, kidney, and colon and cerebrovascular disease
Cadmium	Welding, PVC products, pesticide, electroplating, fertilizer, Cd, and Ni batteries, nuclear fission plant, and cigarette smoking	Respiratory irritation, chronic lung disease, prostate cancer, kidney failure, renal dysfunction, tubular damage, bone defects (osteomalacia, osteoporosis), skeletal damage, bronchitis, gastrointestinal disorder, and cancer
Lead	Food cans, paint, ceramic ware, antiknock motor vehicle fuel additives, fossil fuel, paint, and pesticide	Symptoms related to the nervous system, acute psychosis, disturbance of hemoglobin synthesis, behavioral, and developmental effects, congenital paralysis and sensory neural deafness
Mercury	Thermal volatilization, cosmetics, pesticides, batteries, and paper industry	Hearing loss, neurological disorder, lung damage, kidney damage, coronary heart disease, acrodynia characterized by pink hands, feet, and spontaneous abortion
Chromium	Mines and mineral sources	Damage to the nervous system, fatigue, and irritability
Copper	Mining, pesticide production, chemical industry, and metal piping	Anemia, liver, and kidney damage, stomach, and intestinal irritation

causes abortion and weight of newborn child will be retarded. Heavy metals cause both acute and chronic disease. In particularly, arsenic, cadmium, lead, mercury, chromium, and copper vastly harmful to the human body (Hu, 2002; Jarup, 2003; Prasher, 2009; Singh et al., 2011).

10.3 BIOSORPTION

Removal of metals from the environment is very important research area which carried out in various fields like environmental engineering, environmental microbiology, and environmental biotechnology. Before two decades, there are so many conventional methods were used to remove the heavy metals from the industrial effluents which include chemical precipitation, membrane filtration, solvent extraction, lime coagulation, reverse osmosis, ion exchange, and adsorption. These methods have both merits and limitations. Conventional methods also have some demerits like high reagent and energy requirements, incomplete metal removal, generation of noxious sludge or other waste products (Abbas et al., 2014). Apart from conventional approaches, various biological methods have so far been used to remove the heavy metals such as trickling filter, biosorption, activated sludge process and a variety of anaerobic methods. Compared to the conventional methods, biological methods for heavy metal removal are very economical, effective, and eco-friendly to the nature (Dhokpande and Kaware, 2013).

Nowadays, microbial biomass has emerged as an alternative strategy for economic and eco-friendly effluent and wastewater treatment to control the metal pollution in the environment. Biosorption is a cost-effective strategy, which is used to remove the metal ions using plants, algae, moss, fungi, and bacteria (Wang and Chen, 2009). Biosorption is a technique, which is used for the removal of heavy metals or unwanted substances (organic and inorganic, and in soluble or insoluble forms) from the waste material using biological material. Biosorption is a physicochemical process and includes mechanisms such as adsorption, ion exchange, surface complexation and precipitation (Gadd, 2009) (Figure 10.1).

The greatest advantage of using dead biomass as absorbent are as follows: (1) growth independent nonliving biomass is not subject to toxicity limitation by cells; (2) waste from fermentation industry can be used as a cheap source; (3) this method is not controlled by physiological constraints of microbial cells; (4) this process is very rapid; it requires few minutes to few hours because dead biomass acts as an ion exchanger; and (5) the desorption and recovery process is rapid (Ahalya et al., 2003; Babak et al., 2012).

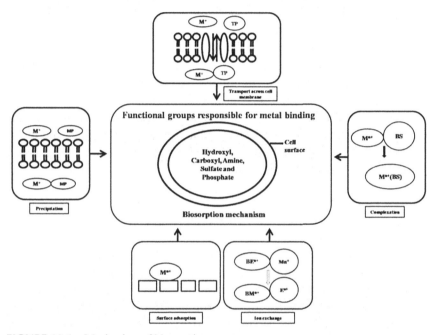

FIGURE 10.1   Mechanism of biosorption.

Before entering into the cells, the metal ions should contact or bind with the surface of the cell wall of organism. The structure of the cell wall comprises polysaccharides, proteins, and lipids. These acts as an active site for binding of metal ions. The macromolecules present in the cell wall composed of many functional groups such as carboxylates, amines, imidazoles, phosphates, sulfhydryls, sulfates, and hydroxyls (Crist, 1981; Najaa et al., 2005; Choi and Yun, 2006). Negatively charged cell wall surface easily attracted by positively charged metal ions present in the solution without metabolic functions of the cell (Swiatek and Krzywonos, 2014). Bacterial cell wall is made up of peptidoglycan layer. Carboxyl groups present in the glutamic acid are the metal binding site of Gram-positive bacterial cell walls with phosphate groups contributing significantly in Gram-negative species (Beveridge, 1989; Wang and Chen, 2009). Cyanobacteria were come under Gram-negative group because it has a similar type of cell wall like Gram-negative bacteria. There are three types of polysaccharides present in the cyanobacteria which mainly possess the metal binding sites which include sheath, capsular, and secreted polysaccharide or soluble polysaccharides. Sulfated polysaccharide may possess the high metal removal efficacy with carboxyl and hydroxyl group (Li et al., 2001). The common component present in the cell wall of plant was cellulose (Davis et al., 2003). Other than cellulose, the following components provides the binding sites for amino, amine, hydroxyl, imidazole, phosphate, and sulfate groups such as polysaccharides like mannan, alginic acid, xylans, as well as proteins (Crist, 1981). Phosphate and carboxyl groups on uranic acids and proteins, and nitrogen-containing ligands on protein and chitin or chitosan present in the fungal cell wall posses the metal binding ability (Gadd 1993; Sag and Kutsal, 2001).

10.3.1 MECHANISM OF BIOSORPTION

General mechanism of heavy metal removal in microorganisms was very much complicated compared to the conventional methods. There are different types of biosorption mechanisms exist, but they are not fully understood. The mechanism of biosorption was elaborated by Veglio and Beolchini (1997). Based on the dependence of cell metabolism (Figure 10.2), biosorption mechanism can be divided into metabolism dependent and non-metabolism dependent (Veglio and Beolchini, 1997).

According to the location, biosorption may be classified as extracellular accumulation/precipitation, cell surface sorption/precipitation and intracellular accumulation. Intracellular accumulation of metals, transport across the cell membrane strictly depends on the cell metabolism. This kind of metabolism only occurs in living cell. In this case, biosorption is not rapid; it requires some time for reaction of the microbes. In this case, dead biomass is used as a biosorbent in biosorption process (Kuyucak and Volesky, 1988). Due to the physiochemical interactions, the metals binds with functional group present in the dead biomass, based on physical adsorption, ion exchange, and complexation. These cell surface sorption mechanism metals can interact with the carboxylate, hydroxyl, sulfate, phosphate, and amino metal binding groups

(Hemambika et al., 2011). This physicochemical phenomenon of metal biosorption, non-metabolism dependent, is relatively rapid and can be reversible (Kuyucak and Volesky, 1988).

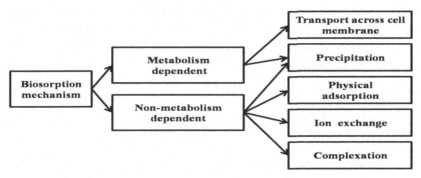

FIGURE 10.2 Biosorption mechanism - based on the dependence of cell metabolism.

10.3.1.1 TRANSPORT ACROSS CELL MEMBRANE

Potassium, magnesium, and sodium are the essential ions which maintain the ionic balance of the cell and across the cell membrane during active and passive transport mechanism. This same mechanism has been used in the transport of heavy metals through the cell membrane. Sometime, the metal transport system may become confused by the presence of heavy metal ions of the same charge and ionic radius (Brierley, 1990). This type of mechanism was not linked with the metabolic activity of the cell. Depending on the microbial species the active or passive transport mechanism turn on during metal uptake. The passive uptake of metal was independent of cellular metabolism and physical conditions such as pH and ionic strength; also this passive uptake was rapid and reversible. Compared to the passive uptake, active uptake was slow and it dependents on the cellular metabolism. In this process metals forms complexes with specific proteins like metallothioneins and both processes can occur simultaneously (Pandey et al., 2001).

The biosorption by living microorganisms comprises two basic steps. Firstly, the independent metabolism binding to cell wall and secondly, metabolism-dependent intracellular uptake, whereby metal ions are transported across the cell membrane into the cell (Gourdon et al., 1990; Tchounwou et al., 2014). Plant pathogen, *Pseudomonas syringae* was resistant against copper, the accumulation and compartmentalization of copper occurs in the cell periplasm and the outer membrane. The copper resistant mechanism in *P. syringae* was due to four types of proteins (CopA, CopB, CopC, and CopD). Bacterial plasmid containing *cop* operon and the proteins are found in the periplasm (CopA and CopC), the outer membrane (CopB), and the inner membrane work together to compartmentalize copper away from bacterial cells (Cooksey, 1993). In *E. coli*, the efflux proteins are expressed by plasmid-borne PCO genes, which are in turn rely upon the expression of chromosomal cut genes. There are two *cut* genes (*cutC* and *cutF*) encode a copper binding protein and an outer membrane lipoprotein. Many bacterial species have normally acquired environmental protective mechanism by environmental stresses (Issazadeh et al., 2013).

10.3.1.2 PHYSICAL ADSORPTION

Physical adsorption is combined with the presence of van der Waals' forces (Crowell, 1966). Uranium, cadmium, zinc, copper, and cobalt biosorption was carried out by dead biomass of algae, fungi, and yeasts, due to the electrostatic interactions between ions in solution and cell walls of biomass (Kuyucak and Volesky, 1988). Thorium and uranium biosorption by fungal biomass of *Rhizopus arrhizus* is also based on physical adsorption in the cell-wall chitin structure (Tsezos and Volesky, 1982). Electrostatic interactions have been demonstrated to be responsible for copper biosorption by bacterium *Zoogloea ramigera,* and algae *Chiarella vulgaris* (Aksu et al., 1992) and chromium biosorption have been demonstrated in fungi like *Ganoderma lucidum* and *Aspergillus niger* (Bang and Pazirandeh, 1999; Tchounwou et al., 2014).

Microbial cell surface possess the heterogeneity of specific metal adsorption. Microbial cell surface display is a technique that expresses a heterologous protein/peptide of interest as a fusion protein with various cell-surface proteins, which are usually cell surface proteins or their fragments (Lee et al., 2003). In addition to using existing proteins/peptides, protein engineering that creates novel proteins/peptides or introduces random mutations or directed evolution to improve the specific metal-binding abilities of known proteins/peptides will contribute to the molecular breeding of biosorbents with specific adsorption properties.

Metalloregulatory proteins like MerR family proteins (Hobman, 2007) which have been employed to create metal specific biosorbents (Bae et al., 2003; Qin et al., 2006) and biosensors (Hakkila et al., 2011; Tao et al., 2013) owing to their highly selective recognition and binding ability to particular metal, might be a excellent choice (Li and Tao, 2013).

10.3.1.3   ION EXCHANGE

Ion exchange is defined as the replacement of an ion in a solid phase with a solution by another ion. Especially, it is the replacement of an absorbed, readily-exchangeable ion by another ion (Sposito, 1989). The Gram-positive bacteria have metal binding capacity because of their negative charge density, mainly in members of the genus *Bacillus*. Bacterial cell wall made up of peptidoglycan layer which consists of teichoic and teichuronic acids. The phosphodiesters of teichoic acid and the carboxylate groups of the teichuronic acid contributes for ion-exchange capacity to the cell wall (Wase and Forster, 1997). Microorganism contains biopolymers or polysaccharides which possess the cationic binding. Biopolymers or polysaccharide implies the functional groups like carboxyl, organic phosphate and organic sulfates. Carboxylic acid mainly dispersed in polymers and they act as side chain with constituents of proteins, the uronic, neuraminic, and muramic acids were associated with monosaccharides of polysaccharides. Phosphodiester and monoester phosphate groups are found in the polysaccharides and related macromolecules. Phosphodiester present as a lipid moiety in lipoprotein and lipopolysaccharides. Sulfate esterified to carbohydrate hydroxyl groups is common in connective tissue and algal polysaccharides and is for most purposes likely to provide the greatest negative charge density, at very low pKa, among the charged biopolymers. Ester sulfate and phosphomonoester groups also occur in proteins. Fungal biomass pretreated with solutions of Ca^{2+} and Na^{2+} released cations of these two light metals into the solution while sorbing Zn^{2+} and Pb^{2+}. Weakly acidic carboxyl groups R-COOH of seaweed, algal, and fungal cell wall constituents as the probable sites of ion exchange (Bloom and McBride, 1979; Fourest et al., 1996).

Organic nitrogen based groups involved in an anion exchange. In proteins, amino (lysyl side chain and N-terminal), imidazole (histidyl) and guanidine (arginyl) groups are common centers of positive charge

(Tsezos et al., 2006). An anion exchange significantly affected by physical condition like pH. For example, anionic species like TcO_4^-, $PtCl_4^{3-}$, CrO_4^{2-}, SeO_4^{2-} and Au $(CN)^-$ exhibit improved biosorption at low pH values (Garnham et al., 1994; Garnham, 1997). Chitin is the notable example for a proportion of glucosamine residue. The amino group of R_2-NH and chitosan R-NH_2 was also subject to examinations. The presence of the former was detected in algae *Sargassum* sp. and *Cladophora* sp. (Crist et al., 1981).

10.3.1.4 COMPLEXATION

The metal removal from solution may also take place through complex formation on the cell surface after interaction between the metal and active groups. Metal ions can bind to unidentate (single) ligands or through chelation (Cabral, 1992a,b). According to Pearson's classification, "hard acids," metals such as Ca, Mg, Na, and K are essential nutrients for microbial growth, bind preferentially to oxygen-containing "hard bases," ligands such as OH^-, HPO_4^{2-}, CO_3^{2-}, R-COO-, and =C=O. Soft acids, metals such as Au, Ag, Pd, and Pt are bound covalently to the cell wall by "soft bases," ligands containing nitrogen or sulfur (Beveridge and Murray, 1980; Tsezos et al., 2006). Ionic natures of surfactant also entrap the metals in the solution by complexation and increase its metal solubilization. Metal bioavailability can also be influenced by common metabolic by-products that results in metal reduction resulting in the formation of less soluble metal salts including sulfide and phosphate precipitates (Banik et al., 2013). Metal detoxification in resistant microorganisms can take place through complexation by binding with bacterial cell envelopes, exopolysaccharides, metal reduction and metal efflux (Filali et al., 1999).

Biosorption of thorium and uranium in *R. arrhizus* was carried out by complex formation. Thorium and uranium bind with nitrogen which is present in the chitin cell wall of fungi (Aksu et al., 1992). Cabral et al. (1992a,b) reported that complexation was found to be the only mechanism responsible for magnesium, mercury, calcium, and cadmium, copper, and zinc accumulation by *P. syringae*. Some cases metal complexes are more toxic than free metals. The combined use of 8-OH-quinoline and ferrous ions was inhibited the growth of *Staphylococcus aureus*. When they are used separately no toxic effect was observed. Although certain metal complexes are more toxic than the free metal, they are often volatile and may disappear from an environment. This is the case with methylated derivatives of mercury. The amount of heavy metal present in wastewater increases with the metal ions which forms complexes with microbes resulting in a lower demand of dissolved oxygen. Complexation of heavy metals with microbes causes death to the bacteria. When the bacteria dead, the biological oxygen demand (BOD) will be increased results in threaten for the microorganisms.

10.3.1.5 PRECIPITATION

Precipitation of metals was dependent on the cellular metabolism or independent of it. In this type, metal removal from solution is often linked with the defense mechanism of microorganisms. Microorganisms produce some compounds which favors the precipitation reaction during removal process (Ahalya et al., 2003). Precipitation leads to formation of insoluble inorganic precipitates (Remoundaki et al., 2003). For example, EPS can entrap or adsorb the precipitated metal sulfides and oxides (Flemming, 1995; Vieira and Melo, 1995). Microbial biofilms have metal binding capability, and it acts as matrices for precipitation of insoluble mineral phases. Microorganisms can produce many kinds of excreted metal-binding metabolites (Bridge et al., 1999; Gadd, 2000). Increasing resistant mechanism to metals in microorganisms, precipitation of metals as metal sulfides or phosphates is another way. In dissimilarly sulfate reduction, sulfate-reducing bacteria (SRB) couple the oxidation of organic compounds or molecular H_2 with the reduction of sulfate as an external electron acceptor under anaerobic conditions (Barton and Tomei, 1995). The following are the general reactions used to describe the sulfate reduction process (Cao et al., 2009):

$$\text{Organic matter (C, H, O)} + SO_4^{2-} \rightarrow HS^- + HCO_3^-$$

$$Me^{2+} + HS^- \rightarrow MeS \downarrow + H^+ \ (Me^{2+} \text{- Metal cation})$$

Detoxification systems of *Arthrobacter* and *Pseudomonas* eliminate the cadmium from the solution and precipitate the cadmium on the cell surface. This mechanism does not require the cell metabolism; the precipitation occurs due to the chemical interaction between the metal and the cell surface (Scott and Palmer, 1990). Membrane-bound acidic phosphatase, precipitate uranium, and other heavy metals in *Citrobacter* sp. (Macaskie et al., 1992). In *M. oxydans*, uranium was not accumulated in the cytoplasm of the cell. Uranium binds with the polyphosphate bodies present in the peptidoglycan layer. Nonspecific acidic phosphatase involved in the precipitation of phosphate groups in and at the cell surface (Boswell et al., 2001).

10.3.2 FACTORS AFFECTING BIOSORPTION OF METALS

Many factors can affect the biosorption of metals. Some of the factors are related to biomass and metal, and the others are related to environmental conditions such as pH and temperature.

10.3.2.1 pH

Metal biosorption in bacteria, cyanobacteria, algae, and fungi was strongly dependent on pH. It will affect the metal solution and also affects the functional group present in the biomass and the metallic ions (Friis and Myers-Keith, 1986; Galun et al., 1987). The number of binding sites present in the cell wall strongly depends on the pH. At low pH, the binding sites of cell surface binds with the hydrogen cation present in the solution. This will leads to the limitation of active site which causes low level of metal absorption. When temperature increases the active sites of cell wall can be increased which leads to increased amount of metal binding (Babak et al., 2012). When the pH is increased, deprotonation was started on functional groups such as carboxyl, sulfhydryl, hydroxyl, and amino groups, behave as negatively charged ions and start attract to the positively charged ions. On the other side when pH is lowered, the cell surface functional group become positive, it will inhibit the positive ion binding (Gupta et al., 2010; Luo et al., 2010). Sometimes extreme pH values, may damage the structure of the biosorbent material. Also, showed damage to the cells, significant weight loss and reduces the sorption capacity (Kuyucak and Volesky, 1988). The speciation of the metal in solution is pH dependent. During low pH, metals are in aqueous solutions occurring as hydrated cations in salvation shells and hydroxides may form at higher pH (Morel, 1983).

10.3.2.2   TEMPERATURE

The stability of metal ion-cell complex and metal ion in solution mainly depends on temperature. The temperature range between 20 to 35°C does not affect the biosorption. Higher temperature permits the higher metal sorption capacity of biomass and also it can affect the microbial living cells and metal uptake (Goyal, 2003; Chojnacka, 2009). Temperature change during biosorption will affect the number of factors which include stability of the ion species initially placed in a solution, stability of microorganism–metal complex depending on the biosorption sites, affect the cell wall configuration and the ionization of chemical moieties on the cell wall (Kumar et al., 2009). The temperature can affect the adsorption process in two ways. One is that increasing the temperature; liquid viscosity was decreased by the rate of adsorbate diffusion particle across cell boundaries and in the internal pores of the adsorbate particles. And another one is it affect the equilibrium capacity of the adsorbate depending on whether the process is exothermic or endothermic (Al-Qodah, 2006). The maximum biosorption capacity for Ni and Pb by *S. cerevisiae* was obtained at 25°C and found to decrease as the temperature was increased to 40°C (White et al., 1997). Greene and Darnall (1998) found that the distribution ratio (metal bound/metal in solution) for biosorption of Cd, Zn, Pb, Ni, and Cu to *Spirulina* algae increased by only ~20% when the temperature was raised from 4°C to 55°C.

10.3.2.3   BIOMASS CONCENTRATION

Concentration of biomass in solution affects the specific uptake during biosorption (Modak and Natarajan, 1995). Biomass adsorbs the more metal ions at low concentration of biomass compared to the high concentration at equilibrium concentration (Gourdon et al., 1990). At the higher concentration of biomass the adsorption rate was increased due to the availability of more empty binding sites as compared to lower dosage in the adsorbate solution (Chong et al., 2013). The electrostatic interaction between the metal-biomass can play a major role in metal uptake (Malkoc and Nuhoglu, 2005). For example, when biomass concentration was increased from 0.5 to 3g, the removal efficiency was also increased which includes the following chemicals such as cadmium from 37.61 to 61.11%, for copper from 23.36 to 50.28%, for nickel 12.04 to 27.54% and for zinc 9.2 to 36.28% (Kumar et al., 2009).

10.3.2.4   INITIAL METAL ION CONCENTRATION

The efficiency of biosorption was not only depends on the physical parameters and biomass concentration, it also depends on the initial metal ion concentration. This plays an important role as a driving force to overcome all mass transfer resistance of metal between the aqueous and solid phases (Dang et al., 2009). When increasing initial concentration of metal ions results in increased amount of metal adsorption because it gives driving force to overcome mass transfer resistance between the biosorbent and biosorption medium. Metal uptake increases with initial concentration with given biomass (Pahlavanzadeh et al., 2010; Abbas et al., 2014). The efficiency of metal removal by biomass was increased with the initial metal ion concentration. At low ion concentration, all metal ions interacts with the binding site of biosorbent leads to 100% metal removal efficiency. On the other hand, elevated concentration, more ions left over in the solution in un-adsorbed state due to the saturation of binding sites (Naiya et al., 2009; Abdel-Ghani and El-Chaghaby, 2014). For example, the removal of cadmium by *H. valentiae* was higher at lower concentration of cadmium solution; a maximum cadmium removal of 86.8% has been obtained for *H. valentiae* biomass (Rathinam et al., 2010). It has been reported that the small difference in the initial metal ion concentration will affect the metal adsorption in low concentration.

10.3.2.5   CONTACT TIME

Contact time also affects the biosorption of metals between biomaterial and metal solution. Most metal were absorbed fast at the beginning of process. The equilibrium concentration was reached at the couple of minutes starting from the exposure of the solution (Chojnacka, 2009). Identify the optimum contact time required to obtain an increasing removal of metal ions. In waste water system equilibrium is one of the parameter for metal removal (Abdel-Ghani et al., 2007). The rate of metal adsorption is rapid, but it slowly decrease with time until it reaches equilibrium. The rate of percentage of metal uptake is higher in the beginning due to a larger surface area of the adsorbent being available for the adsorption of the metals (Abdel-Ghani and El-Chaghaby, 2014). For example, removal efficiency increases for cadmium from 39.12 to

52.31%, for copper 31.24 to 82.13%, for nickel 13.13 to 27.36% and zinc is 10.26 to 40.74% as the contact time increases from 15 to 240 min (Kumar et al., 2009).

10.3.2.6   PRESENCE OF OTHER IONS IN SOLUTION

Presence of more than one metal in the water may be inhibited by the presence of other metal ions. Other ions present in the solution may compete with the own interest metal for binding sites. Uranium uptake by *R. arrhizus* was influenced by the presence of Fe^{2+} and Zn^{2+} in the solution (Tsezos and Volesky, 1982). Cobalt uptake by different microorganisms seemed to be completely inhibited by the presence of uranium, lead, mercury, and copper (Sakaguchi and Nakajima, 1991).

10.3.2.7   DESORPTION OF METALS

Desorption is the process used to regenerate the metal ion from the sorbed biomass, and the same biomass will be useful for further biosorption process. It will reduce the cost of biosorption process. For this purpose, it is desirable to desorb the sorbed metals and to regenerate the biosorbent material for another cycle of application. The desorption process have following characteristics: it will revive the metals in concentrated form, and restore the original metal uptake compared to the first usage without damaging the biosorbent physical state (Alluri et al., 2007). During metal regeneration process, the biosorbent should be washed by the appropriate solution to remove or unbind the metals from the biosorbent. This choice of solution is based on how the metal will interact with the sorbent, whether weak or strong. For desorption of metals, the diluted mineral acids were sulphuric acid, hydrochloric acid, nitric acid and acetic acid (Zhou and Kiff, 1991; Bai and Abraham, 2003). For instance, *Aspergillus* used for the removal of iron, calcium, and nickel from the loaded waste. Desorption takes place due to increasing concentration of HCl. The results showed that desorption was increased at 5M HCl, complete elimination of calcium and iron would be achieved while about 78% nickel would be desorbed (Chandrashekar et al., 1998).

10.4   BIOSORPTION ISOTHERM MODELS

Biosorption isotherms are considered to be a significant method to access the biosorption capacity (Volesky, 2005). Biosorption isotherm depicts the relationship between the mass of the adsorbed component per biosorbent mass and the concentration of this component in the solution (Krowiak et al., 2011). Preliminary examination of solid, liquid sorption system is based on two types of investigations: (a) equilibrium batch sorption tests and (b) dynamic continuous flow sorption studies. The equilibrium of the biosorption process is often described by fitting the experimental points with models (Gadd et al., 1998). This study is very useful to define a single biosorbant system and metal. In this category, biosorption often examined under various parameters such as pH, biomass density, metal concentration and presence of competing cations (Gadd, 2009). This thermo dynamic approach of physicochemical parameters will provide an insight into the adsorption mechanism, surface properties as well as the degree of affinity of the adsorbents (Bulut et al., 2008). This equilibrium parameter gives insight to future design of adsorption systems (Abdel-Ghani and El-Chaghaby, 2014). There are several equilibrium isotherm models applied to fit the biosorption experimental data in order to study the nature of adsorption process. These includes two parameter isotherm (Langmuir, Freundlich, Temkin, Dubinin, Radushkevich, Flory–Huggins, Elovich, Fowler–Guggenheim, Kiselev, Hill–de Boer, Jovanovic, Halsey, and Harkin-Jura), three parameter isotherm (Redlich–Peterson, Hill, Toth, Jossens, Fritz–Schlunder, Sips, Koble–Corrigan, Khan, Radke–Prausnitz, Frumkin, and Liu), four parameter isotherm (Weber-van Vliet, Fritz–Schlunder and Baudu) and five parameter isotherm (Fritz–Schlunder) (Foo and Hameed, 2010). Table 10.1 shows some of the isotherm models and their brief description.

10.5   BIOSORBENTS

A variety of biomass has been used for the biosorption studies under various conditions which includes agricultural products such as dried plant leaves, wool, cottonseed hulls, waste coffee powder, waste tea, cork biomass, rice straw and coconut husks (Orhan and Buyukgungor, 1993). Compared to the agricultural wastes, biological material has high affinity towards the metals and other pollutants. All kinds of microbes, plants, animal biomass, and derived products, have investigated in a variety of forms, and in relation to a variety of substances (Volesky, 2005). Especially microorganisms like, bacteria, fungi, yeast, and algae from their natural habitats are exceptional sources of biosorbents (Zouboulis et al., 1997). A good biosorbent can effectively remove the metals from diluted complex solution rapidly and efficiently (Kapoor and Viraraghavan, 1998). It should be cost-effective and provide new opportunities for pollution control, element recovery, and recycling (Gadd, 2008).

10.5.1 BACTERIAL BIOSORBENTS

Bacteria are predominant biosorbent used in metal sequestration. Because of their small size, ubiquity, and ability to grow under controlled conditions, and their resistance against a wide range of varying environmental conditions (Urrutia, 1997). Based on their cell wall components bacteria classified into two types such as Gram-positive and Gram-negative bacteria. The Gram-positive cell wall made up of two functional layers. Negatively charged teichoic acid (acidic polysaccharides) and teichuronic acid are attached to the cell wall. The cell wall of Gram-positive made up of peptidoglycan which is surrounded on the cytoplasmic membrane. In Gram-positive bacteria, teichoic acid give the Gram-positive cell wall, but an overall negative charge due to the presence of phosphodiester bonds between the teichoic acid monomers. In Gram-positive cell wall, carboxyl groups are the main binding site for metal cations. But in Gram-negative species, phosphate groups play a vital role in metal binding (Beveridge, 1989; McLean et al., 2002). There are some other metal binding components present in the bacteria which include proteinaceous S-layer, and sheaths largely composed of polymeric materials including proteins and polysaccharides (Camargo et al., 2003).

Johnson et al. (2007) studied the impact of metabolic state on Cd adsorption onto both Gram-positive and Gram-negative bacterial cells. They demonstrated that compared to the Gram-negative, activity of Gram-positive bacteria has a significant impact on Cd adsorption onto cell wall, functional groups. Metabolizing Gram-positive cells adsorbed significantly less than non-metabolizing Gram-positive cells. Conversely, metabolizing, and non-metabolizing Gram-negative cells exhibited roughly similar extents of Cd adsorption. The effect of bacterial metabolism Cd adsorption onto Gram-negative cells suggests that Cd binding occurs at greater distance from the inner plasma membrane than occurs within the cell wall of Gram-positive cell. They recommended Gram-positive species for passive metal adsorption of Cd. Metal tolerant *Bacillus thuringiensis* OSM29 isolated from rhizosphere of cauliflower grown in soil irrigated consistently with industrial effluents. The biomass of this strain success-fully removed the metals such as Cd, Cr, Cu, Pb, and Ni from aqueous solution. Metal adsorption of *B. thuringiensis* OSM29, due to the presence of metal binding cell wall functional groups such as amino, carboxyl, hydroxyl, and carbonyl groups. The maximum biosorption of heavy metals by *B. thuringiensis* OSM29 occurred at pH 6 within 30 min at $32 \pm 2^{\circ}C$. Ozdemir et al. (2004) reported that a species of Gram-negative bacterium, *Pantoea* TEM 18 exhibited greatest Cu tolerance which is isolated from wastewater treatment of petrochemical industry. Due to the presence of lipopolysaccharides, phospholipid, and proteins, it shows the possibility of metal binding.

P. aeruginosa biofilms and biomass showed the metal adsoption capacity in various metals like Pb, Cu, and Cd (Chang et al., 1997; Teitzel and Parsek, 2003; Chien et al., 2013). Sulfate-reducing bacteria isolated from natural and constructed wetlands are useful to remove metal ions from wetland treatment. The production of H_2S is the main factor in the precipitation of metals as sulfites (Amacher et al., 1993; Webb et al., 1998). Rani et al. (2010) comparatively assessed the heavy metal removal by immobilized and dead bacterial cells. They concluded that the adsorption capacity of the immobilized bacterial cells was greater than that of dead bacterial cells by 0.002, 0.023, and 0.004 in *Bacillus* sp., *Pseudomonas* sp. and *Micrococcus* sp. respectively. Because compared to the immobilized cells, dead bacterial cell have small particles with low density, reduced mechanical strength, and little inflexibility. An indigenous bacterial isolate *Enterobacter* sp. J1 exhibited good metal uptake capacity and high resis-tance to various heavy metals. It was able to absorb Pb, Cu, and Cd with a capacity of 50.9, 32.5 and 46.2 mg/g dry cell, respectively. During Pb uptake, both cell-surface binding and intracellular accumulation seemed to be involved. For metal recovery, hydrochloric acid was used. The pH adjustment with HCl achieved over 90% recovery of Pb, Cu, and Cd from metal-loaded biomass.

Biosorption of heavy metals by a novel acidic polysaccharide like bioflocculant produced from microorganisms have been reported (Kong et al., 1998). Bioflocculant which is produced by *P. fluorescens* removed 70% mercury, 30% zinc and 45% cadmium. Bioflocculant from the three thermo tolerant isolates, *E. agglomerrans* SM 38, *B. subtilis* WD 90 and *B. subtilis* SM 29, adsorbed nickel and cadmium up to 90 and 85% respectively (Kaewchai and Prasertsan, 2002; Noghabi et al., 2007). EPS produced by the halophilic bacterium *Halomonas almeriensis* exhibited both emulsifying and metal binding (Tchounwou et al., 2014). EPs producing *Ochrobactrum anthropi* isolated from activated sludge showed maximum adsorption capacity to various meals like Cr(VI), Cd(II), Cu(II) about removal of 86.2 mg/g, 37.3 mg/g and 32.6 mg/g respectively (Ozdemir et al., 2003). Geddie and Sutherland (1993) studied the acetylation and de-acetylation of polysaccharide which influence the uptake of ions. The native acetylated polymers showed selectivity for $Ca^{2+}>Mg^{2+}>$monovalent cations, whereas samples missing acetyl groups showed a selectivity for monovalent cations$>Mg^{2+}>Ca^{2+}$. The enterobacterium *Klebsiella oxytoca* strain BAS–10 was isolated from sediments under an iron mat formed in a stream receiving leached waters from pyrite mine tailings. The Fe(III) binds strongly to oxygen donors of OH^- moieties of sugars in polysaccharide which is produced from the *K. oxytoca*. This iron gel co-precipitated toxic metals with high concentration (0.1 mM and 1.0 mM) of Cd-, Pb-, and Zn-acetate, which co-precipitated in the following order: Pb>Cd>Zn (Baldi et al., 2001).

Kinoshita et al. (2013) isolated Hg (II) binding protein from *Lactobacillus oris* using the N-terminal amino acid sequencing. This 14 kDa hypothetical protein has a CXXC motif. CXXC motif is also known as a heavy metal binding motif to Cd (II), Co (II), Cu(II), and Zn(II) ions. *S. aureus* have domain CXXC motifs containing CopA, CopZ, and McsA proteins which are capable of binding various types of heavy metals (Boonyodying et al., 2012; Sitthisak et al., 2012). This supported that 14 kDa protein may bind to Hg (II) through the CXXC motif.

10.5.2 FUNGAL BIOSORBENTS

Fungi are the large and diverse group of eukaryotic microorganisms, and there are three groups of fungi have major practical importance such as molds, yeasts, and mushrooms. Fungi are everywhere in natural environments and important in many industrial processes. Fungi have variety of morphology based on their genera, which have unicellular yeasts to poly-morphic and filamentous fungi; many of these contain complex fruiting bodies. The vital role of fungus is decomposing of organic material. Also, it produces some economically important substances such as ethanol, citric acid, antibiotics, polysaccharides, enzymes, and vitamins (Gadd, 1993). The cell wall composition of fungus exhibits metal binding property. The important constituents present in the fungal cell wall which binds with metals ions includes carbohydrates chitin (3–39%), chitosan (5–33%), polyuronide, and polyphosphates (2–12%), lipids (2–7%) and proteins (0.5–2.5%). These properties vary between the different taxonomic groups (Javaid et al., 2011).

A. niger NCIM–501, A. oryzae NCIM–637, R. arrhizus NCIM–997 and R. nigricans NCIM–880 dead biomass were used to remove the Cr(VI) from aqueous solution in batch mode. Bai and Abraham (1998) optimized the basic parameters such pH (2.0–8.0), initial metal ion concentration (100–500mgl^{-1}), contact time (2–24h) and varying biomass concentration (0.5–3.0g). R. nigricans and R. arrhizus possessed good specific uptake of 11 mg Cr (VI)g^{-1} of biomass at the pH range of 2.0–7.0. Metal uptake capacity was in the order of R. nigricans>R. arrhizus>A. oryzae>Aspergillus niger. Seventy-six fungal isolates were isolated from sewage, sludge, and industrial effluents. Phanerochaete chrysosporium, A. awamori, Aspergillus flavus, and Trichoderma viride were having metal tolerance capacity. There was a significant higher uptake of Pb (0.89 mg/g), Cd (0.16 mg/g) by Trichoderma viride, Cr (0.76 mg/g) by A. niger and Ni (0.66 mg/g) by A. flavus. Yan and Viraraghavan (2003) explored the NaOH pretreated Mucor rouxii showed higher metal adsorption capability for the removal of zinc, nickel, lead, and cadmium from aqueous solution.

Non-living, dead biomass of Streptoverticillium cinnamoneum was used for recovery of Pb and Zn at optimum pH 5.0–6.0 and 3.5–4.5, respectively. The metals were significantly desorbed with dilute HCl, nitric acid and 0.1 M ethylene diamine tetra acetic acid (EDTA) by Puranik and Paknikar (1997). The biosorption of heavy metal may be pH dependent. Penicillium removed the uranium at the pH of 2.5 to 9.5, but the removal was more or less same (Galun et al., 1983). Simonescu and Ferdes (2012) comparatively studied the copper removal capacity of following fungal strains: Aspergillus oryzae ATCC 20423, A. niger ATTC 15475, Fusarium oxisporum MUCL 791 and Polyporus squamosus. The highest copper removal capacity was obtained in Aspergillus oryzae ATCC 20423. The optimum pH for metal removal depending on the fungal species and the optimum temperature for biosorption were 30°C. FT-IR spectra revealed that the hydroxyl, amino, carboxyl, and carbonyl functional groups are present in the cell wall of fungus. The uptake of copper by fungal biomass involved the copper interaction between these functional groups.

Among yeasts, most significant yeasts are the baker's and brewer's yeasts, which are members of the genus *Saccharomyces*. The habitats of this yeast were fruits and fruit juices. The commercial yeast is different from wild type because they have been genetically improved and manipulated through the years and these yeasts are excellent models for the study of many important problems in eukaryotic biology (Madigan et al., 1997). *S. cerevisiae* is a medicore biosorbent which have metal removal capacity from aqueous solutions (Wang and Chen, 2006). For example, metal removal done through living or dead cells (Horikoshi et al., 1981), immobilized cell or free cell and flocculent or non-flocculent cell (Marques et al., 1999) of *S. cerevisiae*.

Many yeast genera have been used for the biosorbent of heavy metal ions such as *Saccharomyces, Candida,* and *Pichia* (Podgorskii et al., 2004). Zinc and uranium was removed from dead cells of *Saccharomyces cerevisiae*, about 40%. Uranium was deposited as fine needle-like crystals inside the cells and on the outer cell surface. The metal biosorption rapidly reached 60% of the final uptake value within 15 min (Volesky and May-Phillips, 1995). Goyal et al. (2003) explained that the increasing temperature would lead to higher affinity of sites for metal or binding sites on the yeast. The metal uptake was increased during the increasing temperature in the range of 25–45°C. The energy of the system facilitates Cr (VI) attachment on the cell surface to some extent. When the temperature is too high, there is a decrease in metal sorption due to distortion of some sites of the cell surface available for metal biosorption.

Pretreatment of the *S. cerevisiae* by using 10–20mM glucose increased the removal efficiency for Cd^{2+}, Cr^{3+}, Cu^{2+}, Pb^{2+} and Zn^{2+} by 30–40%, but the pretreatment by using 60mM glucose decreased the removal of Cr^{6+} by almost 50%. The mechanism of Cr^{6+} uptake may be different from other metal ions (Mapolelo and Torto, 2004). Compared to the intact cell walls of yeast, isolated components of cell wall like mannans, glycans, and chitin highly accumulate the metal ions. Wherelse 30% reduction in metal-accumulating capacity occurred after cell wall protein removed by enzymatic digestion. This result suggests that the outer mannan-protein layer of the cell wall is more important than the inner glucan-chitin layer in heavy metal cation accumulation (Brady et al., 1994).

Mushrooms are grown in natural habitat which is capable to accumulate metal pollutants present in the soil. Accumulation takes place in mycelia and sporocarps up taken by fruiting bodies of mushroom. Nowadays, not so much reports in metal absorption by mushrooms (Das et al., 2008). Accumulation of Cd in non-edible mycorrhizal mushroom cell wall and vascular compartment has been reported by Blaudez et al. (2000). Stihi et al. (2011) was worked to determine the heavy metal content of the young fruiting bodies of *Lycoperdon perlatum* and *Pleurotus ostreatus* and their substrate collected at various distances from a metal smelter in Dambovita County, Romania. *Lycoperdon perlatum* accumulate more Zn and Pb, and *Pleurotus ostreatus* accumulate more Mn and Fe. *Lycoperdon perlatum* accumulate a large amount of Pb, and that fact can be one of the reasons of the non-edibility of this species at the maturity. The concentrations obtained for heavy metals in *Pleurotus ostreatus* species seems to be acceptable for human consumption and nourishment value. Fruiting bodies of non-edible mushrooms were screened for copper uptake (Muraleetharan et al., 1995). Highest accumulation of cadmium and lead in edible mushroom *Psalliota campestris* cause various health risks in Spanish consumers.

10.5.3 ALGAL BIOSORBENTS

The name algae coined to a large and diverse group of organisms that contain chlorophyll and carry out oxygenic photosynthesis (Davis et al., 2003). Algae have grown in low nutrient requirements, being autotrophic, they produce a large biomass, and unlike other microbial biomass, such as bacteria and fungi. They generally do not produce toxic substances. There are three types of algae present in environment which includes macro-algae (brown or marine algae), micro-algae (green and freshwater algae) and red algae (Abbas et al., 2014). Algae considered to be one of the most promising types of biosorbents because it has both economic and eco-friendly nature and also readily available in a large quantities in seas and oceans. It has a high metal recovery potential and regeneration capacity and also acts as a cost-effective absorbent in wastewater management (Rincon et al., 2005; Brinza et al., 2007; Pahlavanzadeh et al., 2010). Electrostatic attraction and complexation of the cell wall play major role in biosorption of algae (Ashkenazy et al., 1997). The major cell wall components like proteins and polysaccharides responsible for the biosorption. In algae, biosorption takes plays with the help of covalent binding of metal amino and carboxyl groups which contains high content of mannuronic and gulucouronic acids. Ionic interaction between carboxyl and sulfate groups with metals leads to biosorption in algae (Petersen et al., 2005; Ibrahim, 2011). Cell wall of brown algae contains cellulose, alginic acid (a polymer of mannuronic and gulucuronic acids) and sulfated polysaccharides. Red algae also contain cellulose and made up of galactanes which includes agar and carragenates. In green algae, cell wall glycoproteins are bonded with polysaccharides (Romera et al., 2006).

Chan et al. (2014) used *Chlorella vulgaris* and *Spirulina maxima* for the treatment of secondary effluent from wastewater plants. They used *Scenedesmus* sp. as a mixture and concluded that the mixture and *S. maxima* strains achieved highest removal efficiencies for Cu and Zn. From this data, the microalgae were able to remove between 69.9% and 81.7% of the Cu and between 0.7% and 94.1% of the Zn in the untreated trial depending on the strain. In the autoclaved trial, removal was higher from 71.7% to 84.4% and from 93.1% to 96.3% for Cu and Zn, respectively, depending on the strain. Chromium sorption by non-living biomass of *Chlorella vulgaris* and *Cladophora crispata* along with bacteria and fungi was studied, and it was observed that pH affected metal uptake capacity (Nourbakhsh et al., 1994). Dinesh Kumar et al. (2015) studied that the *C. marina* which was inoculated in aquaculture wastewater reduced phosphate from 0.010 µmol/L to 0.0003 µmol/L, nitrate reduced from 1.10 µmol/L to 0.05 µmol/L, nitrate reduced from 2.30 µmol/L to 0.54 µmol/L and ammonia was reduced from 29.33 µmol/L to 9.65 µmol/L. And also *C. vulgaris* removed the 95.3% and 96% of nitrogen and phosphorus from 25% secondarily treated swine wastewater after four days of incubation (Kim et al., 1998). Pond micro-algae in anaerobic fixed-bed reactor able to remove the obtained 90.2%, 84.1%, and 85.5% organic nitrogen, ammonia, and total phosphorus removal, respectively from the treated distillery wastewater (Travieso et al., 1996).

In cyanobacteria, metal binding ability is based on the following factors such as the chemical and morphological features of microbial cells, the chemical characteristics of the EPS surrounding the cells and present in solution, the chemical and physical properties of metals, their interactions with the other compounds in solution and the operational conditions utilized in the treatment. Cyanobacteria grown in metal-polluted environments had proven the ability to tolerate the high concentration of metals ions Cu, Cd, and Zn. Marine cyanobacterium *Oscillatoria* sp. was efficiently removed the heavy metal chromium from the wastewater. Heat-killed cells of *Oscillatoria* sp. was able to adsorb 10 µg/mL and 5 µg/mL of chromium. This results in the reduction of cellular contents like chlorophyll a, carotenoids, phycocyanin, allophycocyanin, sugars, free amino acids and proteins (Jayashree et al., 2012).

Lake harvested water bloom cyanobacterium *Microcystis aeruginosa* adsorbed the uranium by powered biomass (Li et al., 2004). Dried biomass of *Microcystis* sp. has the potential to remove Cu, Cd, and Cr in bimetallic, as well as multi-metallic conditions in Bold's basal medium. In this study, EDTA supplemented with Bold basal medium has the dual property of binding the heavy metals of medium when present in excess, and making available the localized (bound metal ions) or low concentration metal ions as free ions (Rai and Tripathi, 2007). In *Synechocystis* sp. BASO671 removal of metals like Cr(VI), Cd(II) and Cr(VI)+Cd(II) greatly influenced by carboxyl group of uronic acid such as glucuronic acid and galacturonic acid content of EPS (Ozturk et al., 2014). Balakiran and Thanasekaran (2012) used Box–Behnken model for experimental design to access the cobalt adsorption in *Lyngbya putealis* isolated from an electroplating industry. Optimal conditions for cobalt adsorption were found to be pH 6–6.5 in initial metal concentration. Higher concentration of cobalt removed from industrial effluents up to 50 mg l^{-1} by *L. putealis*. The composition of CPS can affect its ability to remove heavy metals. In *Gloeocapsa gelatinosa*, pH of solution is higher than this pKa value. In this condition, carboxyl group on the cyanobacterial cells deprotonates and is presented in the negatively charged form (COO$^-$) which could bind to positively charged Pb^{2+} ions (Raungsomboon et al., 2006).

Nine pre-treated marine algal biomass were evaluated for their metal uptake potential. Those nine cultures were *Ascophyllum nodosum, Durvillaea potatorum, Ecklonia radiata, E. maxima, Laminaria hyperbola, Laminaria japonica, Lessonia avicans,* and *Lessonia nigresense.* The metal uptake capacity of these algal biomass for lead (II), copper(II) and cadmium(II) ions were in the ranges of 1.0 to 1.6, 1.0 to 1.2 and 0.8 to 1.2 mmol/g, respectively (Yu et al., 1999). Kaewsarn (2002) studied the copper (II) biosorbent capacity of *Padina* sp. in wastewater streams. The metal uptake capacity of *Padina* sp. was pH dependent. The maximum adsorption capacity was obtained to be 0.80 mmol/g at a solution pH of about 5. The kinetics of adsorption by this biomass was rapid with 90% of total adsorption occurring within 15 min.

Commercial ion exchange resins were used to compare the heavy metal adsorption capacities of *D. potatorum* and *E. radiata*. Metal uptake was higher in low initial concentration. Some of the ion exchange resins, the ions of hard water (Ca^{2+} and $Mg^{2+)}$ or monovalent cations (Na^+ and K^+) do not significantly interfere with the binding of heavy metal ions. The biosorption of cadmium (II) and lead (II) ions from aqueous solution was evaluated using the algae nonliving biomass (*Rivularia bulata*) and the maximum adsorption capacities for Cd (II) and Pb(II) were found to be 26.36 and 34.30 mg/g (Kizilkaya et al., 2013). *S. muticum*, brown algae is widely used for biosorption studies. This algae efficiently adsorbed the hexavalent chromium from the aqueous solutions with maximum biosorption capacity equal to 196.1 mg/g. Similarly, copper and nickel was significantly removed by the *Sargassum filipendula* with maximum capacities of 1.324 and 1.070 mmol/g, respectively (Kleinübing et al., 2011). In *Sargassum* during nickel biosorption, carboxyl group play a major role in metal binding. Likewise, functional groups containing nitrogen and sulfur, such as amino/amido and sulfonate/thiol, were also involved in the adsorption of the heavy metals like lead and cadmium (Raize et al., 2004).

10.6 CONCLUSION

Biosorption is a cost-effective and promising process for metal removal in industrial effluents and wastewater treatment. Recovery and regeneration of biomass were more attractive and alternative process when compared to the conventional methods. Commercialization of biosorbent material is very much limited compared to the other biological products. A good biosorbents material may be useful to the society for removing heavy metals which accumulated by their own activities. Researchers will tend to find new strategy for improving biomass metal removing capability. Researchers may concentrate on the gene-specific metal binding studies which can give another insight for metal biosorption.

KEYWORDS

- microorganism
- exopolysaccharides
- heavy metals
- human health
- bacteria
- fungi
- algae
- biosorption
- environment

REFERENCES

Abbas, S. H., Ismail, I. M., Mostafa, T. M., & Sulaymon, A. H., (2014). Biosorption of heavy metals: A review. *J. Chem. Sci. Technol., 3*(4), 74–102.

Abdel-Ghani, N. T., & El-Chaghaby, G. A., (2014). Biosorption, for metal ions removal from aqueous solutions: A review of recent studies. *Int. J. L. Res. Sci. Technol., 3*(1), 24–42.

Abdel-Ghani, N., Hefny, M., & El-Chaghaby, G., (2007). Removal of lead from aqueous solution using low cost abundantly available adsorbents. *Int. J. Environ. Sci. Tech., 4*(1), 67–73.

Ahalya, N., Ramachandra, T. V., & Kanamadi, R. D., (2003). Biosorption of heavy metals. *Res. J. Chem. Environ., 7*, 71–79.

Ahluwalia, S. S., & Goyal, D., (2007). Microbial and plant derived biomass for removal of heavy metals from waste water. *Bioresour. Technol., 98*(12), 2243–2257.

Aksu, Z., Sag, Y., & Kutsal, T., (1992). The biosorption of copper (II) by *C. vulgaris* and *Z. ramigera. Environ. Technol., 13*(6), 579–586.

Al-Jaboobi, M., Zouahril, A., Tijane, M., Housni, A. E., Mennane, Z., Yachoul, H., & Bouksaim, M., (2014). Evaluation of heavy metals pollution in groundwater, soil and some vegetables irrigated with wastewater in the Skhirat region "Morocco." *J. Mater. Environ. Sci., 5*(3), 961–966.

Alkarkhi, A. F., Norli, M., Ahmad, I., & Easa, A. M., (2009). Analysis of heavy metal concentrations in sediments of selected estuaries of Malaysia - a statistical assessment. *Environ. Monit. Assess., 153*, 179–185.

Alluri, H. K., Ronda, S. R., Settalluri, V. S., Bondili, J. S., Suryanarayana, V., & Venkateshwar, P., (2007). Biosorption: An eco-friendly alternative for heavy metal removal. *Afr. J. Biotechnol., 6*(25), 2924–2931.

Al-Qodah., (2006). Biosorption of heavy metal ions from aqueous solutions by activated sludge. *Desalination, 196*, 164–176.

Al-Thagafi, Z., Arida, H., & Hassan, H., (2014). Trace toxic metal levels in canned and fresh food: A comparative study. *Int. J. Innov. Res. Sci. Eng. Technol., 3*(2), 8978–8989.

Altmann, L., Weinsberg, F., Sveinsson, K., Lilienthal, H., Wiegand, H., & Winneke, G., (1993). Impairment of long-term potentiation and learning following chronic lead exposure. *Toxicol. Lett., 66*(1), 105–112.

Amacher, M. C., Brown, R. W., Kotuby-Amacher, J., & Willis, A., (1993). Adding sodium hydroxide to study metal removal in a stream affected by acid mine drainage. *US Department of Agriculture Forest Service Research Paper, 465*, 1–28.

Anna, J., Piotr, K., Krzystof, S., & Jacek, N., (2011). Bioaccumulation of metals in tissues of marine animals, Part 1: The role and impact of heavy metals on organisms. *Pol. J. Environ. Stud., 20*(5), 1117–1125.

Ashkenazy, R., Gottlieb, L., & Yannai, S., (1997). Characterization of acetone-washed yeast biomass functional groups involved in lead biosorption. *Biotechnol. Bioeng., 55*(1), 1–10.

Atchison, G. J., Henry, M. G., & Sandheinrich, M. B., (1987). Effects of metals on fish behavior: A review. *Environ. Biol. Fishes, 18*(1), 11–25.

Ayandiran, T. A., Fawole, O. O., Adewoye, S. O., & Ogundiran, M. A., (2009). Bioconcentration of metals in the body muscle and gut of *Clarias gariepinus* exposed to sublethal concentrations of soap and detergent effluent. *J. Cell Anim. Biol., 3*, 113–118.

Baatrup, E., (1991). Structural and functional effects of heavy metals on the nervous system, including sense organs, of fish. *Comp. Biochem. Physiol., 100*(1 & 2), 253–257.

Babak, L., Supinova, P., Zichova, M., Burdychova, R., & Vitova, E., (2012). Biosorption of Cu, Zn, and Pb by thermophilic bacteria – effect of biomass concentration on biosorption capacity. *Acta. Univ. Agric. Silvic. Mendelianae Brun., 60*, 9–18.

Bae, W., Wu, C. H., Kostal, J., Mulchandani, A., & Chen, W., (2003). Enhanced mercury biosorption by bacterial cells with surface-displayed MerR. *Appl. Environ. Microbiol., 69*(6), 3176–3180.

Bai, R. S., & Abraham, T. E., (2003). Studies on chromium (VI) adsorption-desorption using immobilized fungal biomass. *Bioresour. Technol., 87*(1), 17–26.

Bai, S. R., & Abraham, T. E., (1998). Studies on biosorption of chromium (VI) by dead fungal biomass. *J. Sci. Ind. Res., 57*, 821–824.

Balakiran, K., & Thanasekaran, K., (2012). An indigenous cyanobacterium, *Lyngbya putealis*, as biosorbent: Optimization based on statistical model. *Ecol. Eng., 42*, 232–236.

Baldi, F., Minacci, A., Pepi, M., & Scozzafava, A., (2001). Gel sequestration of heavy metals by *Klebsiella oxytoca* isolated from iron mat. *FEMS Microbiol. Ecol., 36*(2 & 3), 169–174.

Bang, S. S., & Pazirandeh, M., (1999). Physical properties and heavy metal uptake of encapsulated *Escherichia coli* expressing metal binding gene (NCP). *J. Micro Encapsulation, 16*(4), 489–499.

Banik, S., Das, K. C., Islam, M. S., & Salimullah, M., (2013). Recent advancements and challenges in microbial bioremediation of heavy metals contamination. *Biotechnol. Bioeng., 2*(1), 1035.

Barton, L. L., & Tomei, F. A., (1995). *Characteristics and Activities of Sulfate-Reducing Bacteria* (pp. 1–17). Plenum Press, New York.

Bertocchi, C., Navarini, L., & Ceshro, A., (1990). Polysaccharides from Cyanobacteria. *Carbohydr. Polym., 12*(2), 127–153.

Beveridge, T. J., & Murray, R. G. E., (1980). Sites of metal deposition in the cell wall of *Bacillus subtilis. J. Bacteriol., 141*(2), 876–883.

Beveridge, T. J., (1989). The role of cellular design in bacterial metal accumulation and mineralization. *Ann. Rev. Microbiol., 43*, 147–171.

Blaudez, D., Botton, B., & Chalot, M., (2000). Cadmium uptake and subcellular comparmentation in the ectomycorrhizal fungus *Paxillus involutus. Microbiology, 146*, 1109–1117.

Bloom, P. R., & McBride, M. B., (1979). Metal ion binding and exchange with hydrogen ions acid-washed peat. *Soil Sci. Soc. Am., 43*, 687–692.

Boening, D. W., (2000). Ecological effects, transport, and fate of mercury: A general review. *Chemosphere, 40*(12), 1335–1351.

Boonyodying, K., Watcharasupat, T., Yotpanya, W., Kitti, T., Kawang, W., Kunthalert, D., & Sitthisak, S., (2012). Factors affecting the binding of a recombinant heavy metal-binding domain (CXXC motif) protein to heavy metals. *Environment Asia, 5*(2), 70–75.

Boswell, C. D., Dick, R. E., Eccles, H., & Macaskie, L. E., (2001). Phosphate uptake and release by *Acinetobacter johnsonii* in continuous culture and coupling of phosphate release to heavy metal accumulation. *J. Ind. Microbiol. Biotechnol., 26*(6), 333–340.

Bradl, H., (2002). *Heavy Metals in the Environment: Origin, Interaction and Remediation* (pp. 1–27). Academic Press, London.

Brady, D., & Duncan, J. R., (1994). Binding of heavy metals by the cell walls of *Saccharomyces cerevisiae. Enzyme Microb. Technol., 16*(7), 633–638.

Bridge, T. A. M., White, C., & Gadd, G. M., (1999). Extracellular metal-binding activity of the sulphate-reducing bacterium *Desulfococcus multivorans. Microbiol., 145*(10), 2987–2995.

Brierley, C. L., (1990). Bioremediation of metal-contaminated surfaces and groundwater. *Geomicrobiol. J., 8*(3 & 4), 201–223.

Brinza, L., Dring, M. J., & Gavrilescu, M., (2007). Marine micro- and macro-algal species as biosorbents for heavy metals. *Environ. Eng. Manage J., 6*, 237–251.

Bryan, G. W., & Langston, W. J., (1992). Bioavailability, accumulation and effects of heavy metals in sediments with special reference to United Kingdom estuaries: A review. *Environ. Pollut., 76*, 89–131.

Bubb, J. M., & Lester, J. N., (1994). Anthropogenic heavy metals inputs to low land river systems, a case study. The River Stour UK. *Water, Air, Soil Pollut., 78*, 279–296.

Bulut, E., Ozacar, M., & Sengil, I. A., (2008). Adsorption of malachite green onto bentonite: Equilibrium and kinetic studies and process design. *Micropor. Mesopor. Mater, 115*(3), 234–246.

Cabral, J. P. S., (1992a). Mode of antibacterial action of dodine (dodecylguanidine monoacetate) in *Pseudomonas syringae. Can. J. Microbiol., 38*(2), 115–123.

Cabral, J. P. S., (1992b). Selective binding of metal ions to *Pseudomonas syringae* cells. *Microbios., 71*, 47–53.

Camargo, F. A., Okeke, B. C., Bento, F. M., & Frankenberger, W. T., (2003). *In vitro* reduction of hexavalent chromium by a cell-free extract of *Bacillus* sp. ES 29 stimulated by Cu^{2+}. *Appl. Microbiol. Biotechnol., 62*(5), 569–573.

Cao, J., Zhang, G., Mao, Z., Fang, Z., & Yang, C., (2009). Precipitation of valuable metals from bioleaching solution by biogenic sulfides. *Miner. Eng., 22*(3), 289–295.

Chakraborty, P., Ramteke, D., Chakraborty, S., & Nagendernath, B., (2014). Changes in metal contamination levels in estuarine sediments around India – an assessment. *Mar. Pollut. Bull., 78*(1 & 2), 15–25.

Chan, A., Salsali, H., & McBean, E., (2014). Heavy metal removal (copper and zinc) in secondary effluent from wastewater treatment plants by microalgae. *ACS Sustainable Chem. Eng., 2*(2), 130–137.

Chandrashekar, R., Curtis, K. C., Lu, W. H., & Weil, G. J., (1998). Molecular cloning of an enzymatically active thioredoxin peroxidase from *Onchocerca volvulus. Mol. Biochem. Parasitol., 93*(2), 309–312.

Chang, C. Y., Yu, H. Y., Chen, J. J., Li, F. B., Zhang, H. H., & Liu, C. P., (2014). Accumulation of heavy metals in leaf vegetables from agricultural soils and associated potential health risks in the Pearl River Delta, South China. *Environ. Monit. Assess, 186*(3), 1547–1560.

Chang, J. S., Law, R., & Chang, C. C., (1997). Biosorption of lead, copper and cadmium by biomass of *Pseudomonas aeruginosa* PU21. *Water Research, 31*(7), 1651–1658.

Chien, C. C., Lin, B. C., & Wu, C. H., (2013). Biofilm formation and heavy metal resistance by an environmental *Pseudomonas* sp. *Biochem. Eng. J., 78*, 132–137.

Choi, S. B., & Yun, Y. S., (2006). Biosorption of cadmium by various types of dried sludge: An equilibrium study and investigation of mechanisms. *J. Hazard Mater, 138*(2), 378–383.

Chojnacka, K., (2009). *Biosorption and Bioaccumulation in Practice* (pp. 1–75). Nova Science Publishers Inc., New York.

Chong, H. L. H., Chia, P. S., & Ahmad, M. N., (2013). The adsorption of heavy metal by Bornean oil palm shell and its potential application as constructed wetland media. *Bioresour. Technol., 130*, 181–186.

Chronopoulos, J., Haidouti, C., Chronopoulou, A., & Massas, I., (1997). Variations in plant and soil lead and cadmium content in urban parks in Athens, Greece. *Sci. Total Environ., 196*(1), 91–98.

Cooksey, D. A., (1993). Copper uptake and resistance in bacteria. *Mol. Microbiol., 7*(1), 1–5.

Crist, R. H., Oberholser, K., Shank, N., & Nguyen, M., (1981). Nature of bonding between metallic ions and algal cell walls. *Environ. Sci. Technol., 15*(10), 1212–1217.

Crounse, R. G., Pories, W. J., Bray, J. T., & Mauger, R. L., (1983). Geochemistry and man: Health and disease; 1. essential elements, 2. elements possibly essential, those toxic and others. In: Thornton, I., (ed.), *Applied Environmental Geochemistry* (pp. 267–333). Academic Press, London.

Crowell, A. D., (1966). Surface forces and solid-gas interface. In: Alison, F., (ed.), *The Solid-Gas Interface.* Marcel Dekker, New York, (pp. 175-202).

Dang, V. B. H., Doan, H. D., Dang-Vu, T., & Loh, A., (2009). Equilibrium and kinetics of biosorption of cadmium(II) and copper(II) ions by wheat straw. *Bioresour. Technol., 100*(1), 211–219.

Das, N., Vimala, R., & Karthika, P., (2008). Biosorption of heavy metals – an overview. *Indian J. Biotechnol., 7*, 159–169.

Davis, T. A., Mucci, A., & Volesky, B., (2003). A review of the biochemistry of heavy metal biosorption by brown algae. *Water Res., 37*(18), 4311–4330.

Dhokpande, R., & Kaware, P., (2013). Biological methods for heavy metal removal - a review. *Int. J. Eng. Sci. Innov. Technol., 2*(5), 304–309.

Dinesh, K. S., Santhanam, P., Jayalakshmi, T., Nandakumar, R., Ananth, S., & Shenbaga, D. A., (2015). Excessive nutrients and heavy metals removal from diverse waste waters using marine microalgae *Chlorella marina* (Butcher). *Indian J. Mar. Sci., 44*(1), 97–103.

Duffus, J. H., (2002). Heavy metals - a meaningless term, IUPAC technical report. *International Union of Pure and Applied Chemistry, 74*(5), 793–807.

Ernst, E., (2002). Heavy metals in traditional Indian remedies. *Eur. J. Clin. Pharmacol., 57*, 891–896.

Ezemonye, L. I. N., & Kadiri, M. O., (2000). Biorestoration of the aquatic ecosystem: The African perspective. *Environ. Rev., 3*, 137–147.

Farombi, E. O., Adelowo, O. A., & Ajimoko, Y. R., (2007). Biomarkers of oxidative stress and heavy metal levels as indicators of environmental pollution in African Cat fish (*Clarias gariepinus*) from Nigeria ogun river. *Int. J. Environ. Res. Public Health, 4*(2), 158–165.

Fendorf, S., La Force, M. J., & Li, G. C., (2004). Temporal changes in soil partitioning and bioaccessibility of arsenic, chromium, and lead. *J. Environ. Qual., 33*(6), 2049–2055.

Filali, B. K., Taoufik, J., Zeroual, Y., Dzairi, F. A. Z., Talbi, M., & Blaghen, M., (1999). Waste water bacterial isolates resistant to heavy metals and antibiotics. *Curr. Microbiol., 41*(3), 151–156.

Flemming, H. K., (1995). Sorption sites in biofilms. *Water Sci. Technol., 32*(8), 27–33.

Foo, K. Y., & Hameed, B. H., (2010). Insights into the modeling of adsorption isotherm systems. *Biochem. Eng. J., 156*, 2–10.

Fourest, E., Serre, A., & Roux, J. C., (1996). Contribution of carboxyl groups to heavy metal binding sites in fungal wall. *Toxicol. Environ. Chem., 54*(1–4), 1–10.

Friis, N., & Myers-Keith, P., (1986). Biosorption of uranium and lead by *Streptomyces longwoodensis. Biotechnol. Bioeng., 28*(1), 21–28.

Gadd, G. M., (1993). Interaction of fungi with toxic metals. *Phytologist, 124*(1), 25–60.

Gadd, G. M., (2000). Bioremedial potential of microbial mechanisms of metal mobilization and immobilization. *Curr. Opinion Biotechnol., 11*(3), 271–279.

Gadd, G. M., (2009). Biosorption: Critical review of scientific rationale, environmental importance and significance for pollution treatment. *J. Chem. Technol. Biotechnol., 84*(1), 13–28.

Gadd, G. M., White, C., & De Rome, L., (1998). Heavy metal and radionuclide by fungi and yeasts. In: Norris, P. R., Kelly, D. P., & Rowe, A., (eds.), *Biohydrometallurgy* (pp. 421–435). Wilts, UK.

Galun, M., Galun, E., Siegel, B. Z., Keller, P., Lehr, H., & Siege, S. M., (1987). Removal of metal ions from aqueous solutions by Pencilium biomass: Kinetic and uptake parameters. *Water Air Soil Pollut., 33*(3), 359–371.

Galun, M., Keller, P., Malki, D., Feldstein, H., Galun, E., Seigel, S. M., & Seigel, B. Z., (1983). Removal of Uranium (IV) from solution by fungal biomass and fungal wall related polymers. *Science, 219*(4582), 285–286.

Garnham, G. W., (1997). The use of algae as metal biosorbents. In: Wase, J., & Forster, C., (eds.), *Biosorbents for Metal Ions* (pp. 11–37). Taylor and Francis, London.

Garnham, G. W., Avery, S. V., Codd, G. A., & Gadd, G. M., (1994). Interactions of microalgae and cyanobacteria with toxic metals and radionuclides: Physiology and environmental implications. In: Dyer, K. R., & Orth, R., (eds.), *Changes in Fluxes in Estuaries – Implications From Science to Management* (pp. 289–293). Olsen and Olsen, Fredensborg, Denmark.

Geddie, J. L., & Sutherland, I. W., (1993). Uptake of metals by bacterial polysaccharides. *J. Appl. Bacteriol., 74*(4), 467–472.

Ghosh, A. K., Bhattacharyya, P., & Pal, R., (2004). Effect of arsenic contamination on microbial biomass and its activities in arsenic contaminated soils of Gangetic West Bengal, India. *Environ. Int., 30*(4), 491–499.

Glanze, W. D., (1996). *Mosby Medical Encyclopedia.* C. V. Mosby Co., St. Louis, USA.

Gourdon, R., Bhende, S., Rus, E., & Sofer, S. S., (1990). Comparison of cadmium biosorption by Gram-positive and Gram-negative bacteria from activated sludge. *Biotechnol. Lett., 12*(11), 839–842.

Govind, P., & Madhuri, S., (2014). Heavy metals causing toxicity in animals and fishes. *Research Journal of Animal Veterinary and Fishery Sciences, 2*(2), 17–23.

Goyal, N., Jain, S. C., & Banerjee, U. C., (2003). Comparative studies on the microbial adsorption of heavy metals. *Adv. Environ. Res., 7*(2), 311–319.

Goyer, R. A., (2001). Toxic effects of metals. In: Klaassen, C. D., & Doull, C., (eds.), *Toxicology: The Basic Science of Poisons* (pp. 811–867). McGraw-Hill Publisher, New York.

Greene, B., & Darnall, D. W., (1998). Temperature dependence of metal ions sorption by *Spirulina. Biorecovery, 1*, 27–41.

Guerra, F., Trevizam, A. R., Muraoka, T., Marcante, N. C., & Canniatti-Brazaca, S. G., (2012). Heavy metals in vegetables and potential risk for human health. *Scientia Agricola, 69*(1), 54–60.

Gupta, V. K., Rastogi, A., & Nayak, A., (2010). Biosorption of nickel onto treated alga (*Oedogonium hatei*): Application of isotherm and kinetic models. *J. Colloid Interface Sci., 342*(2), 533–539.

Hakkila, K. M., Nikander, P. A., Junttila, S. M., Lamminmäki, U. J., & Virta, M. P., (2011). Cd-specific mutants of mercury-sensing regulatory protein MerR, generated by directed evolution. *Appl. Environ. Microbiol., 77*(17), 6215–6224.

Hale, W. G., & Margham, J. P., (1998). *Collins Dictionary of Biology*, Collins, Glasgow, (p. 518).

He, Z. L., Yang, X. E., & Stoffella, P. J., (2005). Trace elements in agroecosystems and impacts on the environment. *J. Trace Elem. Med. Biol., 19*, 125–140.

Hemambika, B., Rani, M. J., & Kannan, V. R., (2011). Biosorption of heavy metals by immobilized and dead fungal cells: A comparative assessment. *J. Ecol. Nat. Environ., 3*(5), 168–175.

Hobman, J. L., (2007). MerR family transcription activators: Similar designs, different specificities. *Mol. Microbiol., 63*(5), 1275–1278.

Horokoshi, T., Nakajima, A., & Sakaguchi, T., (1981). Studies on the accumulation of heavy metal elements in biological systems: Accumulation of uranium by microorganisms. *Eur. J. Appl. Microbiol. Biotechnol., 12*(2), 90–96.

Hu, H., (2002). In: McCally, M., (ed.), *Human Health and Heavy Metal Exposure, Life Support: The Environment and Human Health* (pp. 1–12). MIT Press, USA.

Ibrahim, D., Froberg, B., Wolf, A., & Rusyniak, D. E., (2006). Heavy metal poisoning: Clinical presentations and pathophysiology. *Clin. Lab. Med., 26*(1), 67–97.

Ibrahim, W. M., (2011). Biosorption of heavy metal ions from aqueous solution by red macroalgae. *J. Hazard Mater, 192*(3), 1827–1835.

Issazadeh, K., Jahanpour, N., Pourghorbanali, F., Raeisi, G., & Faekhondeh, J., (2013). Heavy metals resistance by bacterial strains. *Ann. Biol. Res., 4*(2), 60–63.

Jaishankar, M., Tseten, T., Anbalagan, N., Mathew, B. B., & Beeregowda, K. N., (2014). Toxicity, mechanism and health effects of some heavy metals. *Interdiscip. Toxicol., 7*(2), 60–72.

Jarup, L., (2003). Hazards of heavy metal contamination. *Br Med. Bull., 68*(1), 167–182.

Javaid, A., Bajwa, R., Shafique, U., & Anwar, J., (2011). Removal of heavy metals by adsorption on *Pleurotus ostreatus*. *Biomass Bioenergy, 35*(5), 1675–1682.

Jayashree, S., Thangaraju, N., & Joel, G. J., (2012). Toxic effects of chromium on the aquatic cyanobacterium *Oscillatoria* sp. and removal of chromium by biosorption. *Journal of Experimental Sciences, 3*(5), 28–34.

Jezierska, B., Ługowska, K., & Witeska, M., (2009). The effects of heavy metals on embryonic development of fish (a review). *Fish Physiol. Biochem., 35*(4), 625–640.

Jiwan, S., & Ajay, S. K., (2011). Effects of heavy metals on soil, plants, human health and aquatic life. *Int. J. Res. Chem. Environ., 1*(2), 15–21.

Johnson, K. J., Ams, D. A., Wedel, A. N., Szymanowski, J. E. S., Weber, D. L., Schneegurt, M. A., & Fein, B., (2007). The impact of metabolic state on Cd adsorption onto bacterial cells. *Geobiology, 5*, 211–218.

Kaewchai, S., & Prasertsan, P., (2002). Biosorption of heavy metal by thermotolerant polymer-producing bacterial cells and the bioflocculant. *Songklanakarin J. Sci. Technol., 24*(3), 421–430.

Kaewsarn, P., (2002). Biosorption of copper (II) from aqueous solutions by pre-treated biomass of marine algae *Padina* sp. *Chemosphere, 47*(10), 1081–1085.

Kapoor, A., & Viraraghavan, T., (1998). Biosorption of heavy metals on *Aspergillus niger*: Effect of pretreatment. *Biores. Technol., 63*(2), 109–113.

Khan, A., Hussain, H. I., Sattar, A., Khan, M. Z., & Abbas, R. Z., (2014). Toxico-pathological aspects of arsenic in birds and mammals: A review. *Int. J. Agric. Biol., 16*(6), 1213–1224.

Khan, M. S., Zaidi, A., Wani, P. A., & Oves, M., (2009). Role of plant growth promoting rhizobacteria in the remediation of metal contaminated soils. *Environ. Chem. Lett., 7*(1), 1–19.

Kim, S. B., Lee, S. J., Kim, C. K., Kwon, G. S., Yoon, B. D., & Oh, H. M., (1998). Selection of microalgae for advanced treatment of swine wastewater and optimization of treatment condition. *Korean Journal of Applied Microbiology and Biotechnology, 26*(1), 76–82.

Kinoshita, H., Sohma, Y., Ohtake, F., Ishida, M., Kawai, Y., Saito, T., & Kitazawa, K., (2013). Biosorption of heavy metals by lactic acid bacteria and identification of mercury binding protein. *Res. Microbiol., 164*(7), 701–709.

Kizilkaya, B., Akgül, R., & Turker, G., (2013). Utilization on the removal Cd(II) and Pb(II) ions from aqueous solution using nonliving *Rivularia bulata* algae. *J. Dispers. Sci. Technol., 34*(9), 1257–1264.

Kleinübing, S. J., Da Silva E. A., Da Silva, M. G. C., & Guibal, E., (2011). Equilibrium of Cu(II) and Ni(II) biosorption by marine alga *Sargassum filipendula* in a dynamic system: Competitiveness and selectivity. *Bioresour. Technol., 102*(7), 4610–4617.

Kong, J. Y., Lee, H. W., Hong, J. W., Kang, Y. S., Kim, J. D., Chang, M. W., & Bae, S. K., (1998). Utilization of a cell-bound polysaccharide produced by the marine bacterium *Zoogloea* sp.: New biomaterial for metal adsorption and enzyme immobilization. *J. Marine Biotechnol., 6*(2), 99–103.

Krowiak, A. W., Szafran, R. G., & Modelski, S., (2011). Biosorption of heavy metals from aqueous solutions onto peanut shell as a low-cost biosorbent. *Desalination, 265*(1–3), 126–134.

Kumar, A. V., Naif, A. D., & Hilal, N., (2009). Study of various parameters in the biosorption of heavy metals on activated sludge. *World Appl. Sci. J., 5*, 32–40.

Kuyucak, N., & Volesky, B., (1988). Biosorbents for recovery of metals from industrial solutions. *Biotechnol. Lett., 10*(2), 137–142.

Lee, S. Y., Choi, J. H., & Xu, Z. H., (2003). Microbial cell-surface display. *Trends Biotechnol., 21*(1), 45–52.

Li, P. F., Mao, Z. Y., Rao, S. J., Wang, X. M., Min, M. Z., Qiu, L. W., & Liu, Z. L., (2004). Biosorption of uranium by lake harvested biomass from cyanobacterium bloom. *Bioresour. Technol., 94*(2), 193–195.

Li, P. S., & Tao, H. C., (2013). Cell surface engineering of microorganisms towards adsorption of heavy metals. *Crit. Rev. Microbiol., 41*(2), 140–149.

Li, P., Harding, S. E., & Liu, Z., (2001). Cyanobacterial polysaccharides: Their nature and potential biotechnological applications. *Biotechnol. Gene Engin. Reviews, 18*, 375–404.

Liang, L. N., He, B., Jiang, G. B., Chen, D. Y., & Yao, Z. W., (2004). Evaluation of mollusks as biomonitors to investigate heavy metal contaminations along the Chinese Bohai Sea. *Sci. Total Environ., 324*, 105–113.

Luo, J., Xiao, X., & Luo, S. L., (2010). Biosorption of cadmium(II) from aqueous solutions by industrial fungus *Rhizopus cohnii. Trans Nonferrous Met. Soc. China, 20*(6), 1104–1111.

Macaskie, L. E., Empson, R. M., Cheetham, A. K., Grey, C. P., & Skarnulis, A. J., (1992). Uranium bioaccumulation by a *Citrobacter* sp. as a result of enzymatically mediated growth of polycrystalline HUO_2PO_4. *Science, 257*(5071), 782–784.

Madigan, M. T., Martinko, J. M., & Parker, J., (1997). *Brock Biology of Microorganisms* (8th edn.), Prentice Hall, Upper Saddle River, NJ.

Mafuyai, G. M., Eneji, I. S., & Shaato, R., (2014). Concentration of heavy metals in respirable dust in Jos Metropolitan Area, Nigeria. *Open Journal of Air Pollution, 3*(1), 10–19.

Malkoc, E., & Nuhoglu, Y., (2005). Investigations of Nickel (II) removal from aqueous solutions using tea factory waste. *J. Hazard Mater, 127*(1–3), 120–128.

Mansour, R. S., (2014). The pollution of tree leaves with heavy metal in Syria. *Int. J. Chem. Tech. Res., 6*(4), 2283–2290.

Mapolelo, M., & Torto, N., (2004). Trace enrichment of metal ions in aquatic environments by *Saccharomyces cerevisiae. Talanta, 64*(1), 39–47.

Marques, P. A., Pinheiro, H. M., Teixeira, J. A., & Rosa, M. F., (1999). Removal efficiency of Cu^{2+}, Cd^{2+} and Pb^{2+} by waste brewery biomass: pH and cation association effects. *Desalination, 124*(1–3), 137–144.

McLean, J. S., Lee, J. U., & Beveridge, T. J., (2002). Interactions of bacteria and environmental metals, fine grained mineral development, and bioremediation strategies. In: Huang, P. M., Bollag, J. M., & Senesi, N., (eds.), *Interactions Between Soil Particles and Microorganisms* (pp. 228–261). John Wiley and Sons, New York.

Mendie, U., (2005). The nature of water. In: *The Theory and Practice of Clean Water Production for Domestic and Industrial Use* (pp. 1–21). Lacto-Medals Publishers, Lagos.

Modak, J. M., & Natarajan, K. A., (1995). Biosorption of metals using nonliving biomass - a review. *Miner Metall. Proc., 12*(4), 189–196.

Momodu, M. A., & Anyakora, C. A., (2010). Heavy metal contamination of ground water: The Surulere case study. *Res. J. Environ. Earth Sci., 2*(1), 39–43.

Morel, F. M. M., (1983). *Principles of Aquatic Chemistry* (pp. 237–266). John Wiley and Sons, New York.

Morira, M. T., Echeverria, J., & Garrido, J., (2002). Bioavailability of heavy metals in soil amended with sewage sledge. *Can. J. Soil Sci., 82*(4), 433–438.

Muraleetharan, T. R., Iyengar, L., & Venkobachar, C., (1995). Screening of tropical wood rotting mushrooms for copper biosorption. *Appl. Environ. Microbiol., 61*(9), 3507–3508.

Naiya, T. K., Bhattacharya, A. K., Mandal, S., & Das, S. K., (2009). The sorption of lead (II) ions on rice husk ash. *J. Hazard Mater, 163*(2 & 3), 1254–1264.

Naja, G., Mustin, C., Volesky, B., & Berthelin, J., (2005). A high-resolution titrator: A new approach to studying binding sites of microbial biosorbents. *Water Res., 39*(4), 579–588.

Neeti, K., & Prakash, T., (2013). Effects of heavy metal poisoning during pregnancy. *Int. Res. J. Environment Sci., 2*(1), 88–92.

Noghabi, K. A., Zahir, H. S., & Yoon, S. C., (2007). The production of a cold-induced extracellular biopolymer by *Pseudomonas fluorescents* BM07 under various growth conditions and its role in heavy metals adsorption. *Process Biochem., 42*(5), 847–855.

Nourbakhsh, M., Sag, Y., Ozer, D., Aksu, Z., Katsal, T., & Calgar, A., (1994). A comparative study of various biosorbents for removal of chromium (VI) ions from industrial wastewater. *Process Biochem., 29*(1), 1–5.

Nriagu, J. O., (1989). A global assessment of natural sources of atmospheric trace metals. *Nature, 338*, 47–49.

Orhan, Y., & Buyukgungor, H., (1993). The removal of heavy metals by using agriculture wastes. *Wat. Sci. Tech., 28*(2), 247–255.

Ozdemir, G., Ceyhan, N., Ozturk, T., Akirmak, F., & Cosar, T., (2004). Biosorption of chromium (VI), cadmium (II) and copper (II) by *Pantoea* sp. TEM 18. *Chem. Eng. J., 102*(3), 249–253.

Ozdemir, G., Ozturk, T., Ceyhan, N., Isler, R., & Cosar, T., (2003). Heavy metal biosorption by biomass of *Ochrobactrum anthropi* producing exopolysaccharide in activated sludge. *Bioresour. Technol., 90*(1), 71–74.

Ozturk, S., Aslim, B., Suludere, Z., & Tan, S., (2014). Metal removal of cyanobacterial exopolysaccharides by uronic acid content and monosaccharide composition. *Carbohydr. Polym., 101*, 265–271.

Pacyna, J. M., (1996). Monitoring and assessment of metal contaminants in the air. In: Chang, L. W., Magos, L., & Suzul, T., (eds.), *Toxicology of Metals* (pp. 9–28). CRC Press, Boca Raton, FL.

Pahlavanzadeh, H., Keshtkar, R., Safdari, J., & Abadi, Z., (2010). Biosorption of nickel (II) from aqueous solution by brown algae: Equilibrium, dynamic and thermodynamic studies. *J. Hazard Mater., 175*(1–3), 304–310.

Pandey, A., Nigam, P., & Singh, D., (2001). Biotechnological treatment of pollutants. *Chem. Ind. Digest, 14*, 93–95.

Petersen, F., Aldrich, C., Esau, A., & Qi, B. C., (2005). Biosorption of heavy metals from aqueous solutions. *WRC Report No.,* 1259/1/05.

Phipps, D. A., (1981). Chemistry and biochemistry of trace metals in biological systems. In: Lepp, N. W., (ed.), *Effect of Heavy Metal Pollution on Plants: Effects of Trace Metals on Plant Function* (pp. 1–54). Elsevier Applied Science Publ., New Jersey.

Podgorskii, V. S., Kasatkina, T. P., & Lozovaiva, O. G., (2004). Yeasts - biosorbents of heavy metals. *Mikrobiol. Z., 66*(1), 91–103.

Prasher, D., (2009). Heavy metals and noise exposure: Health effects. *Noise Health, 11*(44), 141–144.

Prescott, L. M., Harley, J. P., & Klein, D. A., (2008). *Microbiology* (pp. 57–59). McGraw Hill Publishers, Boston.

Puranik, P. R., & Paknikar, K. M., (1997). Biosorption of lead and zinc from solutions using *Streptoverticilium cinnamoneun* waste biomass. *J. Biotechnol., 55*(2), 113–124.

Qin, J., Song, L. Y., Brim, H., Daly, M. J., & Summers, A. O., (2006). Hg (II) sequestration and protection by the MerR metal-binding domain (MBD). *Microbiol., 152*(3), 709–719.

Rai, P. K., & Tripathi, B. D., (2007). Removal of heavy metals by the nuisance cyanobacteria *Microcystis* in continuous cultures: An eco-sustainable technology. *Environmental Sciences, 4*(1), 53–59.

Raize, O., Argaman, Y., & Yannai, S., (2004). Mechanisms of biosorption of different heavy metals by brown marine macroalgae. *Biotechnol. Bioeng., 87*(4), 451–458.

Rajeshwari, R. T., & Sailaja, N., (2014). Impact of heavy metals on environmental pollution. *J. Chem. Pharm. Sci., 3S,* 70–72.

Rangabhashiyam, N., Anu, M. S., Giri, N. M. N., & Selvaraju, N., (2014). Relevance of isotherm models in biosorption of pollutants by agricultural byproducts. *Journal of Environmental Chemical Engineering, 2*(1), 398–414.

Rani, J. M., Hemambika, B., Hemapriya, J., & Rajesh, K. V., (2010). Comparative assessment of heavy metal removal by immobilized and dead bacterial cells: A biosorption approach. *Afr. J. Environ. Sci. Technol., 4*(2), 77–83.

Rathinam, A., Maharshi, B., Janardhanan, S. K., Jonnalagadda, R. R., & Nair, B. U., (2010). Biosorption of cadmium metal ion from simulated wastewaters using *Hypnea valentiae* biomass: A kinetic and thermodynamic study. *Bioresour. Technol., 101*(5), 1466–1470.

Raungsomboon, S., Chidthaisong, A., Bunnag, B., Inthorn, D., & Harvey, N. W., (2006). Production, composition and Pb^{2+} adsorption characteristics of capsular polysaccharides extracted from cyanobacterium *Gloeocapsa gelatinosa. Water Res., 40*(20), 3759–3766.

Remoundaki, E., Hatzikioseyian, A., Kousi, P., & Tsezos, M., (2003). The mechanism of metals precipitation by biologically generated alkalinity in biofilm reactors. *Water Res., 37,* 3843–3854.

Rich, G., & Cherry, K., (1987). *Hazardous Waste Treatment Technologies.* Pudvan Publishers, Northbrook, IL.

Rincon, J. F., Gonzalez, F., Ballester, A., Blazquez, M. L., & Munoz, J. A., (2005). Biosorption of heavy metals by chemically-activated alga *Fucus vesiculosus. J. Chem. Technol. Biotechnol., 80*(12), 1403–1407.

Romera, E., Gonzalez, F., Ballester, A., Blazquez, M. L., & Munoz, J. A., (2006). Biosorption with algae: A statistical review. *Crit. Rev. Biotechnol., 26*(4), 223–235.

Sabat, S., Kavitha, R. V., Shantha, S. L., Gopika, N., Megha, G., & Niranjana, C., (2012). Biosorption: An eco-friendly technique for the removal of heavy metals. *Indian J. Appl. Res., 2*(3), 2249–2255.

Sag, Y., & Kutsal, T., (2001). Recent trends in the biosorption of heavy metals: A review. *Biotechnol. Bioprocess Eng., 6*, 376–385.

Saha, J. K., Selladurai, R., Coumar, M. V., Dotaniya, M. L., Kundu, S., & Patra, A. K., (2017). *Soil Pollution - An Emerging Threat to Agriculture* (pp. 75–104). Springer, Singapore.

Sakaguchi, T., & Nakajima, A., (1991). Accumulation of heavy metals such as uranium and thorium by microorganisms. In: Smith, R. W., & Misra, M., (eds.), *Mineral Bioprocessing*. The Minerals, Metals and Materials Society, Warrendale, Pennsylvania, USA.

Scott, J. A., & Palmer, S. J., (1990). Sites of cadmium uptake in bacteria used for biosorption. *Appl. Microbiol. Biotechnol., 33*(2), 221–225.

Shallari, S., Schwartz, C., Hasko, A., & Morel, J. L., (1998). Heavy metals in soils and plants of serpentine and industrial sites of Albania. *Sci. Total Environ., 209*(2 & 3), 133–142.

Sharma, B., Singh, S., & Siddiqi, N. J., (2014). Biomedical implications of heavy metals induced imbalances in redox systems. *Biomed. Res. Int.*, 640754, doi: 10.1155/2014/640754.

Siebert, A., Bruns, I., Krauss, G. J., Miersch, J., & Markert, B., (1996). The use of the aquatic moss *Fontinalis antipyretica* L. ex Hedw. as a bioindicator for heavy metals: 1. fundamental investigations into heavy metal accumulation in *Fontinalis antipyretica* L. ex Hedw. *Science of the Total Environment, 177*(1–3), 137–144.

Simonescu, C. M., & Ferdes, M., (2012). Fungal biomass for Cu (II) uptake from aqueous system. *Pol. J. Environ. Stud., 21*(6), 1831–1939.

Singh, R. P., & Agarwal, M., (2008). Potential benefits and risks of land application of sewage sledge. *Waste Manage, 28*(2), 347–358.

Singh, R., Gautam, N., Mishra, A., & Gupta, R., (2011). Heavy metals and living systems: An overview. *Indian J. Pharmacol., 43*(3), 246–253.

Sitthisak, S., Kitti, T., Boonyonying, K., Wozniak, D., Mongkolsuk, S., & Jayaswal, R. K., (2012). McsA and the roles of metal-binding motif in *Staphylococcus aureus*. *FEMS Microbiol. Lett., 327*(2), 126–133.

Song, Q., & Li, J., (2014). Environmental effects of heavy metals derived from the e-waste recycling activities in China: A systematic review. *Waste Manage, 32*(12), 2587–2594.

Soundararajan, M., & Veeraiyan, G., (2014). Effect of heavy metal arsenic on haematological parameters of fresh water fish, *Tilapia Mossambic. Int. J. Modn. Res. Revs., 2*, 132–135.

Sposito, G., (1989). *The Chemistry of Soils*. Oxford University Press, New York.

Stihi, C., Radulescu, C., Busuioc, G., Popescu, I. V., Gheboianu, A., & Ene, A., (2011). Studies on accumulation of heavy metals from substrate to edible wild mushrooms. *Rom. Journ. Phys., 56*(1 & 2), 257–264.

Su, C., Jiang, L. Q., & Zhang, W. J., (2014). A review on heavy metal contamination in the soil worldwide: Situation, impact and remediation techniques. *Environmental Skeptics Critics, 3*(2), 24–38.

Swiatek, M. Z., & Krzywonos, M., (2014). Potentials of biosorption and bioaccumulation process for heavy metal removal. *Pol. J. Environ. Stud., 23*(2), 551–561.

Tao, H. C., Peng, Z. W., Li, P. S., Yu, T. A., & Su, J., (2013). Optimizing cadmium and mercury specificity of CadR-based *E. coli* biosensors by redesign of CadR. *Biotechnol. Lett., 35*(8), 1253–1258.

Tchounwou, P. B., Yedjou, C. G., Patlolla, A. K., & Sutton, D. J., (2014). Heavy metals toxicity and the environment. *NIH Public Access, 101*, 133–164.

Teitzel, G. M., & Parsek, M. R., (2003). Heavy metal resistance of biofilm and planktonic *Pseudomonas aeruginosa. Appl. Environ. Microbiol., 69*(4), 2313–2320.

Travieso, L., Benitez, F., Weiland, P., Sanchez, E., Dupeyron, R., & Dominguez, A. R., (1996). Experiments on immobilization of micro-algae for nutrient removal in wastewater treatments. *Bioresour. Technol., 55*(3), 181–186.

Tsezos, M., & Volesky, B., (1982). The mechanism of thorium biosorption by *Rhizopus arrhizus. Biotechnol. Bioeng., 24*(4), 955–969.

Tsezos, M., Remoundaki, E., & Hatzikioseyian, A., (2006). Biosorption - principles and applications for metal immobilization from waste-water streams, *EU Asia Workshop on Clean Production and Nano Technologies*. Seoul, Korea, http: //www.metal.ntua.gr/uploads/2748/299/LIBRO_SEOUL.pdf., pp. 23–33.

Urrutia, M. M., (1997). General bacterial sorption processes. In: Wase, J., & Forster, C., (eds.), *Biosorbents for Metal Ions* (pp. 39–66). CRC Press, London.

Vanloon, G. W., & Duffy, S. J., (2005). The Hydrosphere. In: *Environmental Chemistry: A Global Perspective* (2nd edn.). Oxford University Press, New York.

Veglio, F., & Beolchini, F., (1997). Removal of metals by biosorption: A review. *Hydrometallurgy, 44*(3), 301–316.

Vieira, M. J., & Melo, L. F., (1995). Effect of clay particles on the behavior of biofilms formed by *Pseudomonas fluorescens. Water Sci. Technol., 32*(8), 45–52.

Vinodhini, R., & Narayanan, M., (2009). The impact of toxic heavy metals on the hematological parameters in common carp (*Cyprinus carpio* L.). *Iran J. Environ. Health Sci. Eng., 6*(1), 23–28.

Volesky, B., & Holant, Z. R., (1995). Biosorption of heavy metals. *Biotechnol. Prog., 11*(3), 235–250.

Volesky, B., & May-Phillips, H. A., (1995). Biosorption of heavy metals by *Saccharomyces cerevisiae. Appl. Microbiol. Biotechnol., 42*(5), 797–806.

Volesky, B., & Naja, G., (2007). Biosorption technology: Starting up an enterprise. *Int. J. Technol. Transf. Commer., 6*(2–4), 196–211.

Volesky, B., (1990). *Biosorption of Heavy Metals* (pp. 139–172). CRC Press, USA.

Volesky, B., (2005). Detoxification of metal-bearing effluents: Biosorption for the next century. *Hydrometallurgy, 59*(2–3), 203–216.

Volesky, B., Sears, M., Neufeld, R. J., & Tsezos, M., (1983). Recovery of strategic elements by biosorption. *Annals. NY. Acad. Sc., 413*, 310–312.

Vosylienė, M. Z., (1999). The effect of heavy metals on haematological indices of fish (survey). *Acta. Zoologica. Lituanica., 9*(2), 76–82.

Wang, J. L., & Chen, C., (2009). Biosorbents for heavy metals removal and their future and review. *Biotechnol. Adv., 27*(2), 195–226.

Wang, J., & Chen, C., (2006). Biosorption of heavy metals by *Saccharomyces cerevisiae*: A review. *Biotechnol. Adv., 24*(5), 427–451.

Wang, X., Sato, T., Xing, B., & Tao, S., (2005). Health risks of heavy metals to the general public in Tianjin, China via consumption of vegetables and fish. *Sci. Total Environ., 350*(1–3), 28–37.

Wang, Z., Zeng, X., Geng, M., Chen, C., Cai, J., Yu, X., et al., (2015). Heath risk of heavy metal uptake by crop grown in a sewage irrigation Area in China. *Pol. J. Env. Stud., 24*(3), 1379–1386.

Wase, J., & Forster, C., (1997). *Biosorbents for Metal Ions* (pp. 67–50). Taylor and Francis Ltd., UK.

Webb, J. S., McGinness, S., & Lappin-Scott, H. M., (1998). Metal removal by sulphate-reducing bacteria from natural and constructed wetlands. *J. Appl. Microbiol., 84*(2), 240–248.

Wei, B., & Yang, L., (2009). A review of heavy metal contaminations in urban soils road dusts and agricultural soils from China. *Microchem. J., 94*(2), 99–107.

White, C., Sayer, J. A., & Gadd, G. M., (1997). Microbial solubilization and immobilization of toxic metals: Key biogeochemical processes for treatment of contamination. *FEMS Microbiol. Rev., 20*(3 & 4), 503–516.

Yan, G., & Viraraghavan, T. A. K., (2003). Heavy-metal removal from aqueous solution by fungus *Mucor rouxii*. *Water Res., 37*(18), 4486–4496.

Yu, Q., Matheickal, J. T., Yin, P., & Kaewsarn, P., (1999). Heavy metal uptake capacities of common marine macro algal biomass. *Wat. Res., 33*(6), 1534–1537.

Zakil, M. S., & Hammam, A. M., (2014). Aquatic pollutants and bioremediations. *Life Science Journal, 11*(2), 362–369.

Zhou, J. L., & Kiff, R. J., (1991). The uptake of copper from aqueous solution by immobilized fungal biomass. *J. Chem. Technol. Biotechnol., 52*(3), 317–330.

Zhou, Q., Zhang, J., Fu, J., Shi, J., & Jiang, G., (2008). Biomonitoring: An appealing tool for assessment of metal pollution in the aquatic ecosystem. *Analytica Chimica Acta., 606*, 135–150.

Zouboulis, A. L., Rousou, E. G., Matis, K. A., & Hancock, I. C., (1997). Removal of toxic metals from aqueous mixtures, Part 1, biosorption. *J. Chem. Technol. Biotechnol., 74*(5), 429–436.

CHAPTER 11

BIOREMEDIATION PROCESSES AND MOLECULAR TOOLS FOR DETOXIFICATION OF HEAVY METALS AND REHABILITATION OF CONTAMINATED ENVIRONMENTS USING MICROBIAL GENETIC AND GENOMIC RESOURCES

JEYABALAN SANGEETHA[1], DEVARAJAN THANGADURAI[2], RAVICHANDRA HOSPET[2], PURUSHOTHAM PRATHIMA[2], SHRINIVAS JADHAV[3], and MUNISWAMY DAVID[3]

[1]Department of Environmental Science, Central University of Kerala, Kasaragod 671316, Kerala, India

[2]Department of Botany, Karnatak University, Dharwad, 580003, Karnataka, India

[3]Department of Zoology, Karnatak University, Dharwad, 580003, Karnataka, India

11.1 INTRODUCTION

Heavy metals are one among the contaminants of the environment. Heavy metal contamination has become crucial problem in the environment, in which human activities also contribute for producing heavy metals with serious side effects. The heavy metals containing sewage sludges migrating as dust or leachates from contaminant to non-contaminant areas are the few examples of contamination to the ecosystem (Gaur and Adholeya, 2004). Based on its metallic properties and the element having atomic number greater than 20 is typically defined as the heavy metals. Few heavy metals contaminate environments are Cd, Cr, Cu, Hg, Pb, and Zn. Metals are the significant natural elements of soil (Lasat, 2000). Few micronutrients are known for growth of the plants, such as Zn, Cu, Co, Mn, Ni, and while other metals are known to possess unknown biological reactions such as Pb, Hg, and Cd (Gaur and Adholeya, 2004). The existence of heavy metals in soil for many years possesses numerous health hazards to the higher organisms. Moreover, it also affects plant growth metabolism, ground cover area and has a significant negative influence on the soil microflora (Roy et al., 2005). Heavy metals are not easily degraded chemically, but it can be easily converted to non-toxic substances by physical methods (Gaur and Adholeya, 2004). Soil environment is the major sink releasing heavy metals into the ecosystem by several anthropogenic activities. Microbes oxidize the organic contaminants into carbon (IV). Most of metals do not possess either chemical or the microbial degradations (Kirpichtchikova et al., 2006) in which total concentration of soil to holds for long time even after it has been introduced (Adriano, 2001). Presence of toxic metal substances in soil can inhibit the biodegradation of organic contaminants (Maslin and Maier, 2000). Direct ingestion or else contact with that of the contaminated soil, contamination in groundwater, the reduction of food quality and land usability in agricultural production and land tenure problems may pose hazards to the human and ecosystem due to heavy metal contamination in soil (Ling et al., 2007).

Naturally, soil consists of heavy metal in trace amount from pedogenetic processes of the weathering of parent material (<1000 mg kg^{-1}), which rarely possess toxicity (Bolan, 2008). Biogeochemical cycles were disturbed naturally by gradual gathering of heavy metals in soil environment by man which causes risks to plants, animals, ecosystem, and human health (D'Amore, 2005). A variety of human activities like as metal mine tailings, leaded gasoline and lead-based paints, disposal of high metal wastes, overuse or misuse of chemical pesticides, biosolids, coal combustion residues, manures, petrochemicals, and the atmospheric deposition liberates the metal bearing solids into contaminated sites (Basta, 2005). Effect of heavy metals on the human health was extensively studied and revealed by certain international organizations such as WHO. Heavy metals were extensively used by mankind for the thousands of years. Even though it pose various health effects for long period of time, exposure to the heavy metals continues, and it is still increasing in few parts of the world, particularly in underdeveloped countries, though percentage of emissions have reduced in the most advanced countries over last century.

11.2 HEAVY METAL POLLUTION: SOURCES AND RISKS

Heavy metals are extensively used in different regions of the world since thousands of years. Lead is the oldest metal though it has been using from at least 5000 years, the applications of it include building materials, pigments used for glazing ceramics, and pipes for conveying water. Prolonged exposure of the heavy metals like cadmium, lead, arsenic, and mercury pose threat to human life. Currently, cadmium compounds are used mainly in rechargeable nickel-cadmium batteries. During 20th century emission of cadmium increased due to rare recycling in cadmium-containing substances that often dumped with household wastes. The key source of cadmium exposure is smoke released from cigarette, while for non-smokers food will be the major source of cadmium exposure. Recent data reveals that adverse effect of the cadmium exposure on health may even occur at less exposure to the chemical substance than earlier anticipated, primarily it damages kidney, and effects on bone and it may also paralyzes the organ. In many developed countries like Europe, it already exceeded the exposure level, and it extends to large group of individual. In order to prevent the adverse effects on health proper measures have to be taken to minimize cadmium exposure.

Generally, human population exposed to the mercury through food; fish is the key source of mercury methyl and the dental amalgam. Most individuals will not face any major health risk from methylmercury, thus certain group of individuals consuming more fish attain blood levels related with low risk of nervous disorder to adults. Pregnant women are restricted for high intake of few fish such as tuna, shark, swordfish, and other fishes such as bass, pike, and walleye those are taken from the polluted water which causes disturbance to fetus.

Lead is exposed to the individuals through food and air more or less in equal proportions. Due to lead emissions from the petrol and petroleum products, from several decades, emission to ambient air causes significant pollution to the ecosystem. Children are more susceptible to continuous exposure of lead which leads to high uptake of substance in gastrointestinal tract and permeable blood-brain barrier. Usage of lead-based paints has to be abandoned, and lead should be avoided for food containers; public must have awareness of glazed food containers, which leaches the lead into food.

The main source of arsenic is through intake of food and drinking water. Exposure to the arsenic for long time period increases risk to skin, which may also lead to skin cancers, as well as skin lesions and changes in pigmentation. Arsenic is abundantly found as natural source on Earth's crust, generally obtained as arsenic sulfide or as metal arsenates and arsenides. Generally, it is released into environment as trioxide due to high-temperature process. It is mainly absorbed as particles which in turn dispersed by winds and tends to deposits on land and water. Arsenic were released to environment by various ways such as natural activities as dissolution of mineral substances, volcanic activity wind-blown dust, exudates from vegetation; human activities like as mining, smelting of metals, combustion of fossil fuels, use of pesticides, timber treatment with that of preservatives, remobilization of historical sources such as mine drainage and drilling of tube wells for mobilization of drinking water from geological deposits. Arsenic combines with oxygen, sulfur, and chlorine in nature to form the inorganic arsenic compounds, which is extensively used in the preservation of wood. Thus organic arsenic is used as the pesticides for cotton plants. Generally, compared to organic arsenic compounds, inorganic arsenic compounds are highly toxic, in which trivalent compounds are harmful than pentavalent compounds (Ampiah-Bonney et al., 2007; Vaclavikova et al., 2008; Chutia et al., 2009). The pentavalent compounds possess minimal toxicity level when compared to that of trivalent compounds. Even though As^{5+} tends to be less toxic to that of As^{3+}, it is a major source of groundwater contamination because it is thermodynamically more stable under normal conditions (Chutia et al., 2009). Pentavalent compound of arsenate (As^{5+}) is toxic and considered to be carcinogenic to human (Yusof and Malek, 2009). It has been generally accepted that inorganic arsenite and arsenate are predominantly obtained in most of the species in environment; organic ones are found to be rarely (Andrianisa et al., 2008). Generally, inorganic species, arsenite [As^{3+}] and arsenate [As^{5+}] were predominant in most of the environments. Thus organic ones may also be present (Andrianisa et al., 2008).

11.3   INTERNATIONAL STATUS OF HEAVY METAL POLLUTION

Industrial mining is a large sector of the global economy. Global metal value has been raised from US$214 billion in 2000 to US$644 billion by 2010 due to the increased output of metals. Disposal of mineral wastes from industrial mines is the main source of heavy metal pollution. Industrial mining, especially in open pit mining forms two solid wastes as waste rock and tailing are produced in large quantity which contains potentially harmful metals, and other elements affect environmental health (Dudka and Adriano, 1997). These mineral containing dusts are transported by wind and contaminate the nearby ecosystems and bioaccumulate in the organisms, plants, and other food sources. Higher amount of lead is contaminated in a town located in Armenia (Petrosyan, 2004). U.S. Environmental Protection Agency's national priority list (NPL) reported that 40% of hazardous waste sites are co-contaminated with other harmful heavy metal and organic pollutants (Benin et al., 1999; Sandrin and Maier, 2002; Cheng, 2003). These pollutants are showing severe health impact to higher organisms, decreasing crop productivity and ground cover, water contamination and disturbing soil microflora (Garbisu and Alkorta, 2001; McGrath et al., 2001; Zhang et al., 2009).

Cadmium content in Atlantic and Pacific Oceans range from 0.1 to 1.0 mg/Kg; marine manganese nodules and marine zinc-bearing phosphorites are also containing high cadmium levels as 60 to 340 mg Cd Kg^{-1} (Page et al., 1987). Globally, no farmland is available without toxic compounds, especially in the developing countries. Mining activities, unlimited use of agrochemicals, disposal of toxic wastes in landfills and other developmental activities are contributing toxic chemical contamination in large area of arable lands. National Environment Protection Council listed contaminated sites in globally, 80,000 in Australia, 30,000 in North South Wales and Queensland, 10,000 in Victoria, 4000 each in South and Western Australia. About 40,000 contaminated sites in USA, 55,000 in six European Countries, 7,800 in New Zealand and 3 million sites in Asia Pacific. Generally, closed, and discarded mine sites and related landfills contain heavy metals and chlorinated compounds (UNEP, 1996; Eshwaran et al., 2001; Sinha et al., 2009).

During 1990's, groundwater of USA, Taiwan, Argentina, China, Vietnam, and Ganges Plains were contaminated with naturally occurring arsenic (Smedley and Kinniburgh, 2002). According to the data on China

National Census of Pollution, sites of heavy metal exposure in China is more than 1.5 million, and from 2005 to 2011, the volume of discharged heavy metal as wastewater, polluted air, and solid waste are 9,00,000 tonnes per year. Based on the statistics in China, in the last four years, more than forty serious incidents have happened due to heavy metal pollution (Hu et al., 2014). Heavy metal pollution in Pakistan is now the main concern in that country, as major sites of the country are polluted by heavy metal. Punjab and Sindh are the provinces mainly contaminated by the arsenic metal in water sources and soil samples. Hayatabad Industrial Estate and Khyber Pakhtunkhwa are possessing cadmium contaminated groundwater (Manzoor et al., 2006). Bhara River in Nowshera district is significantly contaminated with lead (Nazif et al., 2006). Various coastal regions of Arabian Sea along the Karachi and Lyari River of Pakistan are also highly affected by lead contamination (Mashiatullah et al., 2013).

In India, Karnataka is one of the places where the metal is found. It has two gold mines which show severe pollution of arsenic in water, soil, and food samples. Bengal delta basin is formed by the rivers Bhagirathi and Padma, is also polluted in the groundwater by arsenic which is because of the geochemical reactions from aquifer sediments. Nearly, 5 to 6 million population is drinking the arsenic contaminated water only in West Bengal. The above-stated delta is also situated in major parts of districts of Bangladesh which is contiguous to the West Bengal affected area, are polluted by arsenic pollutant. Generally, for drinking purposes, less than 10% of the groundwater source is used, and 90% of the source is mainly used for agricultural purposes. Arsenic is entering in food chain through groundwater, animal tissues, and agricultural products. In bauxite mining and aluminum production India ranks sixth and eighth places respectively. The state of Orissa is affected by aluminum worst. In Tamil Nadu, Ranipet is polluted by chromium due to tanneries.

11.4   MICROBIAL REMEDIATION OF HEAVY METAL CONTAMINATION

Non-biodegradable heavy metal remains in soil ecosystem long after the process which released the metals has ceased. These toxic substances are biomagnified in the organisms which are exposed to the contamination and entered into the food chain ultimately leads to severe environmental and health problems. Remediation of these metal contaminations through any physicochemical techniques is unsuitable and not effective. There are other methods available for heavy metal remediation such as excavation, *in situ* vitrification, phytoremediation, and microbial remediation. Bioremediation of the heavy metal pollution offers an effective reclamation solution for contaminated soil. A wide range of fungal, bacterial, and algal species are generally used as biosorbents and cause metal transition to detoxify the heavy metals through assimilation, oxidization, and precipitation (Johnson and Hallberg, 2004; Umrania, 2006; Suciu et al., 2008; Iram, 2009).

Microorganisms are playing an important role in the biotransformation of the heavy metals into more nontoxic forms; understanding of accumulation of metal in microbial species in molecular level and mechanisms involved has numerous biotechnological applications for reclamation of heavy metal contaminated sites. Remediation of the heavy metal pollution by microbial species is commonly based on available concentration, and it is complex remediation process which is influenced by several factors (Goblenz et al., 1994). Microorganisms have developed tolerance and resistant for survival in the metal contaminated systems and these organisms adopt various detoxifying mechanisms as bioaccumulation, biosorption, biomineralization, and biotransformation (Gadd, 2000; Lin and Lin, 2005). Microorganisms are not interacting with some oxyanions of metals, and these metals are remediated by catalyzed redox conversion to insoluble forms by enzymatic activity. The genetic level understanding of such microbial metal accumulators leads the research to transfer traits to other microbes through microarrays for betterment of environmental health (Dixit et al., 2015).

11.5 MECHANISM OF BIOSORPTION AND BIOACCUMULATION OF HEAVY METALS

Microbial biosorption of heavy metal is carried out by dead or live microorganisms. Biosorption is classified by the location of non-specific metal ions binding to the polysaccharides and proteins associated with extracellular (precipitation) or cell surface or intracellular accumulation (Vasudevan et al., 2001). According to dependence on cellular metabolism, it can also be classified as metabolism dependent active process and non-metabolism dependent, passive process. Microbial species may uptake metals through active or passive or both processes depending on the species. Passive uptake of heavy metal is independent of cellular metabolism and environmental conditions; it is nonspecific to metal species which is rapid and irreversible process. In contrast, the active process is slow and cellular metabolism dependent, heavy metal forms complexes with the specific proteins such as metallothioneins (Pandey et al., 2001; Issazadeh et al., 2013). In addition to cell wall content such as chitosan and glucans, microbes with chitin cell wall content act effectively in biosorbent process of heavy metals. Likewise, Gram-positive bacterial cell wall attaches more easily with heavy metals than Gram-negative (Rani and Goel, 2009; Issazadeh et al., 2013). During passive and reversible biosorption, uptake of metal is undertaken between metal and functional group like acetamido group of chitin, structural polysaccharide of fungi, amine (amino and peptidoglycosides), sulfhydryl, and carboxyl groups in protein, phosphodiester (teichoic acid), phosphate, hydroxyl in polysaccharides, present on cell surface by physicochemical interaction as physical adsorption, ion exchange and chemical sorption, which is not dependent on cells metabolism. The mechanism of heavy metal biosorption is given in Figure 11.1.

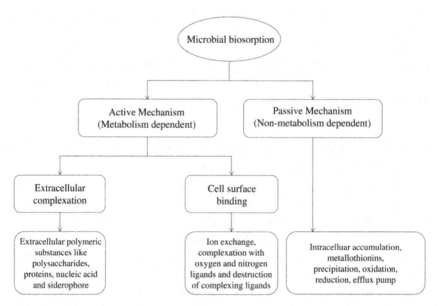

FIGURE 11.1   General mechanism of microbial biosorption of heavy metals.

Bioaccumulation is the process which transports the metals across cell membrane for intracellular accumulation. Ion channels, ion pumps, carrier-mediated transport; complex permeation, endocytosis, and lipid permeation are mediating the transport and bioaccumulation of heavy metals across cell membrane. Precipitation process may take place both in the solution and on the cell surface (Ercole et al., 1995). Furthermore, it is partially cell metabolism dependent process; microorganisms produce substances which enhance precipitation process in presence of toxic metals and also, it may not be a cell metabolism dependent, if the process occurs after the chemical interaction of metal and cell surface (Issazadeh et al., 2013). Table 11.1 presents microbial species capable of bioaccumulation of heavy metals. Generally, heavy metals are not destroyed or degraded by microorganism, but these metals are transformed from its oxidation state to another which may become either water soluble or less toxic or less water soluble to form precipitate or volatilized compound.

11.6 FACTORS INFLUENCING HEAVY METAL UPTAKE BY MICROBES

Microorganisms have ability to produce or release certain organic compounds like pigments or siderophores in presence of heavy metals which can directly reduce or increase the mobility of heavy metals. Moreover, microorganisms also produce some inorganic compounds such as sulfur thereby it reduces mobility of many heavy metals. In addition, microorganisms also influence metal mobility; hence they affect pH and redox potential. All these are significant parameters while studying influence of microbes on mobility of metals (Ledin, 2000).

Soil pH is the principal factors which play an eminent role for uptake of heavy metals by several processes. If soil solutions maintains neutral to alkaline nature, it possesses low mobility of all heavy metals. Hence to increase uptake of heavy metals, the soil solution pH should be lowered. In general, sorption increases with the increasing pH. Thus, in the case of low pH, more metal was found in solution, and hence more metal was mobilized. When pH is less than five mobility of a substance increases the proton concentration. If the pH value is above 7, it possesses hydroxyl-complexes which tend to increase the solubility of metal. The uptake will be more at pH 6.5 than at pH 4.5. Sar et al. (1999) have reported that uptake of Cu at pH 7.0 while in case of Ni it persists up to pH 8.0. The optimum

TABLE 11.1 List of Microorganisms Capable of Biosorption of Heavy Metals

Microbial species	Heavy metal	References
Aspergillus parasitica, Cephalosporium aphidicola, Pseudomonas aeruginosa, Saccharomyces cerevisiae, Citrobacter sp.	Lead	Dahmani-Muller et al., 2000; Tunali et al., 2006; Akar et al., 2007; Khan et al., 2009; Dhamodharan et al., 2011
Hymenoscyphus ericae, Neocosmospora vasinfecta, Verticillum terrestre, Pseudomonas sp.	Mercury	De et al., 2003; Kelly et al., 2006
Pseudomonas syringe, Thiobacillus ferrooxidans, Pseudomonas aeruginosa, Saccharomyces cerevisiae, Candida utilis	Copper	Lin and Olson, 1995; Umrania, 2006; Chen and Wang, 2007; Talos et al., 2009; Dhamodharan et al., 2011; Ahmed and Malik, 2012
Pseudomonas putida, Pseudomonas aeruginosa, Bacillus cereus, Rhizopus arrhizus, Cupriavidus taiwanensis, Bacillus safensys, Xanthomonas sp., *Saccharomyces cerevisiae, Citrobacter* sp.	Cadmium	Pandey et al., 2001; Khan et al. 2009; Rani and Goel, 2009; Dhamodharan et al., 2011; Siripornadulsil and Siripornadulsil, 2013; Priyalaxmi et al., 2014
Bacillus subtilis, Pseudomonas aeruginosa, Pseudomonas fluorescens, Aspergillus niger, Acinetobacter sp., *Bacillus cereus*	Chromium	Khan and Ahmad, 2006; Khan et al., 2009; Thakur and Srivastava, 2011; Tripathi et al., 2011
Pseudomonas aeruginosa, Saccharomyces cerevisiae, Candida utilis, Escherichia coli	Zinc	Dioro et al., 1995; Beard et al., 1997; Chen and Wang, 2007; Talos et al., 2009; Dhamodharan et al., 2011; Ahmed and Malik, 2012
Geobacter sp., *Pseudomonas aeruginosa, Desulfovibrio vulgaris, Saccharomyces cerevisiae, Aspergillus niger*	Uranium	Dioro et al., 1995; Beard et al., 1997; Williams et al., 2011; Ahmed and Malik, 2012
Sulfurospirillum arsenophylum, Sulfurospirillum barnesii, Chrysiogenes arsenatis, Bacillus selenitireducens, Alcaligenes faecalis	Arsenic	Oremland and Stolz, 2003; Silver and Phung, 2005; Lloyd and Oremland, 2006

pH for the uptake of Ag by *Saccharomyces cerevisiae* is 5.0. Likewise, for the uptake of lead is persisted by *Rhizopus oligosporus* at pH 5.0. Sar and D'Souza (2001) observed that the sorption is negligible at low pH, but it is maximum at pH 5.0 by *Pseudomonas aeruginosa*. The filamentous fungus *Phanerochaete chrysosporium* shows maximum uptake of certain metals like Cd, Pb, and Cu at pH 6.0. pH 2.0 was found to be optimum for the removal of gold by *Azolla filicutoides*. Bai and Abraham (2001) reported that sorption of Cr(VI) is carried through *Rhizopus nigricans* at pH 2.0, the Co removal by *Rhizopus* spp. Above pH 6.0 the sorption of Cd attributes to form cadmium hydroxide complex which also decreases uptake of ions. Hence Cd ions are absorbed by *Trametes versicolor* between the pH 5.0 to 6.0. Thus the average pH for the biosorption ranges from pH 2 to 8. Several metal ions such as Au, Cr, Hg, and Ag attributes greater tendency as it occurs naturally as negatively charged complexes, for example, $AuCN^-$, $AuCl_4^-$ such compounds will decrease the uptake with increasing in pH, or it will remain insensitive even with the pH change. Several scientists have also proved that arsenic distribution in soils is associated with pH (Akins and Lewis, 1976; Adriano, 2001). Akins and Lewis (1976) reported that effect of pH ranges from 4 to 8 for sorption of arsenic in soil by using sequential fractionation procedure. They also illustrated that at low pH (pH 4) Fe-As is most abundant form which is followed by Al-As whereas at high pH (pH 6 to 8) Ca-As is the predominant form.

Most of chemical reactions are temperature dependent on preceding concentration of the metals present on sewage sludges amended with that of soil. The absorption rate extensively depends on temperature as the rise in temperature helps in bioavailability and mobility of metal in a soil solution as well as in plant root. Sherene (2010) reported that soil organic matter will be degraded if the soil temperature is high, which releases the organic acids, in turn, enhances uptake of heavy metals in soil. Thus it increases the bioavailability. Several reports show the subsequent uptake of metal by microorganism with that of change in the temperature regime (Skowronski, 1986; Mehta and Gaur, 2001; Mehta et al., 2002) in addition to maximum uptake of metals takes place at specific temperature optima. Higher the temperature thus increase the sorption leads to increase in surface activity and kinetic energy of solute (Sag and Kutsal, 2000; Vijayaraghavan and Yun, 2007). There are also reports which show that there is no effect of the temperature on the metal sorption (Norris and Kelly, 1979; Zhao et al., 1994). Chen and Ting (1995) reported that living cells pertains to remove metal ions from the aqueous solution was also influenced by environmental conditions like temperature, pH, and also the biomass concentration. The incubation temperature for the uptake of metal such as Cd^{2+}, Co^{2+}, and Pb^{2+} is 35°C, whereas for Zn^{2+} 25°C was optimum. Sag and Kutsal (1995) reported that optimum temperature for Cu^{2+} by *Zooglea ramigera* could be 25°C, an activated sludge bacterium could absorb copper at 25°C. The best temperature for Ni^{2+} was found to be at 35°C by the bacterium *Bacillus thuringensis*, the uptake confined to metabolism dependent metal accumulation. In addition to this, *Arcanobacterium bernardiae* and *B. amyloliquefaciens* attributes the maximum capacity for lead uptake at 35°C (Aksu et al., 1991; Tohamy et al., 2006; Jackson et al., 2011).

Organic matter plays important role in the retention of heavy metal from soil constituents. Thus it reduces bioavailability and mobility. If the organic matter is dense in soil content, it indicates its fertility. Fertility implies that sufficient amount of nutrients and minerals are available to the plants. Hence the complexation of metals by soluble organic matter in soil results in the release of metals ions from solids to the soil, which exclusively results with uptake of metal ions by plants.

Contact time is of significant contributor for adsorption of metals; it basically constitutes two phases; one is rapid phase where total metal biosorption take place and the second phase contributes relatively low as a sorbent. As uptake of metal increase as that of increase in contact time but it remains constant after certain equilibrium time period (Murugesan et al., 2006). The removal of metal is rapid hence the adsorption taking place within 60 min (Sheng et al., 2004). Chen and Pan (2005) reported that absorption rate of Pb is rapid at initial stage where it takes 0–12 min for the removal of Pb by *Spirulina*. Some heavy metals such as Cu^{2+}, Cr^{3+}, and Cd^{2+} removal from the solutions by *Spirulina* species possess equilibrium after 10 min (Chojnacka et al., 2005). Increase in biosorption of Cu by *Spirogyra* species was observed with increase in contact time from 0 to 120 min, and after that, it becomes constant up to 180 min (Gupta et al., 2006).

Soil microorganisms and mycorrhizae were supposed to grow on the heavy metal contaminated soils (Shetty et al., 1994; Chaudry et al., 1998; Chaudry et al., 1999). Thus symbiotic association among them has significant potentiality to enhance absorption of root and hence stimulate the acquisition of plant nutrients including metal ions (Khan et al., 2000). Arbuscular mycorrhizal fungi are one among the soil microorganisms. It tends to expand the interface between plants and the soil environment and one of the significant contributor for the uptake of nutrients especially P

(Li et al., 1991). It also facilitates the nitrogen uptake (Ames et al., 1983), Cu (Li et al., 1991), and Zn (Burkert and Robson, 1994). The exact mechanism of how the mycorrhizae reduce the toxicity is poorly understood, but it has been evident that AMF can filter out the toxic heavy metals and hence keep them away from plants. Arbuscular mycorrhizal fungi colonize ferns (Sharma, 1998) and play a dominant role in hyperaccumulation of arsenic in association with that of mycorrhizae (Ma et al., 2001).

11.7 MOLECULAR TOOLS FOR IMPROVING MICROBIAL DEGRADATION OF HEAVY METALS

Traditional bioremediation techniques involve utilization of microbial population in environmental cleansing acts. However, these methods face technical drags because of recalcitrant nature of pollutants. Recombinant DNA technology could be utilized to increase efficacy of microorganisms to effectively remove environmental pollutants. One such tool is genetic engineering of microorganisms; where, genetic material of an individual microbe will be altered either by addition or deletion of desired gene (Sayler and Ripp, 2000). Such genetically engineered microbes can withstand high levels of qualitative pollution load. Genetic engineering has even led to the discovery and the development of surveillance device such as 'microbial biosensors' which can detect and measure the extent of contamination at given site precisely. So far, several biosensors have been developed to monitor heavy metal concentrations such as cadmium (Cd), mercury (Hg), copper (Cu), nickel (Ni) and arsenic (As) (Verma and Singh, 2005; Bruschi and Goulhen, 2006). Remediation of the heavy metal contaminated soils critically depends on microbial remediation which leads the new technology for remediation as genetic engineering of endophytes and rhizospheric microbes (Divya and Deepak Kumar, 2011). Some strains of the genetically modified bacteria expressing phytochelatin–20, such as *Escherichia coli* and *Moreaxella* sp. have shown to accumulate 25 times more Cd and Hg on their surface (Bae et al., 2001; Bae et al., 2003). Sustaining in soil with different environmental conditions and competition with indigenous microbial population is the major disadvantage of using genetically modified organisms in unreceptive field (Wu et al., 2006). In addition, molecular techniques have been applied to limited microbial species successfully for the heavy metal remediation.

Hence, this technique needs to be explored with many other microbial species for heavy metal remediation purposes.

11.7.1   GENETIC ENGINEERING FOR METAL REMEDIATION

Clean up of the heavy metals by genetic engineering facilitates introduction of specific genes into the microbial cells (Chen and Wilson, 1997). Krishnasamy and Wilson (2000) constructed an *E. coli* strain that accumulated Ni^{2+} by introducing nix A gene (coding for a nickel transport system) from *Helicobacter pylori* into *E. coli* JM 109 that expressed a glutathione S-transferase-pea metallothionein fusion protein. This recombinant strain had the capability to hold nickel four folds more than that of wild strain. Bang et al. (2000) engineered *E. coli* to produce sulfide by heterologous expression of thiosulphate reductase gene from *Salmonella entenltca*, which enhanced protein synthesis leading to precipitation of cadmium sulfide. Heavy metal tolerant *R. eutropha* isolate altered to process metallothionein capitalized to accumulate more Cd^{2+} in contaminated soil when inoculated. It also provided protection to tobacco plant from Cd^{2+} load by adsorbing it on surface layer (Valls et al., 2000). Studies have shown periplasmic expression of *Neurospora crassa* produced metallothionein in *E. coli* were more superior in holding metal ions (Pazirandeh, 1998). Kotrba et al. (1999) engineered metal binding peptides that contain either histidines (GHHPHG) 2 (HP) or cysteines (GCGCPCGCG) (CP) into Lam B and expressed on surface of *E. coli*.

11.7.2   APPLICATIONS OF NANOTECHNOLOGY AND GENOMICS

Apart from modern genetic engineering strategies, the heavy metal remediation can be achieved by microbial nanotechnology processes. Various nanoparticles or nanomaterials aiding in microbial removal of toxic heavy metal deposits are at its virtue, and the process may be termed as nanobioremediation. This process not only reduces the time barrier but also increases the amount of heavy metals that can be effectively taken up. Bionanotechnology through biotechnology is biofabrication of the nanoobjects or bifunctional macromolecules utilized as tools to manipulate or construct nanoobjects. The physiological diversity, genetic manipulability and controlled culturability make microbial cells novel producers of nanostructures ranging from natural products such as polymers and magnetosomes to engineered proteins such as virus-like proteins (VLP) and tailored metal particles (Sarikaya et al., 2003). *Deinococcus radiodurans*, radioactive-resistant organism has ability to tolerate radiation well beyond naturally occurring level. Thus its application in the radioactive waste clean-up initiatives was funded by US Department of Energy (DOE) (Smith et al., 1998; Brim, 2000). Till date genomics has been extensively studied in microbes and agriculture. This is due to vast appearing potentials in utilizing them into research and development. A major outcome is GMO crops, but now it is an emerging new tool of bioremediation. Knowledge of whole microbial genome provides an understanding of new genes which would further aid in developing new GM organisms for bioremediation purposes.

11.7.3   MICROBIAL GENETICS OF METAL RESISTANCE

Large amounts of anthropogenic activities and subsequent utilization of metal-containing commercial products pose an immediate need for research concerning microbial metal resistance as well as its remediation. Large plasmids (165–250 kb), encoded specific metal conferring resistance to variety of metals including Ag^+, Cd^{2+}, Co^{2+}, CrO, Cu^{2+}, Hg^{2+}, Ni^{2+}, Pb^{2+}, Sb^{3+}, Tl^+, and Zn^{2+}. Roane and Pepper (2000) categorized mechanism of the metal resistance as: (1) general and do not require metal stress; (2) dependent on a specific metal for activation; and (3) general and are activated by metal stress. The plasmid-borne cadA gene encodes a cadmium specific ATPase in several bacterial genera, including *Staphylococcus*, *Pseudomonas*, *Bacillus*, and *E. coli*. This eliminates toxic metal, which is deposited by the uptake of essential divalent cations (e.g., Mg^{2+}, Mn^{2+}).

11.7.4   BIOMASS IMMOBILIZATION FOR METAL REMEDIATION

Biosorption of the heavy metals by microbes in soil, even with low degree of understanding of mechanism involved is crucial in remediation processes. However, a better understanding on mechanism of the biosorption would mean a lot in effective application method (Volesky, 1999). Laboratory applications of free cells in remediation pose limitations in field and industrial applications (Gadd and White, 1993). The use of the freshly suspended suffers a limitation like small particle size, lack of mechanical strength and efficiency in sequestration process. Immobilization of these free cells in physical matrices like sodium alginate beads, polyurethane foam (PUF), agar, and acrylamide provide required efficiency to these cells to remediate heavy metals effectively (Gupta et al., 2000). Ibanez and Umetsu (2002) have demonstrated the ability of protonated alginate beads in removal of chromium, copper, zinc, nickel, and cobalt ions from dilute aqueous solutions. Karna et al. (1999) immobilized *Phormidium valderianum* BDU 30501 in polyvinyl foam and used for removal of Ca^{2+}, CO^{2+}, Cu^{2+}, and Ni^{2+}. Ca-alginate has been found to be the effective in nickel biosorption as immobilizing agent, and inorganic salts were used for nickel desorption from immobilized microbial biomass (Asthana et al., 1995). HCl and EDTA are the most efficient desorption agents for nickel removal (Ramteke, 2000).

11.8 CONCLUSION AND FUTURE PERSPECTIVES

Contaminated soils are the serious environmental issue all over the world. The widespread increase in pollution causes contamination of vast areas of land which become non-arable and harmful for both human and wild-life. Metals are unique in its nature in that way they do not undergo any biological or chemical degradation but it can either alter or reduce their toxicity. Traditional methods are conventional which do not provide any significance in detoxifying metal from soil. For reducing or detoxification microbes may use metal as terminal acceptor for removal of metals from contaminated environments. Microorganisms are not alchemists, however, it may act upon toxic metal for altering, but it cannot be completely destroyed. Degradation of heavy metals by biological methods (change in the nuclear structure of the substance) is not possible hence it can only alter or transform the complex substances to another form. Thus metals show less toxicity in oxidation state. In general, change in microbial populations for the bioremediation of metals is very limited and, thus microbial community still remains as a "black box." Hence the recent advancement in the field of molecular biology is used in studying dynamics and structure of microbial communities without bias introduced by cultivation. Those molecular approaches were certainly used for studying microbial ecology. Bioremediation of heavy metals remains as immature technology hence it is necessary to determine its boundaries in between promise and reality. Successful bioremediation requires an interdisciplinary approach which involves several disciplines such as microbiology, ecology, geology, and chemistry. Thus, interdisciplinary approach is essential because of the complexity encountered in its type and extent of contamination and the enormous social and legal issues in relevance to contaminated sites.

Better understanding of microbial ecology, physiology, evolution, and genetics exploit and improve the microbial metabolism for remediation of xenobiotics. Bioremediation of heavy metals is lagged due to development of *in situ* bioremediation of organics. In recent years, funding opportunity for research in heavy metal remediation to find out novel advances and techniques is increased. Microbial remediation is a potential technology for removal of heavy metal from polluted environment and is approaching commercialization. Bioremediation techniques are well suitable for large scale where other techniques are not practicable and effective. Even though it is not a complex process, expertise in the field is required to design and implement a bioremediation process successfully.

KEYWORDS

- bioaccumulation
- biofabrication
- biomass immobilization
- bionanotechnology
- bioreactor
- biosorption
- biotransformation
- carrier mediated transport
- chemiosmotic proton-antiporter system
- complex permeation
- endocytosis
- fluidized beds
- genetic engineering
- genome shuffling
- heavy metals
- lipid permeation
- magnetosomes
- metagenomics
- metallothioneins
- mycorrhizae
- nanobioremediation
- organic matter
- petrochemicals
- polyurethane foam
- trickle filter
- xenobiotics

REFERENCES

Adriano, D. C., (2001). *Trace Elements in Terrestrial Environments: Biogeochemistry, Bioaccessibility and the Risk of Metals*, Springer, New York.

Ahemad, M., & Malik, A., (2012). Bioaccumulation of heavy metals by zinc resistant bacteria isolated from agricultural soils irrigated with wastewater. *Bacteriology Journal, 2*, 12–21.

Akar, T., Tunali, S., & Cabuk, A., (2007). Study on the characterization of lead (II) biosorption by fungus *Aspergillus parasiticus*. *Appl. Biochem. Biotech., 136*, 389–406.

Akins, M. B., & Lewis, J. R., (1976). Chemical distribution and gaseous evolution of arsenic–74 added to soils as DSMA74-As. *Soil. Sci. Soc. Am., 40*, 655–658.

Aksu, Z., Kutsal, T., Songul, G. N., Haciosmanoglu, N., & Gholaminejad, M., (1991). Investigation of biosorption of Cu(II), Ni(II) and Cr(VI) ions to activated sludge bacteria. *Environ. Technol., 12*, 915–921.

Ames, R. N., Reid, C. P. P., Porter, L. K., & Cambardella, C., (1983). Hyphal uptake and transport of nitrogen from two ^{15}N-labeled sources by *Glomus mossea*, a vesicular-arbuscular mycorrhizal fungus. *New Phytol., 95*, 381–396.

Ampiah-Bonney, R. J., Tyson, J. F., & Lanza, G. R., (2007). Phytoextraction of arsenic from soil by *Leersia oryzoides*. *Int. J. Phytoremediation., 9*(1), 31–40.

Andrianisa, H. A., Ito, A., Sasaki, A., Aizawa, J., & Umita, T., (2008). Biotransformation of arsenic species by activated sludge and removal of bio-oxidized arsenate from wastewater by coagulation with ferric chloride. *Water Research, 42*(19), 4809–4817.

Asthana, R. K., Chatterjee, S., & Singh, P., (1995). Investigations on nickel biosorption and its remobilization. *Process Biochemistry, 30*, 729–734.

Bae, W., Mehra, R. K., Mulchandani, A., & Chen, W., (2001). Genetic engineering of *Escherichia coli* for enhanced uptake and bioaccumulation of mercury. *Appl. Environ. Microbiol., 67*, 5335–5338.

Bae, W., Wu, C. H., Kostal, J., Mulchandani, A., & Chen, W., (2003). Enhanced mercury biosorption by bacterial cells with surface-displayed MerR. *App. Environ. Microbiol. 69*, 3176–3180.

Bai, S., & Abhraham, T. E., (2001). Biosorption of Cr (VI) from aqueous solution by *Rhizopus nigricans. Biores, Technol., 79*, 73–81.

Bang, S. W., Clark, D. S., & Keasling, J. D., (2000). Engineering hydrogen sulfide production and cadmium removal by expression of the thiosulfate reductase gene (phs ABC) from *Salmonella enterica* serovar typhimurium in *Escherichia coli. Appl. Environ. Microbiol., 66*, 3939–3944.

Basta, N. T., Ryan, J. A., & Chaney, R. L., (2005). Trace element chemistry in residual-treated soil: Key concepts and metal bioavailability. *J. Environ. Qual., 34*(1), 49–63.

Beard, S. J., Hashim, R., Hernandez, J., Hughes, M., & Poole, R. K., (1997). Zinc (II) tolerance in *Escherichia coli* K–12: Evidence that the *zntA* gene (o732) encodes a cation transport ATPase. *Mol. Microbiol., 25*(5), 883–891.

Benin, A. L., Sargent, J. D., Dalton, M., & Roda, S., (1999). High concentrations of heavy metals in neighborhoods near ore smelters in Northern Mexico. *Environ. Health Perspect., 107*(4), 279–284.

Bolan, N. S., Ko, B. G., Anderson, C. W. N., & Vogeler, I., (2008). Solute interactions in soils in relation to bioavailability and remediation of the environment. In: *Proceedings of the 5th International Symposium of Interactions of Soil Minerals with Organic Components and Microorganisms*. Pucón, Chile.

Brim, H., McFarlan, S. C., Fredrickson, J. K., Minton, K. W., Zhai, M., Wackett, L. P., & Daly, M. J., (2000). Engineering *Deinococcus radiodurans* for metal remediation in radioactive mixed waste environments. *Nat. Biotechnol., 18*, 85–90.

Bruschi, M., & Goulhen, F., (2006). New bioremediation technologies to remove heavy metals and radionuclides using Fe(III)-sulfate- and sulfur reducing bacteria. In: Singh, S. N., & Tripathi, R. D., (eds.), *Environmental Bioremediation Technologies* (pp. 35–55). Springer, New York.

Burkert, B., & Robson, A., (1994). ^{65}Zn uptake in subterranean clover (*Trifolium subterraneum* L.) by three vesicular-arbuscular mycorrhizal fungi in root-free sandy soil. *Soil Biol. Biochem., 26*, 117–1124.

Chaudry, T. M., Hayes, W. J., Khan, A. J., & Khoo, C. S., (1998). Phytoremediation focusing on accumulator plants that remediate metal contaminated soils. *Aust. J. Ecotoxicol., 4*, 37–51.

Chaudry, T. M., Hill, L., Khan, A. J., & Keuk, C., (1999). Colonization of iron and zinc-contaminated dumped filter cake waste by microbes, plants and associated mycorrhizae. In: Wong, M. W., & Baker, A. J. M., (eds.), *Remediation and Management of Degraded Land* (pp. 275–283). CRC Press, Boca Raton.

Chen, C., & Wang, J. L., (2007). Characteristics of Zn^{2+} biosorption by *Saccharomyces cerevisiae*. *Biomed. Environ. Sci., 20*, 478–482.

Chen, H., & Pan, S. S., (2005). Bioremediation potential of *Spirulina*: Toxicity and biosorption studies of lead. *J. Zhejiang. Univ. Sci. B., 6*(3), 171–174.

Chen, P., & Ting, Y. P., (1995). Effect of heavy metal uptake on the electrokinetic properties of *Saccharomyces cerevisiae*. *Biotechnol. Let., 17*(1), 107–112.

Chen, S., & Wilson, D. B., (1997). Construction and characterization of *Escherchia coli* genetically engineered for bioremediation of Hg^{2+} contaminated environments. *Appl. Environ. Microbiol., 63*, 2442–2445.

Cheng, S. P., (2003). Heavy metal pollution in China: Origin, pattern, and control. *Environ. Sci. Pollut. Res., 10*(3), 192–198.

Chojnacka, K., Chojnacki, A., & Gorecka, H., (2005). Biosorption of Cr^{3+}, Cd^{2+} and Cu^{2+} ions by blue-green algae *Spirulina* sp.: Kinetics, equilibrium and the mechanism of the process. *Chemosphere, 59*(1), 75–84.

Chutia, P., Kato, S., Kojima, T., & Satokawa, S., (2009). Arsenic adsorption from aqueous solution on synthetic zeolites. *J. Hazard Mater, 162*(1) 440–447.

D'Amore, J. J., Al-Abed, S. R., Scheckel, K. G., & Ryan, J. A., (2005). Methods for speciation of metals in soils: A review. *J. Environ. Qual., 34*(5), 1707–1745.

Dahmani-Muller, H., Van Oort, F., Gelie, B., & Balabane, M., (2000). Strategies of heavy metal uptake by three plant species growing near a metal smelter. *Environ. Pollut., 109*, 231–238.

De, J., Ramaiah, N., Mesquita, A., & Verlekar, X. N., (2003). Tolerance to various toxicants by marine bacteria highly resistant to mercury. *Mar. Biotechnol., 5*(2), 185–193.

Dioro, C., Cai, J., Marmor, J., Shinder, R., & Dubow, M. S., (1995). An *Escherichia coli* chromosomal *ars* operon homolog is functional in arsenic detoxification and is conserved in gram negative bacteria. *J. Bacteriol., 177*(8), 2050–2056.

Divya, B., & Deepak, K. M., (2011). Plant-microbe interaction with enhanced bioremediation. *Res. J. Biotechnol., 6*, 72–79.

Dixit, R., Wasiullah, M. D., Pandiyan, K., Singh, U. B., Sahu, A., Shukla, R., Singh, B. P., Rai, J. P., Sharma, K., Lade, H., & Paul, D., (2015). Bioremediation of heavy metals from soil and aquatic environment: An overview of principles and criteria of fundamental processes. *Sustainability, 7*, 2189–2212.

Dudka, S., & Adriano, D. C., (1997). Environmental impacts of metal ore mining and processing: A review. *J. Environ. Qual., 26*, 590–602.

Ercole, C., Veglio, F., Toro, L., Ficara, G., & Lepidi, A., (1995). Immobilization of microbial cells for metal adsorption and desorption. In: *Mineral Bioprocessing II* (pp. 181–188). Minerals, Metals and Materials Society, Snowboard, Utah, USA.

Eswaran, H., Lal, R., & Reich, P., (2001). Land degradation: An overview. In: Bridges, E. M., Hannam, I. D., Oldeman, L. R., Pening de Vries, F. W. T., Scherr, S. J., & Sompatpanit, S., (eds.), *Response to Land Degradation* (pp. 20–35). Science Publishers, Enfield, USA.

Gadd, G. M., & White, C., (1993). Microbial treatment of metal pollution - a working biotechnology? *Trends Biotechnol., 11*(8), 353–359.

Gadd, G. M., (2000). Bioremedial potential of microbial mechanisms of metal mobilization and immobilization. *Curr. Opin. Biotechnol., 11*, 271–279.

Garbisu, C., & Alkorta, I., (2001). Phytoextraction: A cost effective plant-based technology for the removal of metals from the environment. *Biores. Tech., 77*(3), 229–236.

Gaur, A., & Adholeya, A., (2004). Prospects of arbuscular mycorrhizal fungi in phytoremediation of heavy metal contaminated soils. *Curr. Sci., 86*(4), 528–534.

Goblenz, A., Wolf, K., & Bauda, P., (1994). The role of glutathione biosynthesis in heavy metal resistance in the fission yeast *Schizosaccharomyces pombe*. *FEMS Microbiol. Rev., 14*, 303–308.

Gupta, R., Ahuja, P., Khan, S., Saxena, R. K., & Mohapatra, H., (2000). Microbial biosorbents: Meeting challenges of heavy metal pollution in aqueous solution. *Curr. Sci., 78*(8), 967–973.

Gupta, V. K., Rastogi, A., Saini, V. K., & Jain, N., (2006). Biosorption of copper (II) from aqueous solutions by *Spirogyra* species. *J. Colloid Interf. Sci., 296*(1), 59–63.

Hu, H., Jin, Q., & Kavan, P., (2014). A study of heavy metal pollution in China: Current status, pollution – control policies and counter measures. *Sustainability, 6*, 5820–5838.

Ibanez, J. P., & Umctsu, V., (2002). Potential of protonated alginate beads for heavy metals uptake. *Hydrometallurgy, 64*(2), 89–99.

Iram, S., Ahmad, I., & Doris, S., (2009). Analysis of mines and contaminated agricultural soil samples for fungal diversity and tolerance to heavy metals. *Pak. J. Bot., 41*, 885–895.

Issazadeh, K., Jahanpour, N., Pourghorbanali, F., Raeisi, G., & Faekhondeh, J., (2013). Heavy metals resistance by bacterial strains. *Ann. Biol. Res., 4*(2), 60–63.

Jackson, T., West, M. M., & Leppard, G., (2011). Accumulation and partitioning of heavy metals by bacterial cells and associated colloidal minerals, with alteration, neoformation, and selective adsorption of minerals by bacteria, in metal-polluted lake sediment. *Geomicrobiol. J., 28*(1), 23–55.

Johnson, D. B., & Hallberg, K. B., (2004). Acid mine drainage remediation options: a review. *Sci. Total Environ., 338*, 3–14.

Karna, R. R., Uma, L., Subramanian, G., & Mohan, P. M., (1999). Biosorption of toxic metal ions by alkali extracted biomass of marine cyanobacterium *Phormidium valderianum* BDU 30501. *World J. Microbiol. Biotechnol., 15*(6), 729–732.

Kelly, D. J. A., Budd, K., & Lefebvre, D. D., (2006). The biotransformation of mercury in pH-stat cultures of microfungi. *Can. J. Bot., 84*, 254–260.

Khan, A. G., Keuk, C., Chaudry, T. M., Khoo, C. S., & Hayes, W. J., (2000). Role of plants, mycorrhizae, and phytochelators in heavy metal contaminated land remediation. *Chemosphere, 41*, 197–207.

Khan, M. S., Zaidi, A., Wani, P. A., & Oves, M., (2009). Role of plant growth promoting rhizobacteria in the remediation of metal contaminated soils. *Environ. Chem. Lett., 7*, 1–19.

Khan, M. W. A., & Ahmad, M., (2006). Detoxification and bioremediation potential of a *Pseudomonas fluorescens* isolate against the major Indian water pollutants. *J. Environ. Sci., 41*, 659–674.

Kirpichtchikova, T. A., Manceau, A., Spadini, L., Panfili, F., Marcus, M. A., & Jacquet, T., (2006). Speciation and solubility of heavy metals in contaminated soil using X-ray microfluorescence, EXAFS spectroscopy, chemical extraction, and thermodynamic modeling. *Geochimica et Cosmochimica. Acta., 70*(9), 2163–2190.

Kotrba, P., Posisil, P., Lorenzo, V. D., & Ruml, T., (1999). Enhanced metalloprotein of *E. coli* cells due to surface display of beta and alpha-domains of mammalian metallothionein as a fusion to Lam B protein. *J. Recept. Signal. Transduct. Res., 19*, 703–715.

Krishnaswamy, R., & Wilson, D. B., (2000). Construction and characterization of an *E. coli* strain genetically engineered for Ni(II) bioaccumulation. *Appl. Environ. Microbiol., 66*, 5383–5386.

Lasat, M. M., (2000). Phytoextraction of metals from contaminated soil: A review of plant/soil/metal interaction and assessment of pertinent agronomic issues. *J. Hazard. Sub. Res., 2*(5), 1–25.

Ledin, M., (2000). Accumulation of metals by microorganisms - processes and importance for soil systems. *Earth-Science Reviews, 51*, 1–31.

Li, X. L., George, E., & Marschner, H., (1991). Extension of the phosphorous depletion zone in VA-mycorrhizal white clover in a calcareous soil. *Plant and Soil, 136*, 41–48.

Lin, C. C., & Lin, H. L., (2005). Remediation of soil contaminated with the heavy metal (Cd^{2+}). *J. Hazard. Mater., 122*, 7–15.

Lin, C., & Olson, B. H., (1995). The sociality of bioremediation: Hijacking the social lives of microbial populations to clean up heavy metal contamination. *Can. J. Microbiol., 41*, 642–646.

Ling, W., Shen, Q., Gao, Y., Gu, X., & Yang, Z., (2007). Use of bentonite to control the release of copper from contaminated soils. *Aus. J. Soil. Res., 45*(8), 618–623.

Lloyd, J. R., & Oremland, R. S., (2006). Microbial transformations of arsenic in the environment: From soda lakes to aquifers. *Elements, 2*, 85–90.

Ma, L. Q., Komar, K. M., Tu, C., Zhang, W., Cai, Y., & Kenelley, E. D., (2001). A fern that hyper-accumulates arsenic. *Nature, 409*, 579.

Manzoor, S., Shah, M. H., Shaheen, N., Khalique, A., & Jaffar, M., (2006). Multivariate analysis of trace metals in textile effluents in relation to soil and groundwater. *J. Hazard. Mater., 137*(1), 31–37.

Mashiatullah, A., Chaudhary, M. Z., Ahmad, N., Javed, T., & Ghaffar, A., (2013). Metal pollution and ecological risk assessment in marine sediments of Karachi Coast, Pakistan. *Environ. Monit. Assess., 185*(2), 1555–1565.

Maslin, P., & Maier, R. M., (2000). Rhamnolipid-enhanced mineralization of phenanthrene in organic-metal co-contaminated soils. *Biorem. J., 4*(4), 295–308.

McGrath, S. P., Zhao, F. J., & Lombi, E., (2001). Plant and rhizosphere processes involved in phytoremediation of metal contaminated soils. *Plant Soil, 232*(2), 207–214.

Mehta, S. K., & Gaur, J. P., (2001). Characterization and optimization of Ni and Cu sorption from aqueous solution by *Chlorella vulgaris*. *Ecol. Eng., 18*, 1–13.

Mehta, S. K., Singh, A., & Gaur, J., (2002). Kinetics of adsorption and uptake of Cu^{2+} by *Chlorella vulgaris*: Influence of pH, temperature, culture age, and cations. *J. Environ. Sci. Health Part A., 37*, 399–414.

Murugesan, G. S., Satishkumar, M., & Swaminathan, K., (2006). Arsenic from groundwater by pretreated waste tea fungal biomass. *Bioresour. Technol., 97*(3), 483–487.

Nazif, W., Perveen, S., & Shah, S. A., (2006). Evaluation of irrigation water for heavy metals of Akbarpura area. *J. Agric. Biol. Sci., 1*, 51–54.

Norris, P. R., & Kelly, D. P., (1979). Accumulation of cadmium and cobalt by *Saccharomyces cerevisiae*. *J. Gen. Microbiol., 99*, 317–324.

Oremland, R. S., & Stolz, J. F., (2003). The ecology of arsenic. *Science, 300*, 939–944.

Page, A. L., Chang, A. C., & El-Amamy, M., (1987). Cadmium levels in soils and crops in the United States. In: Hutchinson, T. C., & Meema, K. M., (eds.), *Lead, Mercury, Cadmium, and Arsenic in the Environment* (pp. 119–146). John Wiley and Sons Ltd, New York.

Pandey, A., Nigam, P., & Singh, D., (2001). Biotechnological treatment of pollutants. *Chem. Ind. Digest, 14*, 93–95.

Pazirandeh, M., Wells, B. M., & Ryan, R. L., (1998). Development of bacterial based heavy metal biosorbents: Enhanced uptake of cadmium and mercury by *Escherichia coli* expressing a metal binding motif. *Appl. Environ. Microbiol., 64*, 4068–4072.

Petrosyan, V., Orlova, A., Dunlap, C. E., Babayan, E., Farfel, M., & Von Braun, M., (2004). Lead in residential soil and dust in a mining and smelting district in Northern Armenia: A pilot study. *Environ. Res., 94*(3), 297–308.

Priyalaxmi, R., Murugan, A., Raja, P., & Raj, K. D., (2014). Bioremediation of cadmium by *Bacillus safensis* (JX126862), a marine bacterium isolated from mangrove sediments. *Int. J. Curr. Microbiol. App. Sci., 3*(12), 326–335.

Ramteke, P. W., (2000). Biosorption of Nickel (II) by *Pseudomollas stutzeri*. *J. Environ. Biol., 21*, 219–221.

Rani, A., & Goel, R., (2009). *Microbial Strategies for Crop Improvement* (pp. 105–132). Springer, Berlin.

Roane, T. M., & Pepper, I. L., (2000). Microorganisms and metal pollution. In: Maier, R. M., Pepper, I. L., & Gerba, C. B., (eds.), *Environmental Microbiology* (pp. 1–55). Academic Press, London.

Roy, S., Labelle, S., Mehta, P., Mihoc, A., Fortin, N., Masson, C., et al., (2005). Phytoremediation of heavy metal and PAH-contaminated brownfield sites. *Plant and Soil, 272*, 277–290.

Sag, Y., & Kutsal, T., (1995). Copper (II) and nickel (II) adsorption by *Rhizopus arrhizus* in batch stirred reactors in series. *Chem. Eng. J., 58*, 265–273.

Sag, Y., & Kutsal, T., (2000). Determination of the biosorption heats of heavy metal ions on *Zoogloea ramigera* and *Rhizopus arrhizus*. *Biochem. Eng. J., 6*(2), 145–151.

Sandrin, T. R., & Maier, R. M., (2002). Effect of pH on cadmium toxicity, speciation, and accumulation during naphthalene biodegradation. *Environ. Toxicol. Chem., 21*(10), 2075–2079.

Sar, P., & D'Souza, S. F., (2001). Biosorptive uranium uptake by *Pseudomonas* strain: Characterization and equilibrium studies. *J. Chem. Technol. Biotechnol., 76*, 1286–1294.

Sar, P., Kazy, S. K., Asthana, R. K., & Singh, S. P., (1999). Metal adsorption and desorption by lyophilized *Pseudomonas aeruginosa*. *Int. Biodeterior. Biodegrad., 44*, 101–110.

Sarikaya, M., Tamerler, C., Jen, A. K., Schulten, K., & Baneyx, F., (2003). Molecular biomimetics: Nanotechnology through biology. *Nat. Mater., 2*, 577–585.

Sayler, G. S., & Ripp, S., (2000). Field applications of genetically engineered microorganisms for bioremediation process. *Curr. Opin. Biotechnol., 11*, 286–289.

Sharma, B. D., (1998). Fungal associations with *Isoetes* species. *Am. Fern. J., 88*, 138–142.

Sheng, P. X., Ting, Y. P., Chen, J. P., & Hong, L., (2004). Sorption of lead, copper, cadmium, zinc, and nickel by marine algal biomass: Characterization of biosorptive capacity and investigation of mechanism. *Colloid Interface Sci., 275*, 131–141.

Sherene, T., (2010). Mobility and transport of heavy metals in polluted soil environment. *Biological Forum, 2*(2), 112–121.

Shetty, K. G., Hetrick, B. A. D., Figge, D. A. H., & Schwab, A. P., (1994). Effects of mycorrhizae and other soil microbes on revegetation of heavy metal contaminated mine spoil. *Environ. Pollut., 86*, 181–188.

Silver, S., & Phung, L. T., (2005). Genes and enzymes involved in bacterial oxidation and reduction of inorganic arsenic. *Appl. Environ. Microbiol., 71*, 599–608.

Sinha, R. K., Valani, D., Sinha, S., Singh, S., & Heart, S., (2009). Bioremediation of contaminated sites: A low-cost nature's biotechnology for environmental clean up by versatile microbes, plants, and earthworms. In: Faerber, T., & Herzog, J., (eds.), *Solid Waste Management and Environmental Remediation* (pp. 1–72). Nova Science Publishers, New York.

Siripornadulsil, S., & Siripornadulsil, W., (2013). Cadmium-tolerant bacteria reduce the uptake of cadmium in rice: Potential for microbial bioremediation. *Ecotoxicol. Environ. Saf., 94*, 94–103.

Skowroński, T., (1986). Influence of some physico-chemical factors on cadmium uptake by the green alga *Stichococcus bacillaris*. *Appl. Microb. Biotechnol., 24*, 423–425.

Smedley, P. L., & Kinniburgh, D. G., (2002). A review of the source, behavior, and distribution of arsenic in natural waters. *Appl. Geochem., 17*(5), 517–568.

Smith, M. D., Lennon, E., McNeil, L. B., & Minton, K. W., (1998). Duplication insertion of drug resistance determinants in the radioresistant bacterium *Deinococcus radiodurans*. *J. Bacteriol., 170*, 2126–2135.

Suciu, I., Cosma, C., Mihai, T., Bolboacă, S. D., & Jäntschi, L., (2008). Analysis of soil heavy metal pollution and pattern in central Transylvania. *Inter. J. Mol. Sci., 9*, 434–453.

Talos, K., Pager, C., Tonk, S., Majdik, C., Kocsis, B., Kilar, F., & Pernyeszi, T., (2009). Cadmium biosorption on native *Saccharomyces cerevisiae* cells in aqueous suspension. *Acta. Univ. Sapientiae. Agric. Environ., 1*, 20–30.

Thakur, I. S., & Srivastava, S., (2011). Bioremediation and bioconversion of chromium and Pentachlorophenol in tannery effluent by microorganisms. *Inter. J. Technol., 3*, 224–233.

Tohamy, E. Y., Abou, Z. M. A., Hazaa, A. M., & Hassan, R., (2006). Heavy metal biosorption by some bacterial species isolated from drinking water at different sites in Sharqia Governorate. *Arab. Univ. J. Agric. Sci., 14*(1), 147–172.

Tripathi, M., Vikram, S., Jain, R. K., & Garg, S. K., (2011). Isolation and growth characteristics of chromium(VI) and pentachlorophenol tolerant bacterial isolate from treated tannery effluent for its possible use in simultaneous bioremediation. *Indian J. Microbiol., 51*(1), 61–69.

Tunali, S., Akar, T., Oezcan, A. S., Kiran, I., & Oezcan, A., (2006). Equilibrium and kinetics of biosorption of lead(II) from aqueous solutions by *Cephalosporium aphidicola. Sep. Purif. Technol., 47*, 105–112.

Umrania, V. V., (2006). Bioremediation of toxic heavy metals using acidothermophilic autotrophes. *Biores. Technol., 97*, 1237–1242.

UNEP, (1996). *Our Planet*, United Nation Environment Program, Nairobi, Kenya.

Vaclavikova, M., Gallios, G. P., Hredzak, S., & Jakabsky, S., (2008). Removal of arsenic from water streams: An overview of available techniques. *Clean Technol. Environ., 10*(1), 89–95.

Valls, M., Atrian, S., Loveno, V. D., & Fernandez, L. A., (2000). Engineering a mouse metallothionein on the cell surface of *Ralsatonia ellIropha* CH34 for immobilization of heavy metals in soil. *Nat. Biotechnol., 18*, 661–665.

Vasudevan, P., Padmavathy, V., Tewari, N., & Dhingra, S. C., (2001). Biosorption of heavy metal ions. *J. Sci. Ind. Res., 60*, 112–120.

Verma, N., & Singh, M., (2005). Biosensors for heavy metals. *J. Biometals., 18*, 121–129.

Vijayaraghavan, K., & Yun, Y. S., (2007). Utilization of fermentation waste (*Corynebacterium glutamicum*) for biosorption of Reactive Black 5 from aqueous solution. *J. Hazard. Mater., 141*(1), 45–52.

Vishwanathan, B., (2009). *Nanomaterials*. Narosa Publishing House Pvt. Ltd., New Delhi.

Volesky, B., (1999). Biosorption for the next century, *International Biohydrometallurgy Symposium* (pp. 20–23). El Escorial, Spain. http: //www.biosorption.mcgill.ca/publication/BVspain/BVspain.htm

Williams, K. H., Long, P. E., Davis, J. A., Wilkins, M. J., N'Guessan, A. L., Steefel, C. I., et al., (2011). Acetate availability and its influence on sustainable bioremediation of uranium-contaminated groundwater. *Geomicrobiol. J., 28*, 519–539.

Wu, C. H., Wood, T. K., Mulchandani, A., & Chen, W., (2006). Engineering plant-microbe symbiosis for rhizoremediation of heavy metals. *Appl. Environ. Microbiol., 72*, 1129–1134.

Yusof, A. M., & Malek, N. A. N. N., (2009). Removal of Cr(VI) and As(V) from aqueous solutions by HDTMA-modified zeolite Y. *J. Hazard. Mater., 162*, 1019–1024.

Zhang, H., Dang, Z., Zheng, L. C., & Yi, X. Y., (2009). Remediation of soil co-contaminated with pyrene and cadmium by growing maize (*Zea mays* L.). *Int. J. Environ. Sci. Tech.*, *6*(2), 249–258.

Zhao, Y., Hao, Y., & Ramelow, G. J., (1994). Evaluation of treatment techniques for increasing the uptake of metal ions from solution by non-living seaweed algal biomass. *Environ. Monit. Assess.*, *33*, 61–70.

CHAPTER 12

MICROBIAL APPLICATIONS IN NANOTECHNOLOGY INDUSTRY

ESSAM ABDELLATIF MAKKY and MASHITAH MOHD YUSOFF

Faculty of Industrial Sciences and Technology, University Malaysia Pahang, Gambang, 26300 Kuantan, Pahang, Malaysia

12.1 INTRODUCTION

Green nanotechnology (Gnanotech) defined as the use of nanotechnology products for enhancing the sustainability and includes green nano-products to support sustainability. Gnanotech also described as the development of clean technologies to reduce environmental and human risks associated with industry and use of nanotechnology (NanoTech) products, and to replace the existing products to that are more environmentally friendly throughout their lifecycle. In this chapter, the potential use of NanoTech, nanoparticles (NPs) in different important areas and scientific terminology definition to be considered has been discussed as follows.

12.1.1 NANOTECHNOLOGY AND NANOBIOTECHNOLOGY

NanoTech indicates to the manufacture, treatment, and exploitation of submicron matters in the range of 1–100 nm nanosize. Physico-chemical sciences have sophisticated materials and procedures to manufacture nanoscale structures with interested applications in different science areas. In the biomedical, the relevance of NanoTech depend on the specific biophysical properties of nanoscale tools and their specific interaction with living beings such as high diffusion in tissues and organs, high surface-volume ratio, appropriate uptake by mammalian cells and high biological interface influences through mechanotransduction notification (Curtis et al., 2006) and alternative spectrum of cell activities and responses (Jiang et al., 2008).

Bionanotechnology (BioNanoTech), as well as NanoTech, are considered mysterious terms whose overlapping meanings, as their related applications and technologies keep improving. Both are understood as the generation of hybrid substances deriving from biological and chemical synthesis, or bioinspired substances (Taylor, 2007). Also, BioNanoTech can be observed as "nanotechnology through biotechnology" (Sarikaya et al., 2003). The bio-manufacture of nano-materials can be used as tools to manipulate or construct nano-objects. The chronology of nanotechnology was explained in Figure 12.1, and Table 12.1 shows different contaminants and microorganism size in micro and nanometer.

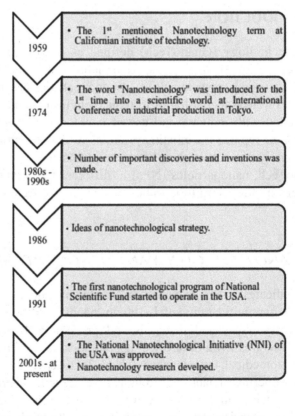

FIGURE 12.1   Chronology of nanotechnology since 1959 to present.

TABLE 12.1   Different Particles and Microorganisms in Relation to Size

Particle	Micrometer (μm)	Nanometer (nm)
One inch	24,500	24,500,000
Bacteria	0.3–60	300–60,000
Mold	3–12	300–12,000
Mold spores	10–30	10,000–30,000
Viruses	0.005–0.3	5–300
Beach sand	10–10000	10,000–10,000,000
Burning wood	0.3–3	300–3,000
Clay	0.1–50	100–50,000
Pollens	10–1000	10,000–1,000,000
Red blood cell	5–10	5000–10,000

12.1.2   MICROBIOLOGY AND INDUSTRIAL MICROBIOLOGY

Microbiology refers to the study of microorganisms, those being consist of single cell (called unicellular organisms), many cells (called multicellular organisms), or lacking cells (called acellular organisms) (Madigan et al., 2006). Microbiology covers many different sub-disciplines such as bacteriology, parasitology, mycology, and virology. Eukaryotic microorganisms such as protists and fungi are containing membrane-bound cell organelles, whereas prokaryotic microorganisms are normally classified as lacking membrane-bound organelles and include archaebacteria and eubacteria. Microbiologists conventionally depend on microbial culture, staining, and microscopy. However, microorganisms present in common environments (about <1%) can be isolated and cultured using particular protocols (Nitesh et al., 2011). Also, they often depend on extraction or detection of nucleic acid, either deoxyribonucleic acid (DNA) or ribonucleic acid (RNA) sequences. Industrial microbiology is an area of applied microbiology which deals with screening, management, improvement, and exploitation of microorganisms for the production of different useful by-products on a large scale.

12.1.3   BIOTECHNOLOGY AND TECHNOLOGY

Biotechnology (BioTech) is the use of living organisms and systems to develop or produce products, or any technological application that uses biological systems, living organisms or derivatives thereof, to make or modify products or processes for specific use. Relying on the objects and applications, it overwhelmingly overlaps with the associated fields of biomedical, bioengineering, and biomanufacturing. For thousands of years, human beings have used BioTech in medicine, food manufacture, and agriculture. The term is largely considered to have been published in 1919 by Károly Ereky. In the late 20th and early 21st century, BioTech has developed to comprise new and varied sciences such as recombinant gene techniques, genomics, applied immunology and development of diagnostic tests and pharmaceutical therapies.

Technology is the gathering of different techniques, methods, skills, and processes used in the manufacture of goods or services or in the achievement of scientific investigation. It can be the scientific information of techniques and processes or embedded in industrial machines, devices, computers, and factories, which can be operated by individuals without detailed information of the operations of such equipment's. The use of technology launched with the transformation of natural resources into simple matters. The ancient discovery of how to control fire and the later Neolithic Revolution increased the available sources of food, and the invention of the wheel helped humans to travel and control their environment. Evolutions in ancient times, including the telephone, printing press, and internet, have minimized physical barriers to communication and allowed humans to interact freely on a global scale. The stable progress of military technology has brought weapons of ever-increasing destructive power, from clubs to nuclear weapons. Technology has various impacts, and it has supported in the development of more advanced economies and has allowed the rise of a leisure class. Many technological processes produce undesirable by-products, like pollution, and minimize natural resources, to the damage of Earth's environment. Many applications of technology impact the significances of a society and new technology often promote new ethical questions. Instances comprise the elevation of the concept of competence in terms of human productivity, a term primarily used only to machines, and the challenge of conventional standards. Philosophical discussions have arisen over the use of technology, with differences over whether technology improves the human condition or worsens it. Neo-Luddism, anarcho-primitivism, and similar reactionary movements fault the universal of technology in the modernistic world, disputing that it damage the environment and alienates people; proponents of ideologies such as transhumanism and techno-progressivism view continued technological progress as beneficial to society and the human condition. The technical term has changed significantly over the last 200 years. Before the 20th century, it was rare in English and usually referred to the description or study of the useful arts (Stratton and Mannix, 2005).

12.1.4 MICROBIAL FERMENTATION AND NANOREMEDIATION

Fermentation occurs due to the microorganisms consume susceptible organic substrate as part of their own metabolic processes. Such microbial interactions are essential to the decay of natural materials and to the extreme return of chemical elements to the soil and air without which life could not be sustained (Table 12.2).

TABLE 12.2 Some Examples of Fermented Products

Product	Detailed information
Beer	Anaerobic fermentation; produced by *Saccharomyces cerevisiae* and maltose can be used as a substrate. Metabolites other than ethanol impart a distinct flavor.
Wine	Occurred under anaerobic fermentation; can be produced by yeast that may be found naturally on the grapes or be added to give better consistency of quality.
Lactic acid, Yoghurt (milk), Sauerkraut (cabbage), Kimchi (vegetables), Salami (meat), Silage	Various bacterial strains such as *Lactobacillus*, *Streptococcus*, and *Leuconostoc* able to ferment sugars at lower pH into lactic acid. It also alters the consistency, texture, and flavor by coagulating proteins.
Antibiotics	Many of antibiotics are produced by fermentation or by enzymes which may be microbial origin under aerobic conditions. Fungicides and bactericides can be used as medicines for human, in agriculture to treat and animals diseases. Some are toxic to eukaryotic cells are used to treat tumor.
Hormones	Various hormones made by genetically modified microorganisms such as human gonadotrophin, insulin, and human growth hormone.
Amino acids	Used in food and beverage industries to enhance flavor, antioxidants, and as sweeteners. Proline is used in cosmetics, and many amino acids can be produced by microorganisms as *Bacillus, Corynebacterium, E. coli, Lactobacillus, Klebsiella*.

TABLE 12.2 *(Continued)*

Product	Detailed information
Vitamins	Microorganisms able to produce various types of vitamins as riboflavin, thiamine, ascorbic acid, and vitamin B12. Different bacterial species able to grow on some substrates as methanol and glucose under anaerobic conditions as *Pseudomonas* and *Propionibacterium* or under aerobic conditions as *Bacillus* and *Streptomyces*.
Enzymes	Microbial rennin used in cheese making. Proteases and lipases have many industrial applications in washing powders, glucose isomerase in fructose syrup production, pectinases in fruit juice extraction and many enzymes are used in the textiles, paper, and animal feed production. Streptokinase is a clot buster, and glucose oxidase is used by diabetics in biosensors to measure blood glucose levels.
Citric acid	Used in the food and drink industries as a flavor enhancer for marmalade, ice cream, and fruit juice. In the pharmaceutical industry, it is used to preserve stored blood, tablets, and ointments. It is also used in cosmetics, as an anti-foaming agent and to soften textiles.
Food	Single cell protein (SCP) can be produced by bacteria from methane, methanol or diesel oil, which can be added to animal feed. Marmite is made from hydrolyzing spent yeast by *Spirulina* is eaten in parts of Africa and Quorn is a mycoprotein made from *Fusarium venenatum*. This is made during aerobic fermentation and is high in both protein and dietary fiber while low in saturated fat.

Environmental pollution can be remediated or cleaned-up using different techniques. Environmental remediation (ER) refers to the field of study that concentrates on exploring the removal or clean-up of contaminants from the environment. ER uses several methods to degrade and/or remove environmental contaminants in groundwater, surface waters, polluted soils, as well as in sediments. The comprehensive objective of ER is to minimize environmental and/or human health risks due to environmental pollution through one or several remediation methods. Some of these cover removing the sediment, contaminated soil, or water from the polluted places and then treating the pollution aboveground recognized as *ex situ* techniques. Other techniques can be used in clean-up the contamination while it is still in the ground called *in situ* methods without the need of off-site treatment. There are many pressures from legislative, financial, and time associated chains that drive the need for continued research into new techniques that provide cheaper, better, and faster ER treatment options. At the latest, the choice of the better remediation technique different from place to place and rely on site-specific conditions such as the contaminant nature and hydrology as well as cost, performance, and environmental effects of the potential clean-up technologies. Recently, the use of engineered materials such as NPs, have acquired increased amount of awareness as potency remediation techniques, often cited as an attractive, cost-effective alternative to traditional approaches. Engineered nanomaterials can be defined as specially designed matters with size range of around 1–100 nm. These nanomaterials often offer various properties compared to their bulk-scale counterparts. Owing to these properties at the nano size, a range of engineered nanomaterials are used in a number of consumer products and other applications, involve ER. "Nanoremediation" is the term used to characterize many methods and techniques to clean-up contaminated sites using engineered nanomaterials. Recently, it is evaluated that around 45 to 70 places all over the world that have used nanoremediation techniques either at pilot or full study scales (Karn et al., 2009; Bardos et al., 2014). Engineered nanomaterials are also being sophisticated for other environmentally associated activities, like the use of anti-fouling, absorbent nanowires for oil spills, nanomaterials for desalination processes using reverse osmosis, as well as photocatalytic nanomaterials that can be used for decontamination of drinking waters or disinfection.

12.2 MICROBIAL NANOTECHNOLOGY APPLICATIONS

12.2.1 PRESERVATION IN FOOD INDUSTRY

The different stages of food manufacture can be achieved using Nano-Tech, i.e., during various processing of labeling, packaging, transporting, and during tracing the food in question. NPs are to be used in foods as composites, or emulsions nanostructured materials and can play a main role in ensuring the food microbial safety and in reservation the quality of the product. Recently, the uses of NanoTech with antimicrobial activity has been offered as a new defense against multiple drug resistant (MDR) infectious organisms and exhibit a different way to damage the functions of microbial cell. Instead of focusing upon specific biochemical processes, as traditional antibiotics, they are probably to damage multiple cellular processes in a less particular style and may make it tough for microbes to promote resistance (Aruguete et al., 2013).

The commercial high demand for available silver NPs products is broad claims due to the power these ingredients, like inhibit and/or kill 99% of bacteria, renders material "permanently antimicrobial and antifungal," "kills approximately 650 kinds of harmful germs and viruses," and "kills bacteria in a short time as 30 minutes, 2–5 times faster than other forms of silver" (Emtiazi et al., 2009). Indeed, silver has a long history of being used as an antimicrobial agent in food and beverage storage applications. Silver was the broad spectrum antimicrobial activity and active disinfecting agent for water treatment in developing countries due to its low cost (Solsona and Méndez, 2003). To date, many mode of actions have been postulated for the antimicrobial susceptibility of silver nanoparticles (AgNPs); adhesion of NPs to the surface changing the membrane properties. AgNPs have been reported to breakdown lipopolysaccharide (LPS) molecules, collecting inside the membrane by forming holes, and largely increasing permeability of the membrane (Sondi and Salopek-Sondi, 2004). AgNPs can enter the bacterial cell causing damage of DNA; dissolution of nano-Ag liberates antimicrobial Ag^+ ions (Morones et al., 2005). Copper nanoparticles have also been reported to be highly toxic against a wide variety of bacteria and fungi due to their high surface-to-volume-ratio, and generally kill cells by various modes of actions, such as blocking biochemical pathways, membrane disruption, complex formation with proteins, and DNA damage (Cioffi et al., 2005; Ruparelia et al., 2008; Ren et al., 2009). However, copper nanoparticles are extremely sensitive to O_2 forming Cu_2O nanoparticles, so it is better to use a matrix that will bind and protect the particles from environmental oxidizing agents (Kaninnen et al., 2008).

Titanium dioxide (TiO_2) is commonly used as semiconductor photocatalyst activated by Ultraviolet-A (UV-A) irradiation. It can be kill both Gram-positive and Gram-negative bacteria, although some Gram-positive bacteria are less sensitive owing to their capability to form spores (Wei et al., 1994). TiO_2 was also reported to be effective against viruses such as poliovirus 1 (Watts et al., 1995), hepatitis B (Zan et al., 2007) and Herpes simplex (Hajkova et al., 2007) through the hydroxyl free radicals and peroxide formulation and also demonstrated to induce *in vivo* side effects in mice like genetic instability and damage of DNA (Trouiller et al., 2009), for these reasons, the strict toxicity testing and precautious is urged (Weir et al., 2012).

12.2.2 AIR DISINFECTION

Many of airborne particles such as bioaerosols are of biological origins involving bacteria, fungi, and viruses, which are able to rising allergenic, toxigenic diseases, or infection. Especially, indoor air bioaerosols were found to accumulate in large quantities on filters of heating, ventilating, and air-conditioning (HVAC) systems (Yoon et al., 2008). It is found that the pollution of outdoor air and insufficient hygiene of HVAC installation often resulted in the low quality of indoor air. Furthermore, the inorganic or organic materials accumulated on the filter medium after air filtration to enhance the microbial growth. About 50% of the biological contamination existent in indoor air comes from different air treatment methods, and harmless microorganism's formation like fungal and bacterial pathogens was found in air filters as reported by the World Health Organization (WHO). The significant note is that most of these pathogens able to produce mycotoxins which are risky to human health. To solve the microbial growth problem in air filters, the combination of antimicrobial AgNPs in these filters has been suggested and developed to reduce the bacterial number, especially which was observed in both Gram-positive strains and Gram-negative of *Pseudomonas luteola, B. subtilis, Micrococcus roseus*, and *Micrococcus luteus*. The obvious reduction in bacterial cells in silver treated filters made the antimicrobial filter technology really necessary for the future of human health.

12.2.3 WATER DISINFECTION

The most important substances on the Earth is the water and essential to all forms of life and covered around 70% of the Earth, but only small percentage (0.6%) is appropriate for human consumption. In many developing countries, the most important social issue and health is the safe drinking water (De Gusseme et al., 2010) and about 1 billion people are not access to safe drinking water according to WHO report. The contaminated drinking water and waterborne diseases are leading of death in many developing countries (Pradeep, 2009). Furthermore, the spectrum and effects of many infectious diseases are increasing globally. Subsequently, there is a tremendous requirement for treatments to prevent the microbial contamination of water and reduce the waterborne diseases. The use of AgNPs for water disinfection is important application, and its impact on microbial societies in wastewater treatment plants was evaluated and developed (Sheng and Liu, 2011). The original of wastewater biofilms are strongly tolerant to AgNPs treatment processes. The bacterial biofilm was successfully measured using heterotrophic plate count (HPC) and reduced when AgNPs of 200 mg L^{-1} was applied after 24 h, and this is may be due to the extracellular polymeric substances (EPS) which play an important role in bacterial physical protection under AgNPs treatment. The sensitivity to AgNPs is different from microorganism to another in the microbial biofilm and provided two suggestions: (a) AgNPs should be effective against the structures of the microbial biofilm of wastewater, depending on the characteristics of each microbial strain, e.g., capability to produce EPS and growth rate, and the interactions within these strains; and (b) the impact of AgNPs on planktonic cells were varied to those on wastewater biofilms.

We can conclude that, the ideal method in water disinfection by using AgNPs treatment and can be combined to core materials and polymeric membranes to disinfect the water contaminated with different microorganisms. The application of silver NPs is very important to prevent the waterborne diseases associated to poor treatment of drinking water. Over the above, the addition of silver NPs could inhibit viral/bacterial attachment and formation of microbial biofilm in filtration medium (Sintubin et al., 2012).

12.2.4   SURFACE DISINFECTION

The bactericidal mechanism has been improved by coatings on the surfaces and attracted increasing the interest to protect the environment and human health. Such AgNPs embedded paints are of specific interest due to their potential bactericidal activity. Kumar et al. (2008) reported that an environmentally friendly chemistry approach to produce metal NPs embedded paint, in one step, from common household paint. The oxidative drying process in oils can be occurred naturally, including the free radical exchange as the basic mechanism for reducing metal salts and dispersing metal NPs in the oil media, without external use of stabilizing or reducing agents. So, NPs in oil dispersions can be used directly on different surfaces such as steel, wood, glass, and various polymers. As results, the surfaces coated with AgNPs paint exhibited excellent antimicrobial properties by inhibiting and/or killing both Gram-positive and Gram-negative bacteria as human pathogens *Staphylococcus aureus* and *E. coli*, respectively.

12.2.5 ANTICANCER ACTIVITY

AgNPs can be synthesized using *Acalypha indica* Linn. and exhibited cell inhibition (about 40%) against human breast cancer cells (MDA-MB–231) (Krishnaraj, 2014). The MCF–7 cells (human breast adenocarcinoma cell line) lose their 50% viability at 5µg/mL for the AgNPs produced by *Dendrophthoe falcata* (Loranthaceae) on lipids, lipoproteins, and lipid-metabolizing enzymes (Sathishkumar et al., 2014). Silver (protein-lipid) nanoparticles prepared using seed extract of *Sterculia foetida* (L.) showed cellular DNA fragmentation against HeLa cancer cell lines (Rajasekharreddy and Rani, 2014). *Datura inoxia* AgNPs inhibited 50% proliferation of breast cancer cell MCF–7 at 20 µg/mL after incubation period of 24 h by inhibiting its growth, decreasing DNA synthesis, and arresting the cell cycle phases to enhance apoptosis process (Gajendran et al., 2014). The cytotoxic assays of *Chrysanthemum indicum*-AgNPs showed no toxicity toward 3T3 mouse embryo fibroblast cells at a concentration of 25 µg/mL (Arokiyaraj et al., 2014).

12.2.6 MICROBIAL BIOSENSORS

An analytical device that combines with a signal transducer determine the biological element and able to convert the response with analytes into a measurable signal known as biosensor (D'Souza, 2001a; Chauhan et al., 2004; Lei et al., 2006; Mohanty and Kougianos, 2006) (Figure 12.2). A biosensor that uses the microorganisms which able to produce various enzymes such as the bioelements called microbial biosensor. These enzymes can produce a response to the analytes selectively and specifically, without neither time-consuming and purification cost nor the negative impacts of the operating environment (D'Souza, 2001a,b; Chauhan et al., 2004; Lei et al., 2006; Su et al., 2011). For better enzyme immobilization, nanotechnology offers an alternative by using some nanomaterials such as NPs and nanotubes, which promote higher stability and reliability of the bioelements (Tuncagil et al., 2011). There are many types of microbial biosensors which are optical biosensor to produce changes in diverse optical properties (Velasco-Garcia, 2009), fluorescent microbial biosensor is commonly used in analysis processes (D'Souza, 2001a,b; Lei et al., 2006). Bioluminescence based microbial biosensor has been extensively used in environmental monitoring for detection of toxicity (Steinberg,

1995) and colorimetric microbial biosensor able to use the color changes in a special compound to determine the target analytes concentration. Lately, microbial fuel cells (MFCs) have been suggested as a new technique for microbial biosensors which depend on optical transducers as a main transducer in the past decade. With the capability to produce potential electricity from biodegradable organic compounds through microbial metabolism, MFCs provide high selective sensing capability and sensitivity (Du et al., 2007; Choi and Chae, 2013) (Figure 12.3).

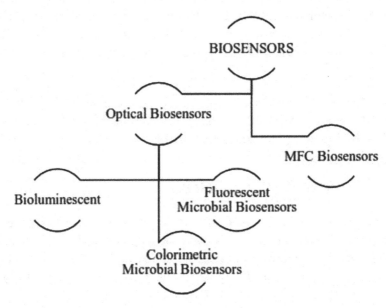

FIGURE 12.2   Types of microbial biosensors.

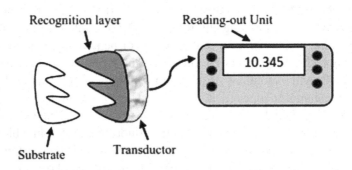

FIGURE 12.3   Biosensor principle diagram.

12.3   APPLICATION OF METAL NANOCOMPOSITES

Metal nanocomposites have drawn great attention owing to their broad antimicrobial applications (Muniz-Miranda and Neto, 2004) and well known that the biocides including metal particles able to penetrate the bacteria cell wall. At present time, the mechanism of AgNPs as antimicrobial agent is unclear, notwithstanding, the theory has been assumed as follows; the AgNPs able to binding the microbial cells and alter their membranes (Sondi and Salopek-Sondi, 2004). AgNPs lead to DNA damage due to the penetrating and dissolution of AgNPs releases antimicrobial Ag^+ ions (Morones et al., 2005). Copper exhibited antifouling properties and has therefore been incorporated into antimicrobial coatings (Acharya et al., 2004). The paint industry has applied Cu and CuO particles into organic matrixes since the 19th century for antifouling coatings in maritime applications (Sedlarik et al., 2010). The antimicrobial mechanism of Cu NPs still unclear, but reported that they have a similar mechanism of action as AgNPs (Ruparelia et al. 2008). TiO_2 photocatalysts have been reported for removing of contaminants from water and purification due to its high photosensitivity, stability, and antimicrobial activity in the form of films and powders (Yu et al., 2003; Maneerat and Hayata, 2006; Xie and Li, 2006). The redox reactions of TiO_2 degrade the microbial cells by damaging their structure or disrupting their biochemistry. These cells are ultimately oxidized to CO_2 and H_2O. The strength of the UV source plays an important role in the bactericidal activity of TiO_2 (Hochmannova and Vytrasova, 2010). Gold NPs can be produced by *in situ* reduction of chloroauric acid ($HAuCl_4$) on the surface of *Rhizopus oryzae* strain. The water contaminated pesticides and *E. coli* can be removed by using gold NPs application, and the antimicrobial mechanism of action included strong adsorption of various organophosphorous pesticides and cell membrane rupturing of bacteria leading to cell death. Gold NPS also reported to be applied in antimicrobial coatings. Zinc oxide (ZnO) NPs can be incorporated into antimicrobial paints and serve as photocatalysts, and its mechanism is not clear, but it has been suggested that H_2O_2, generated through photocatalysis and plays a primary role (Sawai, 2003). Also, inhibits the microbial growth by intracellular accumulation and cell membrane damage (Brayner et al., 2006). Many other NPs applications with antimicrobial properties include NPs based on magnesium oxide (MgO), alginate, antimicrobial peptides, cadmium selenide/telluride, gold, chitosan, fullerenes, as well as carbon nanotubes. Many of these reported are targeted specifically at food or food packaging applications, and a new publication reviewed many classes of nanomaterial antimicrobials targeted for use in drinking water sterilization (Duncan, 2011) (Figure 12.4).

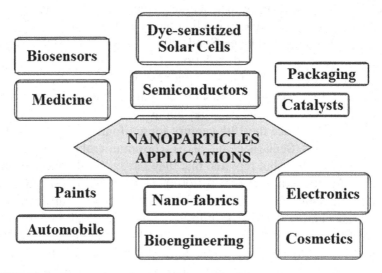

FIGURE 12.4   An overview of nanoparticles applications.

12.3.1 SOLID LIPID AND POLYMERIC NANOPARTICLES

Solid Lipid Nanoparticles (SLNPs) consists of a solid lipid matrix, where the drug is normally incorporated, with an average diameter below 1 μm. Many surfactants can be used to prevent aggregation and stabilize the dispersion have been accepted Generally Recognized as Safe (GRAS) status. SLNPs have been considered as new transfection agents using cationic lipids for the matrix lipid composition. Cationic SLNPs for gene transfer can be formulated using the same cationic lipids as for liposomal transfection agents. In comparison to SLNPs or nanosuspension, polymeric nanoparticle (NsPNP) composed of a biodegradable polymer. Biocompatibility is an important characteristic for many applications such as gene, and drug delivery, tissue engineering, and new vaccination strategies. Most of biodegradable polymers containing synthetic polyesters as polycyanoacrylate and related polymers like poly(lactide-co-glycolide) or poly lactid acid (PLA). Mostly under the term of nanoparticle, nanospheres are understood to be considered as uniformly dispersed matrix system. Besides of these spheric vesicular systems nanocapsules are also known as a polymeric membrane surrounds the drug in a matrix core. The polymer choice and the ability to modify drug release from polymeric NPs have made them ideal candidates for cancer therapy, delivery of vaccines, contraceptives, and delivery of targeted antibiotics. Furthermore, polymeric NPs can be easily incorporated into other activities related to drug delivery such as tissue engineering, and into drug delivery for species other than humans. In future, the challenge field of polymers will be focused on to create new polymers matching lipophilic and hydrophilic properties of new drugs for smart formulation (Kayser et al., 2005).

12.4 CONCLUSION AND FUTURE PERSPECTIVES

In the future, there are many future surprises might be avoided by carefully testing NPs up front to measure their toxicities and to predict how easily they could spread in our environment. The NanoTech applications are almost unlimited when new properties of matter are discovered and known as a platform technology: it easily consolidates and huddles with other technologies. Many scientists said that we are not doing anything radically new; just we are simply continuing our tradition of making tools, investigating how the natural world works and manipulating that world at ever smaller scales. Simply, nanotechnology is about qualifying and exploiting the atoms to serve our needs. Recently, researchers just built the first complete integrated circuit using a single carbon nanotube, which is 50,000 times thinner than a human hair.

According to some reports recently released, we can conclude that the global nanotechnology market should reach $90.5 billion by 2021 from $39.2 billion in 2016 at a compound annual growth rate (CAGR) of 18.2%, from 2016 to 2021. For microsensors, reached nearly $9.9 billion in 2014 and totaled almost $10.9 billion in 2015. The market should reach more than $18.9 billion in 2020, with CAGR of 11.7% between 2015 and

2020. While the global nanotechnology market in environmental applications reached $23.4 billion in 2014. This market is expected to reach about $25.7 billion by 2015 and $41.8 billion by 2020, registering CAGR of 10.2% from 2015 to 2020.

As application, a portable microbial biosensor array system will be promising in reducing the cost, time, and personnel required for detecting the water toxicity while intensive farming and industrialization have led to the release of various toxic compounds in our environment, causing an important pollution of aquatic ecosystems. The use of biological methods in environmental monitoring is essential to complement chemical analysis with information about actual toxicity or genotoxicity of environmental samples. Microorganisms are widely applied test species in many bioassays because of the low costs and their culturing as well as the lack of ethical issues often accompanying the use of higher organisms. With the development of toxicogenomic approaches, the use of microorganisms for environmental monitoring purposes is expected to become even more extensive because of better knowledge about potential analogies in toxicity mechanisms between higher organisms and microbes. One of the main problems in the water treatment industry is biofouling. Ultimately, research has focused on applying NanoTech in antimicrobial coatings and films as alternative techniques to prevent biofouling.

KEYWORDS

- biotechnology
- fermented products
- industrial application
- market size
- microbial size
- microorganisms
- nanoparticles
- nanotechnology
- particle size

REFERENCES

Acharya, V., Prabha, C. R., & Narayanamurthy, C., (2004). Synthesis of metal incorporated low molecular weight polyurethanes from novel aromatic diols, their characterization, and bactericidal properties. *Biomaterials, 25*, 4555–4562.

Arokiyaraj, S., Arasu, M. V., Vincent, S., Udaya, P. N. K., Choi, S. H., Oh, Y. K., et al., (2014). Rapid green synthesis of silver nanoparticles from *Chrysanthemum indicum* land its antibacterial and cytotoxic effects: an *in vitro* study. *International Journal of Nanomedicine, 9*(1), 379–388.

Aruguete, D. M., Kim, B., Hochella, M. F., Ma, Y., Cheng, Y., Hoegh, A., et al., (2013). Antimicrobial nanotechnology: It's potential for the effective management of microbial drug resistance and implications for research needs in microbial nanotoxicology. *Environmental Science: Processes and Impacts, 15*, 93–102.

Bardos, P., Bone, B., Daly, P., Elliott, D., Jones, S., Lowry, G., & Merly, C., (2014). *A Risk/Benefit Appraisal for the Application of Nano-Scale Zero Valent Iron (nZVI) for the Remediation of Contaminated Sites.* www.nanorem.eu/Displaynews.aspx?ID=525.

Brayner, R., Ferrari-Iliou, R., Brivois, N., Djediat, S., Benedetti, M. F., & Fievet, F., (2006). Toxicological impact studies based on *Escherichia coli* bacteria in ultrafine ZnO nanoparticles colloidal medium. *Nano Letters, 6*(4), 866–870.

Chauhan, S., Rai, V., & Singh, H. B., (2004). Biosensors. *Resonance, 9*(12), 33–44.

Choi, S., & Chae, J., (2013). Optimal biofilm formation and power generation in a micro-sized microbial fuel cell (MFC). *Sensors and Actuators A: Physical, 195*, 206–212.

Cioffi, N., Torsi, L., Ditaranto, N., Tantillo, G., Ghibelli, L., Sabbatini, L., et al., (2005). Copper nanoparticle/polymer composites with antifungal and bacteriostatic properties. *Chem. Mater., 17*, 5255–5262.

Curtis, A. S., Dalby, M., & Gadegaard, N., (2006). Cell signaling arising from nanotopography: Implications for nanomedical devices. *Nanomed., 1*, 67–72.

D'Souza, S. F., (2001a). Microbial biosensors. *Biosensor and Bioelectronics, 16*(6), 337–353.

D'Souza, S. F., (2001b). Immobilization and stabilization of biomaterials for biosensor applications. *Appl. Biochem. Biotechnol., 96*(1–3), 225–238.

De Gusseme, B., Sintubin, L., Hennebel, T., Boon, N., Verstraete, W., Baert, L., & Uyttendaele, M., (2010). Inactivation of viruses in water by biogenic silver: Innovative and environmentally friendly disinfection technique. *4th International Conference on Bioinformatics and Biomedical Engineering* (pp. 1–5). Chengdu, China.

Du, Z. W., Li, H. R., & Gu, T. Y., (2007). A State of the art review on microbial fuel cells: A promising technology for wastewater treatment and bioenergy. *Biotechnol. Adv., 25*(5), 464–482.

Duncan, T. V., (2011). Applications of nanotechnology in food packaging and food safety: Barrier materials, antimicrobials, and sensors. *J. Coll. Interface Science, 363*, 1–24.

Emtiazi, G., Hydary, M., & Saleh, T., (2009). Collaboration of *Phanerocheate chrysosporium* and nanofilter for MTBE removal. *The International Conference on Nanotechnology: Science and Applications* (p. 276). Barcelona, Spain.

Gajendran, B., Chinnasamy, A., Durai, P., Raman, J., & Ramar, M., (2014). Biosynthesis and characterization of silver nanoparticles from *Datura inoxia* and its apoptotic effect on human breast cancer cell line MCF7. *Materials Letters, 122*, 98–102.

Hajkova, P., Spatenka, P., Horsky, J., Horska, I., & Kolouch, A., (2007). Photocatalytic effect of TiO₂ films on viruses and bacteria. *Plasma Processes and Polymers, 4*, 397–401.

Hochmannova, L., & Vytrasova, J., (2010). Photocatalytic and antimicrobial effects of interior paints. *Progress in Organic Coatings, 67*(1), 1–5.

Jiang, W., Kim, B. Y., Rutka, J. T., & Chan, W. C., (2008). Nanoparticle-mediated cellular response is size-dependent. *Nature Nanotechnology, 3*, 145–150.

Kaninnen, P., Johans, C., Merta, J., & Kontturi, K., (2008). Influence of ligand structure on the stability and oxidation of copper nanoparticles. *Journal of Colloid and Interface Science, 318*, 88–95.

Karn, B., Kuiken, T., & Otto, M., (2009). Nanotechnology and *in situ* remediation: A review of the benefits and potential risks. *Environmental Health Perspectives, 117*(12), 1813–1831.

Kayser, O., Lemke, A., & Hernández-Trejo, N., (2005). The impact of nanobiotechnology on the development of new drug delivery systems. *Current Pharmaceutical Biotechnology, 6*(1), 3–5.

Krishnaraj, C., Muthukumaran, P., Ramachandran, R., Balakumaran, M., & Kalaichelvan, P., (2014). *Acalypha indica* Linn.: Biogenic synthesis of silver and gold nanoparticles and their cytotoxic effects against MDA-MB–231, human breast cancer cells. *Biotechnology Reports, 4*, 42–49.

Kumar, A., Vemula, P. K., Ajayan, P. M., & John, G., (2008). Silver-nanoparticle-embedded antimicrobial paints based on vegetable oil. *Nature Materials, 7*(3), 236–241.

Lei, Y., Chen, W., & Mulchandani, A., (2006). Microbial biosensors. *Analytica. Chimica. Acta., 568*(1 & 2), 200–210.

Madigan, M. T., Martinko, J., & Brock, T. D., (2006). *Brock Biology of Microorganisms*, Pearson Prentice Hall, Upper Saddle River, NJ, USA, (p. 1096).

Maneerat, C., & Hayata, Y., (2006). Antifungal activity of TiO₂ photocatalysis against *Penicillium expansum in vitro* and in fruit tests. *International Journal of Food Microbiology, 107*(2), 99–103.

Mohanty, S. P., & Kougianos, E., (2006). Biosensosrs: A tutorial review. *IEEE Potentials, 25*(2), 35–40.

Morones, J. R., Elechiguerra, J. L., Camacho, A., Holt, K., Kouri, J. B., Ramirez, J. T., & Yacaman, M. J., (2005). The bactericidal effect of silver nanoparticles. *Nanotechnol., 16*, 2346–2353.

Muniz-Miranda, M., & Neto, N., (2004). Surface-enhanced Raman scattering of π-conjugated "push-pull" molecules: Part I. p-Nitroaniline adsorbed on silver nanoparticles. *Colloids and Surfaces A: Physicochemical and Engineering Aspects, 249*(1–3), 79–84.

Nitesh, R. A. I., Ludwig, W., & Schleifer, K. H., (2011). Phylogenetic identification and *in situ* detection of individual microbial cells without cultivation. *Microbiology Review, 59*(1), 143–169.

Pradeep, T. A., (2009). Noble metal nanoparticles for water purification: A critical review. *Thin. Solid Films, 517*(24), 6441–6478.

Rajasekharreddy, P., & Rani, P. U., (2014). Biofabrication of Ag nanoparticles using *Sterculia foetida* L. seed extract and their toxic potential against mosquito vectors and HeLa cancer cells. *Materials Science and Engineering: C., 39*(1), 203–212.

Ren, G., Hu, D., Cheng, E. W. C., Vargas-Reus, M. A., Reip, P., & Allaker, R. P., (2009). Characterization of copper oxide nanoparticles for antimicrobial applications. *International Journal of Antimicrobial Agents, 33*, 587–590.

Ruparelia, J. P., Chatterjee, A. K., Duttagupta, S. P., & Mukherji, S., (2008). Strain specificity in antimicrobial activity of silver and copper nanoparticles. *Acta Biomaterialia., 4*, 707–716.

Sarikaya, M., Tamerler, C., Jen, A. K., Schulten, K., & Baneyx, F., (2003). Molecular biomimetics: Nanotechnology through biology. *Nature Materials, 2*, 577–585.

Sathishkumar, G., Gobinath, C., Wilson, A., & Sivaramakrishnan, S., (2014). *Dendrophthoe falcata* (L.f) Ettingsh (Neem mistletoe): A potent bioresource to fabricate silver nanoparticles for anticancer effect against human breast cancer cells (MCF–7). *Spectrochimica Acta A: Molecular and Biomolecular Spectroscopy, 128*, 285–290.

Sawai, J., (2003). Quantitative evaluation of antibacterial activities of metallic oxide powders (ZnO, MgO, and CaO) by conductimetric assay. *Journal of Microbiological Methods, 54*(2), 177–182.

Sedlarik, V., Galya, T., Sedlarikova, J., Valasek, P., & Saha, P., (2010). The effect of preparation temperature on the mechanical and antibacterial properties of poly(vinyl alcohol)/silver nitrate films. *Polymer Degradation and Stability, 95*(3), 399–404.

Sheng, Z., & Liu, Y., (2011). Effects of silver nanoparticles on wastewater biofilms. *Water Research, 45*(18), 6039–6050.

Sintubin, L., Awoke, A. A., Wang, Y., Ha, D., & Verstraete, W., (2012). Enhanced disinfection efficiencies of solar irradiation by biogenic silver. *Annals of Microbiology, 62*(1), 187–191.

Solsona, F., & Méndez, J. P., (2003). *Water Disinfection.* Pan American Center for Sanitary Engineering and Environmental Sciences, Pan American Health Organization, Lima, Peru.

Sondi, I., & Salopek-Sondi, B., (2004). Silver nanoparticles as antimicrobial agent: A case study on *E. coli* as a model for Gram-negative bacteria. *Journal of Colloid and Interface Science, 275*, 177–182.

Steinberg, S. M., Poziomek, E. J., Engelmann, W. H., & Rogers, K. R., (1995). A review of environmental applications of bioluminescence measurements. *Chemosphere, 30*(11), 2155–2197.

Stratton, J. A., & Mannix, L. A., (2005). *Mind and Hand: The Birth of MIT* (pp. 190–192). MIT Press, Cambridge.

Su, L., Jia, W., Hou, C., & Lei, Y., (2011). Microbial biosensors: A review. *Biosensors and Bioelectronics, 26*(5), 1788–1799.

Taylor, P. M., (2007). Biological matrices and bionanotechnology. *Philosophical Transactions of the Royal Society of London, Series B, Biological Sciences, 362*, 1313–1320.

Trouiller, B., Reliene, R., Westbrook, A., Solaimani, P., & Schiest, R. H., (2009). Titanium dioxide nanoparticles induce DNA damage and genetic instability *in vivo* in mice. *Cancer Research, 69*, 8784–8789.

Tuncagil, S., Ozdemir, C., Demirkol, D. O., Timur, S., & Toppare, L., (2011). Gold nanoparticle modified conducting polymer of 4-(2,5-di(thiophene–2-yl)–1H-pyrrole–1-l) benzenamine for potential use as a biosensing material. *Food Chemistry, 127*(3), 1317–1322.

Velasco-Garcia, M. N., (2009). Optical biosensors for probing at the cellular level: A review of recent progress and future prospects. *Seminars in Cell and Developmental Biology, 20*(1), 27–33.

Watts, R. J., Kong, S., Orr, M. P., Miller, G. C., & Henry, B. E., (1995). Photocatalytic inactivation of coliform bacteria and viruses in secondary wastewater effluent. *Water Research, 29*, 95–100.

Wei, C., Lin, W. Y., Zainal, Z., Williams, N. E., Zhu, K., Kruzic, A. P., et al., (1994). Bactericidal activity of TiO_2 photocatalyst in aqueous media: Toward a solar-assisted water disinfection system. *Environmental Science and Technology, 28*, 934–938.

Weir, A., Westerhoff, P., Fabricius, L., Hristovski, K., & Von Goetz, N., (2012). Titanium dioxide nanoparticles in food and personal care products. *Environmental Science and Technology, 46*, 2242–2250.

Xie, Y. B., & Li, X. Z., (2006). Degradation of bisphenol A in aqueous solution by H_2O_2-assisted photoelectrocatalytic oxidation. *Journal of Hazardous Materials, 138*(3), 526–533.

Yoon, K. Y., Byeon, J. H., Park, C. W., & Hwang, J., (2008). Antimicrobial effect of silver particles on bacterial contamination of activated carbon fibers. *Environmental Science and Technology, 42*(4), 1251–1255.

Yu, C. J., Zhang, L., Zheng, Z., & Zhao, J., (2003). Synthesis and characterization of phosphated mesoporous titanium dioxide with high photocatalytic activity. *Chemistry of Materials, 15*(11), 2280–2286.

Zan, L., Fa, W., Peng, T., & Gong, Z., (2007). Photocatalysis effect of nanometer TiO_2 and TiO_2-coated ceramic plate on Hepatitis B virus. *Journal of Photochemistry and Photobiology B: Biology, 86*, 165–169.

INDEX

Printed in the United States
by Baker & Taylor Publisher Services